The Last Boy

D0964215

ALSO BY JANE LEAVY

Fiction
Squeeze Play

Nonfiction
Sandy Koufax

The Last Boy

Mickey Mantle and the End of America's Childhood

Jane Leavy

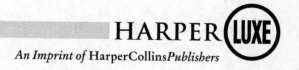

HARPER LUXE

An Imprint of HarperCollins*Publishers*

HarperCollins books may be purchased for educational, business, or sales promotional use. For information please write: Special Markets Department, HarperCollins Publishers, 10 East 53rd Street, New York, NY 10022.

Photo inserts designed by William Ruoto

FIRST HARPERLUXE EDITION

HarperLuxe™ is a trademark of HarperCollins Publishers

Library of Congress Cataloging-in-Publication Data is available upon request.

ISBN: 978-0-06-177488-1

10 11 12 13 14   ID/RRD   10 9 8 7 6 5 4 3 2 1

For Nick,
with love and hope

In memory of my father,
who taught me not to throw like a girl

Played hard, died hard.

—Don Larson

Ya gotta be honest.

—Tony Kubek

CONTENTS

PART TWO
A Round with The Mick, Atlantic City, April 1983

PART THREE
Nightcap, Atlantic City, April 1983

PART FOUR
Dream On, Atlantic City, April 1983

PART FIVE
Riding with The Mick, Atlantic City, April 1983

PREFACE:
MY WEEKEND WITH THE MICK

Mickey Mantle's sweater hangs on the door to my office. I put it there the day I decided to write this book. It is the first thing I see when I sit down at my desk in the morning and the last thing I see when I shut down the computer at night. It has followed me from closet to closet and house to house since he gave it to me twenty-seven years ago. I packed it away in an old garment bag right after I said goodbye to him. I thought I was done with The Mick.

The sweater is as gray as the day we met—Yankee colors, road-trip gray and pinstripe blue. There is a baseball embroidered over the left breast with raised seams and the words MICKEY MANTLE INVITATIONAL GOLF TOURNAMENT stitched around the circumference. On the sweet spot it says, THE CLARIDGE, the hotel where Mantle had accepted a job as Director of

Sports Promotions. I went there to interview him for *The Washington Post* in the spring of 1983.

Forty-eight absurdly tumultuous hours later I headed home flaunting his sweater. How many times have I told this story? How he saw me shivering outside the clubhouse that raw day after nine holes of promotional golf and offered me his sweater. Everybody always said he'd give you the shirt off his back; I had the sweater to prove it. V-necked in the Fifties style that remains a fashion statement chiefly at exclusive American country clubs, Mickey's sweater evokes a time when Mantle was at the top of his game and his game was the only one that mattered.

His gesture was warm, spontaneous, and authentic; the fabric felt like cashmere against my skin. But, like so many of the imperial vestments draped about the shoulders of heroes, this was the product of human engineering: 100 PERCENT VIRGIN ORLON ACRYLIC.

Turned out I had never read the fine print on the manufacturer's label. The warranty got me thinking: Which Mantle would I discover? An authentic human being or a synthetic construct of memory and imagination?

It took five years, far longer than I wished or imagined, to find an answer. Every time I heard a story of baseness and excess, I checked out the sweater in my office. Every time I heard an anecdote infused with kindness, humor, self-deprecation, or irony, I heard his laconic drawl: "Can't we get this girl a *fuckin'* sweater? She's gonna *fuckin'* freeze."

Then, deep into my research, there came in the mail a videotape of Robert Lipsyte's Media Day interview at the Claridge for *CBS Sunday Morning*. There's Mick kicking at the damp sand, his ball buried in a trap and the wind playing havoc with his hair. He seizes his wedge, mutters imprecations at his ball and . . . he's wearing *my* sweater! There I am in the background, hair freshly poodle-permed, all decked out in a brown suede suit, Loehmann's circa 1983. I had gotten all dressed up for The Mick, the way I did when I wore Mary Janes on opening day of the baseball season.

I remembered being underdressed for the occasion. That's why I needed his sweater. I remembered wearing cotton, not suede. Confronted with this disconcerting shard of videographic truth, the story I had told so often began to unravel like the yarn wound around the hard center of an official Major League baseball. I rushed to have another look at the sweater in my office. Somehow I had never noticed the "L" stitched into the collar. Shouldn't Mickey's sweater have been at least an XL or an XXL? If my memories didn't match the evidence, what else might have gotten blurred by the hero worship of childhood? Did he give me the shirt off his back? Or did I invent a kinder, warmer, bigger Mick, the Mick I wanted him to be?

I believe in memory, not memorabilia. Mickey's sweater is the only artifact I kept from my tenure as a sportswriter. Memorabilia is a goal, a get, an end in itself leading nowhere except to the next acquisition. Memory is a process, albeit a faulty one. Mantle began burrowing into American memory the moment he stepped onto the public stage in 1951. Blond and blue-eyed, with a coast-to-coast smile, he was an unwitting antidote to the darkness and danger embodied by that other Fifties icon, Elvis Presley.

Mickey Charles Mantle was born on October 20, 1931. But like the sweater hanging in my office, *Mickey Mantle* is a blend of memory and distortion, fact and fiction, repetition and exaggeration. However far Mantle's home runs traveled, his acolytes remembered them going farther; however great his pain, they remembered it as more disabling. In a life so publicly led, the accretion and reiteration of fable and detail are as thick as fifty years of paint jamming an old windowsill. My challenge was to strip away the layers and let in the air.

As much as anyone, the photographer Ozzie Sweet

was responsible for how Mantle is remembered. Sweet first photographed him in the spring of 1952, when his boyish blemishes still required retouching. With his tripod and his old-fashioned view camera—the same equipment Mathew Brady used in the Civil War—Sweet produced portraits for 1,700 magazine covers but was known best for the confections he created for *SPORT* magazine. His specialty was what he called "simulated action," set pieces choreographed to evoke heroes and hero worship. Every spring, he and Mantle would get together and decide how to expand the Sweet trove of Mantle iconography. "What do you think, Mick?" Sweet would ask. "Can we stage this and make it *look* real?"

He always shot from below, the angle of icons, rendering his subjects larger than life. Clouds and foliage were banished from the frame; nothing was allowed to clutter the image. His photos look as if they could have been taken anywhere, anytime. The context is timelessness.

In 1968, the last spring of his career, Mantle posed before a lavender sky with a Louisville Slugger resting lightly on his shoulder. Only the "Slugger" in the redoubtable trademark is visible beneath his grip on the unsullied wood. Hand on hip in a posture of feigned informality, he gazes over his shoulder into a purple distance with just the barest hint of cumulous cloud on the horizon. Tufts of blond hair on his chiseled forearms shimmer in the glow from an unseen source of light. Beneath the smudged bill of his cap Mantle's brow is puckered with age. The grain of his black leather belt is worn and punctured with asymmetrical holes added to accommodate the passage of time. Sweet says it was one of Mantle's favorite pictures. It was also a template for a "no-dirt-under-the-fingernails" hero whose image would continue to be airbrushed long after his death.

The Mantle postage stamp issued by the U.S. Postal Service in July 2007 was an idealized collage, the head cloned from one photograph and the body from another. Phil Jordan, the art director, rejected the initial design because it didn't look enough like "the image people had of him." He wanted "the soft, baby-faced character, the blondness of him."

He sent the artist back to the drawing board. Lonnie Busch, who outgrew his boyhood fondness for baseball long before he took up his palette, screened Billy Crystal's Mantlecentric HBO movie *61** for inspiration. He painted an effulgent young slugger who fills the batter's box with promise. The sky is an odd orange hue, as if the day could go either way. The green copper frieze of the House that Ruth Built rests squarely on his shoulders. His arms extend beyond the frame, too big to be

contained in a miniaturist's portrait. A better rendering of a person who felt bigger and smaller than life
cannot be found. I asked Busch what he was trying to
capture inside the sticky 1-by-1½-inch rectangle. "The
light within the darkness," he replied.

Who else—besides maybe Elvis—is lodged so
firmly in pop iconography? They were two country
boys fated for unimaginable fame and infamy. Look
at their smiles: Elvis with a dark, brooding forelock
dangling over his brow like an apostrophe and a curled
upper lip. And Mick, with that slight overbite, those
buckteeth (for which he had been sufficiently teased)
pushing the corner of his mouth upward, into an irrepressible grin. "Mantle-esque," the catcher-cum-
broadcaster Tim McCarver called it. "Quite unlike any
other that I remember. It was almost a measure of a
man in his smile."

Elvis served notice: the blatant sexual braggado-
cio was a come-on that came with a caution. Mantle's
darkness was concealed by a sunny, coltish smile and
a euphonious name that makes product branders grin.
Linguistically, "*Mickey Mantle*" is *mmm mmm good*.

"He had a way of putting a smile on your face," his
compatriot, Duke Snider, said.

Even in Joe Torre's Brooklyn neighborhood, where
Snider ruled, you had to wear your hat like Mickey

Mantle. A half century later, the Yankee manager demonstrated how the look was achieved: wetting the NY on his cap with spit, he wrapped the brim around a ball and stuck it in a coffee cup in the manager's office just as he had as a boy in his mother's kitchen. It was called the "Mantle Roll," and it was as popular in the cornfields of Illinois as it was on the sidewalks of Bensonhurst. As Mantle's teammate Jerry Coleman once said, "Everything he was is what everyone wanted to be and couldn't."

With his aura of limitless potential, Mantle was America incarnate. His raw talent, the unprecedented alloy of speed and power, spoke directly to our postwar optimism. His father mined Oklahoma's depths for the lead and zinc that supported the country's infrastructure and spurred its industrial growth. Mutt's boy had honest muscles. His ham-hock forearms were wrought by actual work, not weight machines and steroid injections.

He was proof of America's promise: anyone could grow up to be president or Mickey Mantle—*even* Mickey Mantle. And he recognized it. "I could have ended up buried in a hole in the ground, and I ended up being Mickey Mantle," he mused. "There must be a god somewhere." And, more succinctly, "I guess you could say I'm what this country is all about."

He was numerologically charmed—lucky number 7, *The* Mick, synonymous with the number of planets, wonders, days, sins. "What does seven mean?" asked teammate Clete Boyer. "It means Mickey Mantle."

"His aura had an aura," his teammate Eli Grba said. "The way he walked, the way he ran, and the way he presented himself once he put on the uniform—he was a symphony. Ever hear Beethoven's Ninth? The *Ode to Joy*? You see him hit and then you see him run, and it's like going into the chorale."

He was, in the author Nick Pileggi's felicitous phrase, "a touched guy," and he connected with something in teammates and opponents, men and (lots of) women, baseball fans and baseball illiterates that all of us struggle to explain.

Listen to Randall Swearingen, a software entrepreneur from Houston, who met Mantle at the last fantasy camp he hosted in 1994. Filling his plate on the buffet line, he heard a familiar drawl: "*Umm-hmmm,* those meatballs sure look good." When Mantle reached over his shoulder and helped himself to a meatball, it was like God was eating off his plate. Swearingen devoted much of the next fifteen years and a considerable amount of money to preserving, protecting, and defending The Mick. He developed the official Web site for Mantle's family, authored a Mick encomium,

"A Great Teammate," and assembled in his Houston office one of the largest collections of Mantle memorabilia in the country. An attempt, he says, to re-create "not Mickey Mantle the person but Mickey Mantle the image, the feeling," to "reconstitute what he stood for."

Imagine Swearingen's anguish as a Category 5 hurricane bore down on the office building. The order to evacuate had been issued. But Swearingen was still at his desk, trying to decide which parts of The Mick to preserve and which to leave behind—he finally picked 1962 and 1965 road jerseys, a 1961 warm-up jacket, the 1955 American League Championship ring, and the 1956 Player of the Year award. Then Hurricane Rita made a sharp left. Houston saw barely a drop of rain.

Listen to Cathy McCammon, who recalls, at age ten, returning to her grandstand seat in the Stadium with a tray full of ketchup-saturated hot dogs when she was hit in the head by a Mantle home run, which left a dent in her scalp $\frac{1}{16}$ inch deep. "The food went everywhere, but I caught the ball," she said. "The security guard came to see if I was okay. I said I'd be okay if I could get Mickey to sign the ball. They took me down to the dugout, and Mickey said, 'Are you okay?'

"I said, 'I will be when you sign the ball.'"

She gave the prize to her son. She has The Mick's mark to remember him by.

I hated to point out that her story didn't quite add up. No one signs autographs by the dugout in the middle of a game. She consulted her brother and called back with an update. Okay, she was sixteen. They were sitting in a field box. She was beaned by an errant fungo during batting practice. But neither the impact of the moment nor the ridge in her skull was diminished. "If I put pressure on it, I can feel that sensation of being hit," she said. "It takes me right back to the ballpark."

Listen to Frank Martin, a welder from Pennsylvania, who took a day off from work in order to watch Mantle's funeral live on TV and tape it for posterity. Eight years later, he was at Madison Square Garden for the preview of the Mantle family auction, which he attended knowing he couldn't afford to bid on even the least expensive item. But he had his picture taken with Mantle's second son, David, who looks so much like his father that people stop him on the street and ask, "Aren't you dead?" Martin pressed his memories on his hero's son. "One time my teacher asked, 'Who was the father of our country?' I said, 'Mickey Mantle. He's more famous than Washington—and his card's worth more.'"

Listen to Thad Mumford, the son of an African-American dentist from Washington, D.C., who cut his hair like The Mick and wore his baseball stirrups like

The Mick and spent boyhood afternoons sketching Mantle's legs for fun. Sure, black was beautiful. But Mantle was "modest, graceful. The way he practiced, the way he just stood—it was noble."

One day in the violent summer of 1967, Mumford bought a round-trip ticket on the Eastern Air Lines shuttle ($9.18 with a student discount, he recalls) and flew to New York to interview for a job as a Yankee ball boy. He had just enough cash for a subway token and a hot dog. But he took the wrong train and ended up in Newark, New Jersey, where race riots that summer caused 26 deaths, 700 injuries, and 1,500 arrests. "And I'm trying to find a way to wear the uniform of The Man, the plutocrat Yankees," Mumford said.

He got the job, moving in with Brooklyn relatives in time for the beginning of the 1968 season. "When Martin Luther King was assassinated, I was worried about how many days they were going to postpone opening day by."

One.

Mere statistics do not explain the devotion of Mantle's fans. True, he played in more games than any other Yankee at a time when the Yankees were the most televised franchise in the country. He played in twelve World Series in his first fourteen seasons and still holds World Series records for home runs (18), RBIs (40), runs (42), walks (43), extra-base hits (26), and total bases (123). He became synonymous with Yankee inevitability and hegemony, an institutional entitlement symbolized by the interlocking NY on their caps, designed by Tiffany & Co. The logo was appropriated from a medal presented to a New York City cop shot in the line of duty in the Tenderloin District—while shaking down the owner of a local saloon.

When he retired, Mantle's 536 career home runs placed him third in major league history. Thirteen of them were game-ending homers. His 1964 World Series home run off Barney Schultz, a "walk-off" home run in the current vernacular, broke Babe Ruth's series record. Ten times, he collected more than 100 walks; nine straight seasons, he scored 100 or more runs; four times, he won the American League home run and slugging titles. He collected 2,415 hits, batted .300 or better ten times, won three MVP Awards, and appeared in twenty All-Star Games. He scored more runs than he drove in (1,677 to 1,509).

Those career totals, now regarded as meaningless expressions of longevity, have been supplanted by a dazzling array of new metrics that measure rates of productivity. "By those standards Mantle is actually underrated," said Dave Smith, founder of Retrosheet, the online database that compiles career statistics for every major leaguer and collects box scores for every game ever played. (Retrosheet supplied and verified all the statistics in this book.)

These prodigious numbers belie the pain and suffering it took to accumulate them. Far more than his contemporaries in center field, Willie and The Duke, Mantle fit the classical definition of a tragic hero—he was so gifted, so flawed, so damaged, so beautiful. The traumatic and defining knee injury he suffered catching a spike in an outfield drain during the 1951 World Series attenuated his breathtaking potential after just seven months in the major leagues. His death from alcohol-related cancer in 1995 attenuated eighteen months of belated, hard-earned sobriety. He had so little time to be his best self.

Today, his memory survives in a kind of protective custody, fiercely guarded against the slings and arrows of a tell-all culture by a cohort of aging fanboys. Call it Mantleology—a cultlike following of Baby Boomers unprecedented in modern sports. Al Taxerman, an otherwise rational New York attorney, calls Mantle "my Achilles, part man, part God, giving the divine fits," but he turned down the chance to meet him lest he be confronted with his hero's flaws.

They will invest almost any sum in order to own a piece of him. Billy Crystal paid $239,000 on a game-used glove—with broken webbing. He intended to quit bidding at $120,000 but his wife persisted, finally outlasting the rival bidder, former Yankee pitcher David Wells. Crystal says the glove is more valuable to him than the Picasso hanging in his home. That's a big number on anyone's ledger but far less impressive than the amount spent by an unemployed limo driver on three Mayo Clinic appointment cards ($649) and a 1951 bankbook from the First State Bank of Commerce ($1,888). That's more than Mantle ever had in the account.

The day before I left for the Heroes in Pinstripes fantasy camp run by his comrades Hank Bauer and Bill Skowron, I got a call from the director wanting to make sure I wasn't planning to ask anything that might upset his campers. These are middle-aged men who paid $5,000 for the privilege of pulling their hamstrings on the fields where Mickey once roamed. They prefer the lavender scrim of Ozzie Sweet's staged portraits.

Upon my arrival at ground zero for Mantleologists, Bauer, the stalwart ex-Marine then dying of lung cancer, poked a gnarly forefinger in my chest and croaked, "Nothing negative. Nothing *negative*! Nothing *NEGATIVE*!" His partner, "Moose" Skowron, once refused to participate in a Mantle roast at the Claridge Hotel, asking with incredulity, "You want me to make fun of Mickey Mantle?" After Mantle's death he expressed disappointment with his family's forthright comments about his alcoholism. "He didn't drink that much," Skowron told me. "He didn't hurt nobody."

When Peter Golenbock's novel 7, a so-called fictional biography of Mantle so prurient that this publisher dropped it from its 2007 list, was published, Johnny Blanchard called the Yankees former PR man Marty Appel to ask: "Would it help if me and Moose went down and beat the shit out of him?"

Such is the force field that surrounds Mantle's memory.

I have a photograph of Mantle and his cohorts Billy Martin and Whitey Ford, part of a series by the photographer Fredrich Cantor, a candid shot taken in the dugout at Shea Stadium on Old Timers' Day, 1975. Graying muttonchops spread like crabgrass across Mantle's cheeks, polyester pinstripes gaping at the formerly taut midriff. He's got a goofy, cross-eyed Jerry Lewis look on his face, chapped lips inverting the famous smile. His thick slugger hands pound his chest as if to say, "Look at me!" Whitey, the slick New York kid, looks away. Billy, the hard case sidekick, laughs, egging him on as usual.

The photo is a touchstone of another era: when boys were allowed to be boys and it was okay to laugh at them for being themselves, when it was okay not to know and to forgive what you did know. Cantor caught the essence of Mantle's appeal. He was the Last Boy in the last decade ruled by boys. He was Li'l Abner in a posse of dreamy reprobates: James Dean, Buddy Holly, Frankie Avalon, Dean Martin, Elvis. Women wanted to have them or mother them. Young men aped them, while behind the scenes, elders and handlers tried to tame them. "And the rest of us got bigger and harder under the testosterone shower," Bob Lipsyte said.

Mickey Mantle was the Last Boy venerated by the last generation of Baby Boomer boys, whose unshakable bond with their hero is the obdurate refusal to grow up. Maintaining the fond illusions of adolescence is the ultimate Boomer entitlement. He inspired awe without envy—except perhaps for what he got away with. Pain inoculated him against jealousy and judgment.

If at the beginning Mantle was the incarnation of the strong, silent Fifties—"surly," sportswriters called him—he evolved into a psychobabble raconteur, laying himself on the public couch to recount the particulars of his recurring nightmares. Mantle himself punctured the protective he-man bubble in an April 1969 guest appearance on *The Dick Cavett Show* shortly after he retired, genially describing his boyhood bed-wetting to the inquisitive host and his disconcerted guest, Paul

Simon. Off camera, Mantle asked why Simon hadn't used "Where have you gone, Mickey Mantle?" in his latest hit song—the only time in memory his name wasn't a good fit. Besides, he hadn't been gone long enough for "lonely eyes" to miss.

In the last years of his life, Mantle morphed into an avatar of the confessional Nineties. He became a 12-step prophet pushing the gospel of recovery in the pages of *Sports Illustrated* after he got out of rehab. Less than eighteen months later, in his last public appearance before his death, he was a desiccated figment of the apple-cheeked youth in Ozzie Sweet's oeuvre.

The packaging and repackaging, construction and deconstruction of his memory continues unabated. SABRmetricians such as Bill James, the guru of baseball's new math, reprocess his statistics. Writers recalibrate his memory and recycle recollections, creating what Richard Sandomir of *The New York Times* calls MickLit (biography, retrospective, novel, novella, koan, list, stat).

He led one of twentieth-century America's most re-iterated lives, leaving a paper trail long enough to pave the way from Commerce, Oklahoma, to the Bronx. The army of scriveners had lots of help from friends happy to exaggerate their relationship with greatness and from Mantle himself. He became an enthusiastic contributor to his own legend but was neither a "good custodian of it," as Thad Mumford put it, nor a reliable one. Over time, alcohol corroded his memory. He misremembered a lot—the year he got married, for example. Factual errors were compounded by repetition and by the raconteur's instinct for a good yarn. His life became a solipsistic loop of video clips and sound bites. Much of what he knew about himself was what reporters told him he had once said.

During his lifetime, Mantle wrote or collaborated on at least six different biographies, in addition to inspirational and instructional tomes. They include *All My Octobers, My Favorite Summer 1956, Whitey and Mickey,* and *The Mick.* Since his death, at least twenty new volumes have been added to the canon, including thoughtful posthumous biographies by Tony Castro, David Falkner, and John Hall. His family has authored or authorized three books, including a collection of condolence notes from his fans and a searingly brave confessional, *A Hero All His Life,* detailing their collective descent into alcoholism and drug abuse. Mantle contributed a chapter written from his deathbed. I relied heavily on this account, supplementing it with my own interviews with Mantle and his family.

The aim here is neither voyeuristic nor encyclopedic—don't look for every home run, clutch base hit, disabling injury, or pub crawl. The narrative land-marks of his life are well known and well documented: Mutt, the dying miner father; the osteomyelitis that almost cost him a leg; the crippling encounter with the drain in right center field at Yankee Stadium. Instead, I picked twenty days from his life and career for closer inspection, each pivotal or defining. They represent highs, lows, flash points, and turning points.

I retraced his steps to Commerce and the corrugated metal shed that was the backstop to his father's dream. I checked into fantasy camp in Fort Lauderdale and vis-ited the Yankees' new spring training home in Tampa. I paid a call at Monument Park at the old Stadium, where tourists clad in John Deere baseball caps tram-pled newly planted geraniums in an effort to get close to him. In Carthage, Missouri, I attended the sixtieth reunion of the Class D Kansas-Oklahoma-Missouri (KOM) League, where he played his first games as a professional. I paid my respects at the Mantle shrine in the all-male clubhouse at the Preston Trail Golf Club, his home away from home in Dallas, and at his grave at the Sparkman Hillcrest Memorial Park. I attended the 2003 family auction at Madison Square Garden, where I listened as *über*-fan Frank Martin bragged to David Mantle, "Nobody knows Mickey like I did, I'll challenge him to a trivia contest." David cheerfully explained that he hears lots of things about his father he's never heard before—especially from people like me. In the background, I could hear his brother Danny on his cell phone telling someone, "I really didn't know him until I was sixteen."

People who loved him and loathed him agree he was an uncommonly honest man, a trait he bequeathed to his family. His surviving sons, Danny and David, and their late mother, Merlyn, answered the most intimate questions as honestly and fully as they could, asking only that I try to find his good heart. But his prolonged absences from their lives left gaping holes in their knowledge of their own history.

They spoke about failed marriages and their addictive inheritance and the deaths of two brothers. So I was puzzled to hear sometime later that Danny was upset that I had been questioning the length of a home run his father had hit in a spring training game sixty years ago. It wasn't my intention to diminish the feat. On the contrary, I was asking questions about March 26, 1951, because that was the day Mantle announced himself to the world. Then I realized: calibrations of clout are among the only unsullied measures of the man they have left.

In an effort to calculate how far those baseballs actually traveled, I bought a radar gun and surveyed the

USC campus where that home run was hit. I asked Eric R. Kandel, a Nobel Prize winner for his research into the biochemistry of remembering, to explain muscle memory. I persuaded Preston Peavy, an Atlanta hitting coach, to transform grainy Mantle game films into kinetic diagrams of his swing. I importuned Alan Nathan, a distinguished physicist and Fenway Park partisan, to climb to the roof of Howard University Hospital in Washington, D.C., to gauge the distance of the first Tape Measure Home Run.

In an effort to find his good heart, I spoke with more than five hundred people—friends and family (brother, sister, wife, sons, cousins); opponents and teammates (sandlot, high school, minor and major league); friends and girlfriends; agents and lawyers; writers who covered him and writers who wished they hadn't. I interviewed linguists, coaches, physicians, batboys, and clubhouse men. I asked each of them the question posed by his minor league teammate Cromer Smotherman in reply to my own query: "What's the one thing you'd ask Mickey if you could talk to him today?" After a choked pause, Smotherman replied, "Mickey, what happened? Why did you do it? Why did you choose to live the life you did? Because you were not that kind of person. That was not you."

When Mantle faced the cameras for the last time

a month before his death, he was a husk of a man, shrunken by cancer. The stiff brim of his 1995 All-Star Game cap dwarfed his brow. There was no Mantle Roll. He looked straight into the cameras and told us all, "Don't be like me."

The transformation of The Mick over the course of eighteen years in the majors and forty-four years in the public eye parallels the transformation of American culture from willful innocence to knowing cynicism. To tell his story is to tell ours. And mine.

I saw the best and the worst of The Mick during the weekend I spent with him in Atlantic City but I wrote little of the latter in the piece that appeared in *The Washington Post.* In 1983, it would have been a firing offense to write what had really happened. Today it would be a firing offense *not* to write it—one measure of how much the landscape of public discourse has changed.

I have tried to balance the claims of dignity and fact. I recognize that some of the material in the pages that follow may offend and disappoint. More than once I was tempted to put his sweater back in the closet.

Well into the late innings of my research, I was still unsure what to write, how to write it, and whether I wanted to write anything at all. Many of his intimates offered opinions—scientific, psychological, and spiritual—about how to tackle the job. Bobby Richardson's wife, Betsy, beseeched me not to "glorify the flesh" and "to pray to do justice to the truth without doing injustice to that which breaks God's heart."

It was good advice but I also believe that denial is treacherous and taking refuge in generalities is the same as giving him another pass. Nobody knew the danger of that better than The Mick.

So how do you write about a man you want to love the way you did as a child but whose actions were often unlovable? How do you reclaim a human being from caricature without allowing him to be fully human? How do you find the light within the darkness without examining the dimensions of both?

I decided to abandon an orthodox biographical structure in favor of an approach that could accommodate my stubborn fandom. At intervals throughout the narrative, I revisit Atlantic City and my weekend with The Mick. This time, the account is full and unexpurgated. It chronicles my journey from childhood thrall to adult appreciation.

This is my attempt to understand the person he was and, given who he was, to understand his paradoxical hold on a generation of baseball fans, including myself, who revered The Mick despite himself, who were seduced by that lopsided, bucktoothed grin.

My course charted, I went to the storage unit to un-
earth notebooks and transcripts of taped interviews
from April 1983. There, in the typed pages of onion-
skin, I found warming words not among the sound
bites *CBS Sunday Morning* had aired in Bob Lipsyte's
report.

"Can't we get her a *smaller* one?" Mickey said.
"She's gonna *fuckin'* freeze."

He saw me as I was, cold and small. I needed to see
him as XXL.

PART ONE

Innocence Lost

ATLANTIC CITY, APRIL 1983

I met Mickey Mantle in the Atlantic City hotel where my mother lost her virginity, three weeks after Pearl Harbor. It was the spring of 1983, the year Mantle's hometown of Commerce, Oklahoma, was named one of the most toxic waste sites in America. I was a reporter for The Washington Post *and a devoted second, who had taken up the gauntlet in the endless verbal duels of protracted childhood: "Who's better? Mickey Mantle or Willie Mays?" He was the newly appointed Director of Sports Promotions at the Claridge Hotel and newly banished from baseball because of his affiliation with its casino.*

My parents' honeymoon had been brief, one winter night by the Jersey shore—Christmas Day 1941, the only day they could find a rabbi in the prenuptial rush to commitment prior to his shipping out. After my father received his orders—he was stationed in the Aleutian Islands for four long, bitter years—my mother moved back in with her parents at 751 Walton Avenue, one very long, very loud foul ball from Yankee Stadium.

The building was called the Yankee Arms and featured a leaded stained-glass window in the lobby with bats crossed over a heraldic shield; the colors were home white and pinstripe blue and yellow-gold to evoke the blondness of ash.

Groundbreaking was in the fall of 1927, just after The Babe swatted his sixtieth home run. Such was his clout that a whole new subdivision of luxury buildings—the Neighborhood That Ruth Built—sprang up in his shadow. Bordered on the east by the Beaux Arts mansions of the Grand Concourse and on the west by the Harlem River, the Stadium area embodied upward mobility. For a time, The Babe himself lived on Walton Avenue, ten blocks north of my grandmother's kitchen.

The apartment had all the latest amenities of 1920s construction: windows that swiveled on a pivot for washing; a dumbwaiter that brought groceries up from the basement and took trash away; a refrigeration system that circulated cold water through pipes to keep the groceries cold. The halls smelled of butter and borscht, chicken schmaltz and stuffed cabbage. I could smell my grandmother's sweet-and-sour salmon two floors away.

The lobby attempted dark Tudor elegance: heavy, brocaded furniture and ocher-colored shellacked stucco walls. Shafts of blue and gold light poured through the stained-glass window, pooling on the hard stone floor. I hopscotched from blue to gold, my party shoes clicking like baseball spikes against a concrete runway.

It was there that I fell in love with Mickey Mantle.

My grandmother's apartment, 2A, faced east toward the Concourse, away from the Stadium. During home stands, the roar of the crowd threatened the kibitzing in her parlor, ricocheting off the buildings on 157th Street, past the candy store and the greengrocer on the corner of Gerard Avenue, past Nick, the shoemaker, and Mr. Kerlan, the kosher butcher, and through her double-hung windows. Crouched beneath the grand piano—with a damaged right leg as precarious as The Mick's—I listened to Mel Allen's honeysuckle baritone, punctuated by the crack of the bat. And then the roar came again as the sound waves vibrated up the street. It was my own primitive version of surround sound and it rattled the glass. I turned up the volume when Mickey was on deck.

In my worldview, Celia Zelda Fellenbaum and

Mickey Charles Mantle were linked by something far deeper than mere proximity. Both were stoic in the face of pain and selfless in pursuit of pleasing others. My diabetic grandmother injected her thigh daily with the insulin she kept in the icebox along with the sweets she stocked for me and my cousins: six-packs of Pepsi, platters piled high with homemade rugelach, and her own seven-layer chocolate cake. How different was it, really—Mantle's insistence upon being in the lineup no matter how much he hurt and her risky determination to fast on Yom Kippur? Weren't they both team players?

"Who's better, Dad? Mickey or Willie?"

My father grew up on the other side of the Harlem River in a tenement hovering above Coogan's Bluff. In the winter of 1927, he patrolled the Polo Grounds as a water boy for the New York football Giants. "Willie," he replied firmly, citing the latest box score.

Mickey was my guy. Or: I was a Mickey guy. Either way you put it, the relationship was proprietary and somehow essential. Like Mick, who had to be sent down to the minors three months after his major league debut, I had arrived prematurely. Conceived the week—perhaps the day—he hit his first home run at the Stadium, I was born two months too soon in a Bronx hospital twenty city blocks from where that ball landed. Like Mick, I had a sense of being physically flawed. Other kids practiced his swing; I practiced his limp and aped his grimace.

My grandmother gave me permission to be who I was, a little girl who liked to play boys' games. One fine spring day, opening day of the baseball season, we took the CC train downtown to Saks Fifth Avenue to buy a baseball glove. The cars still had those old straw seats and the bristles caught in my tights and we almost missed the stop while trying to untangle me. I often got tangled up when I tried to be a proper girl.

We bought me a mitt, the only one they had, a Sam Esposito model, which was firmly attached to the glove hand of a mannequin in the Saks Fifth Avenue window. "I'll have that for my granddaughter," she told the flummoxed salesman.

No matter how many times he demurred—"Madam, it's not for sale"—she would not be deterred. I took Sammy home with me and everywhere else until my mother disposed of the glove in an unhappy spring purge. I told my grandmother that Sam was a Yankee. She had no reason to know better. In the twenty-five years she lived at 751 Walton Avenue, she never once felt compelled to cross the threshold of the cathedral of baseball.

She celebrated the Jewish High Holy Days in the

ballroom of the Concourse Plaza Hotel at the corner of 161st Street, where Mickey and Merlyn Mantle spent their first year as newlyweds. No matter what the temperature, she wore her mink coat to shul. It had a shawl collar and no buttons and was big enough to keep her and several grandchildren warm. In fact, her coat was two sizes too large—marked down, wholesale. She didn't wear it to temple on sweltering fall afternoons of prayer to show off. That would have required a mere stole. It was to accommodate me, Sammy, and my red, plastic transistor radio with a tinny gold flower-shaped speaker at its center. She greeted the New Year, waiting for me by a bench in front of Franz Siegel Park, arms spread wide, an expanse of mink catching me in a satin embrace.

Services were held in the sumptuous ballroom of the hotel, which opened for business the same year as Yankee Stadium. With its vast onlookers' balcony, the ballroom was well suited to my grandmother's Conservative congregation, in which men and women worshiped in sacred isolation. The women sat upstairs in the gallery in ballroom chairs facing toward Jerusalem. I faced the opposite direction, called to prayer by the large, green, looming presence of the outfield wall at the bottom of 161st Street. Just down the hill, past Joyce Kilmer Park, where African-American men sold towers of undulating marbleized balloons, past Addie

Vallens, the ice cream parlor where Joe DiMaggio enjoyed an ice cream soda between ends of a double-header. Mickey was so close, and so far away.

While my grandmother listened for the sound of the shofar, I listened to Red Barber inside a cocoon of heavy red velvet drapery that concealed his voice and my apostasy. While she prayed for my future, I prayed that no one would ever humiliate Mickey again, the way Sandy Koufax did in the 1963 World Series.

The 1964 World Series was my last opportunity to pray with her and for him. Mickey got old fast, and so did my grandmother. I was sitting in my parents' maroon-on-black Dodge sedan with the push-button transmission in the parking lot of Montefiore Hospital when she suffered the stroke that precipitated her death at age seventy-four. The night she died, Monday, May 2, 1965, the Yankees did not play.

I didn't go back to Yankee Stadium until September 1968. This time, it was to pay homage to The Mick. It had been an awful year of abrupt and tragic goodbyes. Robert Kennedy and Martin Luther King, Jr., were assassinated. The cover of Time magazine asked if God was dead too. And Mickey Mantle was playing his last season.

The particulars of the game are hazy. Was it a Sunday? A doubleheader against the Senators, perhaps? Memory returns in shards: traffic whizzing by the pigeons loitering on the median dividing the Concourse; the rumble of the D train below tar-patched macadam; a steel girder buttoned with bolts, blocking the view from our seats in the lower deck behind and to the left of home plate. The netting cut the batter's box into tidy rectangles of time and space. I don't remember what Mickey did that day. But then, my view was obstructed.

Just how little I'd really seen of him became apparent when he agreed to meet me for breakfast in Atlantic City fifteen years later. I was sitting at my desk in the sports department at The Washington Post when he called. "Hi, this is Mickey," he drawled. "Mickey Lipschitz."

"I didn't know you were Jewish."

"Let me tell you something a guy told me when I first come to New York," Mickey said. "When you're going good, you're Jewish. When you're going bad, you're Eye-talian."

He said he'd meet me at 11 A.M.

1

March 26, 1951

THE WHOLE WORLD OPENED UP

On March 20, 1951, shortly after arriving in Los Angeles for the beginning of their spring training tour of California, the World Champion New York Yankees visited the lot at MGM where Betty Grable was rehearsing dance numbers for her newest flick. A PR still, later published in *Movie Fan* magazine, was taken to commemorate the occasion. There's Yogi Berra front and center wearing a garish paisley sports shirt as bright as his smile, with a collar as wide as his ears and Grable on his well-tailored arm. There's The Scooter, Phil Rizzuto, the unlikely MVP of the 1950 season, and his double-play partner, second baseman

Jerry Coleman; Johnny Hopp, Johnny Mize, and big Joe Collins offering Grable his right elbow.

In the back row, like a schoolboy who'd wandered into the wrong class picture, stands the rookie Mickey Mantle, his features as unformed as his future. He gazes over Grable's shoulder, his blond hair smartly parted, cowlick neatly slicked, necktie tautly knotted.

Mantle and his roommate, Bob Wiesler, were the only rookies in the bunch, both movie buffs. They couldn't understand why more of the veteran players hadn't jumped at the chance to go to Hollywood. They met Esther Williams, Red Skelton, Howard Keel, and the guy who later played Miss Kitty's bartender on *Gunsmoke*. They saw Debbie Reynolds hurrying down the hall carrying two fur coats and called out, "Hiya, Deb!" Mantle wrote home to his Oklahoma sweetheart about the starlets who returned his hello. "Wasn't any as pretty as you."

It was a big time for Mickey Mantle. His childhood friend from Commerce, Nick Ferguson, who had migrated west after high school, drove up from San Diego in his old '42 Plymouth to show him the California coast. Ferguson wanted his Okie buddy to see the Pacific. They went straight out Wilshire Boulevard to the Santa Monica Pier. It was Mantle's first opportunity to feel the surf and sand between his toes. But he did neither. Baseball was the only thing on his horizon. All he cared about was getting to the ballpark on time.

Del Webb, the Yankees' entrepreneurial co-owner, had contrived that spring to switch training camps with the New York Giants. Webb was a Phoenix real estate developer, ahead of his time in grasping the westward rush of postwar America. *Sporting News* reported that he was considering selling his stake in the Yankees to his partner Dan Topping as part of a plan to extend major league baseball beyond the Mississippi.

Bringing the Yankees to train in Phoenix allowed him to play the big shot in his hometown. Then he sent them barnstorming up and down the California coast in order to showcase Joe DiMaggio in the Clipper's home state and whet the appetite for big league ball. The schedule called for thirteen games in California, mostly against Class AAA Pacific Coast League teams, with stops at Glendale, manager Casey Stengel's hometown; in Oakland, where Stengel had managed the Oaks before being promoted to the Yankees' job; at Seals Stadium in San Francisco, where DiMaggio had made his name; and finally at the University of Southern California against the Trojans, better known for their gridiron exploits. That spring was the last time the Yankees would train anywhere other than Florida.

It also marked the opening act of one of baseball's—

or Broadway's—greatest hits, an SRO psychodrama with a very long run.

Stengel had seen Mantle for the first time a year earlier at a pre–spring training camp held in Phoenix for the top prospects in the Yankees system. The kid, just eighteen, had missed the team bus to the practice field. He was standing with a teammate, Cal Neeman, neither of them knowing what to do, when a taxi pulled up. "Well, hop in, boys," Stengel said, "we'll go to the park."

Neeman recalled, "And we're ridin' along, and he wants to know who's in the car. Well, we really didn't want to tell him. I give him my name. He come to Mickey and says, 'Who are you?' And he says, 'I'm Mickey.' And he says, 'Oh, you're that kid that's all mixed up. You're not supposed to be able to run like that and hit the ball so far.' "

Mantle was all but invisible until the coaches said, "Take your marks . . ." Hank Workman, a prospective first baseman, recalled, "They were timing guys from home to first. Nobody noticed Mantle up to that. He was very quiet and extremely shy. He would pull his cap down so far over his brow that you could hardly see his face. Then he ran. And I swear he was going so fast you could still see the tufts of dust in the air from his footprints a couple of feet back from where he was."

Bunny Mick, one of Stengel's lieutenants, timed him from the left-handed batter's box to first base in 3.1 seconds, a new land-speed record.

Workman also recalled Mantle's debut in intrasquad games: "The first time Mantle came up, he hit one a mile outta that ballpark. About three innings later he comes up again. The pitcher's changed, and he hits one a mile out the other way. And all he does after is, he trots out to shortstop in his non-ostentatious way with his hat pulled way down."

The camp was shut down when Commissioner Happy Chandler got wind of the big league instructors getting a head start on spring training. But Stengel had seen enough to see the future. "Mantle's at shortstop taking ground balls, throwing 'em by the first baseman—and outta the dugout comes Stengel," Workman remembered. "He's got a fungo bat in his hand, and he runs right at Mantle. He starts waving this bat at him, and he shoos him out into the outfield, and turns around and loudly announces to all the coaches and everybody that's assembled that this guy is gonna be a center fielder. 'I'm gonna teach him how to play center field myself, and I don't wanna see him at shortstop again.'"

But that's where he played for the 1950 Joplin Miners. His .383 batting average deflected attention from his 55 errors and he was named the Most Valuable Player of the Western League. In January 1951, *The Sporting News* hailed him as a "Jewel from Mine Country."

"Nineteen-year-old Mickey Mantle, dubbed by some big-time scouts as the No. 1 prospect in the nation, will be off for Phoenix in a few weeks to display the talents that won him such raves from veteran talent hunters," baseball writer Paul Stubblefield declared. And in a special box: *The Sporting News* announced the engagement of the Yankees' big catch—who "didn't cost a nickel"—to Miss Merlyn Louise Johnson of Picher, Oklahoma.

The groom to be was a no-show when rookies reported to spring training camp two weeks later. A reporter and photographer from the Miami *Daily News-Record* found Mantle at the Eagle-Picher motor pool and delivered a message from Yankee farm director Lee MacPhail: "Where are you?"

Helping out the pump crew, came the reply, for $35 a week. The Yankees hadn't sent him a train ticket, and Mantle wasn't a bonus baby like Skowron ($25,000) and Kal Segrist ($50,000). His half brother, Ted Davis, used his Army discharge money to pay for Miss Johnson's engagement ring. The photographer snapped a picture of the overall-clad prospect with the smudged grin leaning against a mining company truck. The next day, Tom Greenwade, who would forever be known as the scout who signed Mickey Mantle, showed up with his fare.

By the time Mantle got off the train in Phoenix, the Yankees' most heralded rookies—Bob Wiesler, Moose Skowron, Gil McDougald, Andy Carey, and Bob Cerv—were working on their baseball tans. They never forgot those early tastes and smells of the big time. For Al Pilarcik, an outfielder from Ohio, it was the scent of orange blossoms from the trees outside the team's motel. For Wiesler, it was the standing rib roast that circulated through the dining room on a rolling cart. Every night, waiters would lift the platter's heavy silver-plated hood and the kids would help themselves to a juicy slab of promise.

The early reports on Mantle were measured in tone. He was being groomed, considered, studied. Neither he nor his employers expected him to play in the big leagues in 1951. Never had anyone in the Yankee system made the leap from Class C to the majors after only two years in professional baseball. And Jackie Jensen, the California golden boy, was also waiting in the wings for DiMaggio to exit stage right.

But Stengel was watching, keeping a close eye on his new kid. "We'd go down through the lobby, Casey would always be sittin' there," Nick Ferguson said. "And he never said anything, but he was eyeballin' me like he knew I wasn't a player, and what was I doin' there with Mickey?"

Spring is a season of profusion, especially for baseball writers, the inevitable consequence of sending grown men to Florida and Arizona with empty notebooks, a per diem—and no wives. Red Smith rued the day he received a wire from Stanley Woodward, his editor at the *New York Herald Tribune*, ordering him to quit "godding up those ballplayers." Nonetheless, in the spring of 1951, Mickey Mantle was elevated and beatified.

In the thin Arizona air, his home runs soared and florid prose burst forth from fallow typewriters like desert wildflowers. Just a week after Mantle arrived in camp, Ben Epstein, the effusive beat writer for the *New York Daily Mirror*, wrote, "Thank the fates for Arizona's ambrosial air. It's practically necessary to fuel one's lungs with the stuff if you want to stay in fashion and carry on about Mickey Mantle. Latest estimates hoisted the Yankee oakie doakie as the eventual successor to Joe DiMaggio."

Bad weather during the first days of camp—snow, even—forced the players indoors. Indolent scribes still had to churn out copy. Encomiums lit up the

Western Union wires: "Rookie of the Eons," "Magnificent Mantle," "Mighty Mickey," "Young Lochinvar," "Commerce Comet," "Oklahoma Kid," "Colossal Kid," "Wonder Boy," "One-Man Platoon," "The Future of Baseball."

Pete Sheehy, the clubhouse man and guardian of Yankee succession, assigned the lockers and the uniform numbers—the Yankees were the first team to do that. Sheehy had gone to work for the club when he was fifteen, summoned to his calling while waiting for the Stadium gates to open one day in 1927—and stayed until his death fifty-nine years later. The Yankee locker room is named after him. He was the institutional memory of the club, who divulged nothing. He fetched hot dogs and bicarb for The Babe and joe for Joe D.; he informed a historically challenged rookie that George Herman Ruth's number 3 was not available, nor was Henry Louis Gehrig's 4. As for 5, everyone knew 5 was still working on immortality. Sheehy gave Mantle 6. "The law of mathematical progression," the Yankees' public relations man Red Patterson called it.

Veterans reported on March 1. Archie Wilson, a pitcher returning from military service, arrived to find both beds in his assigned room taken. Archie's widow, Sybil Wilson, recalled that as her husband put his things down on a roll-away, Mantle rose from his bed and said, "You're not going to sleep on the cot." He took it—Wilson was an Army vet and his senior.

On March 2, Stengel announced that he was moving Mantle to the outfield. The next day, DiMaggio announced that the 1951 season would be his last. His throwing shoulder was sore, his left knee was swollen, and his pride was smarting. All those questions didn't help, either. Louella Parsons, the dominatrix of Hollywood gossip, wanted to know about a possible reconciliation with his estranged wife, Dorothy. The baseball writers wanted the dope on the new kid. So DiMaggio threw them all a curve.

His retirement was on the horizon, but the Yankees had no idea an announcement was coming that day. "What am I supposed to do, get a gun and make him play?" Stengel groused. Overnight, Mantle went from a good story to *the* story. "When they'd go into the hotel lobbies, all the newspaper people would flock to Mickey," Sybil Wilson recalled. "He would get down behind Archie and squat down so they wouldn't see him. He was so scared of them."

Tommy Henrich, Old Reliable, was assigned the task of turning him into an outfielder, teaching him how to gauge the angle of the ball off the bat; how to position his body to catch the ball on his back foot and get rid of it in one smooth motion; how to react to a drive hit straight at him. His arm was plenty strong—legend had it that minor league ballparks refused to sell tickets behind first base when Mantle was patrolling the infield or put chicken wire up to protect the spectators. Delbert Lovelace, a friend from sandlot ball back home, was on the receiving end of more than one errant heave: "One time he let the ball loose, and it looked like surely that ball was goin' to drop into the dirt, and I put my glove down, and it hit me on the wrist above my glove."

The seams of the baseball were engraved in his flesh.

In the outfield, Mantle couldn't hurt anybody—except maybe himself. Out there, he could outrun his mistakes.

When the New Yorkers opened their 1951 Cactus League season against the Cleveland Indians in Tucson, Mickey Mantle was the starting center fielder for the New York Yankees. He got three hits. The next day he was conked on the forehead by a line drive while trying to adjust his sunglasses; he had never worn them before. The Southwest sun was so intense that players slathered black shoe polish under their eyes to minimize the glare radiating off their cheeks. "Still couldn't see," said Al Rosen, the Indians' third baseman.

In the Indians' dugout, Mantle's blunder was greeted with sympathy and laughter. "Here comes a kid, and everybody is talking about how great he is," Rosen said. "First thing, he gets hit with a fly ball. Everybody says, 'Some kind of great.' I never saw him drop another fly ball, by the way."

By the end of spring training, when Mantle threw out a runner at third who unwisely wandered off the bag on a fly ball to right, to complete a rare 9–5 double play, Henrich declared his work done. "Best throw I've ever seen," he said.

Players size up other players. That spring, rookies and veterans alike stopped to watch when Stengel's protégé took batting practice. "It was like he was hittin' golf balls," Yankee pitcher Tommy Byrne said.

"Who in the heck is this kid?" wondered Yogi Berra.

Mantle's talents were unprecedented. Only four switch-hitters played regularly in the majors in 1951, and none of them ever hit more than eighteen home runs. "He has more speed than any slugger and more slug than any speedster—and nobody has ever had more of both of 'em together," Stengel declared. "This kid ain't logical. He's too good. It's very confusing."

Compounding Stengel's befuddlement was the disconnect between Mantle's power and his actual size. At only five feet eleven and maybe 185 pounds, he wasn't big at all. Yankee pitcher Eddie Lopat was first to observe, "That kid gets bigger the more clothes he takes off."

Potential is the most elastic of human qualities. By the time the Yankees boarded the train for California, the dispatches being wired back east were inflated with wonder and speculation: How much more might he grow? And if he filled out, what place in baseball history might he occupy?

Stan Isaacs, writing for the *Daily Compass*, was the lone voice of reason, but he had the advantage of being in New York:

Since the start of spring training, the typewriter keys out of the training camps have been pounding out one name to the people back home. No matter what paper you read, or what day, you'll get Mickey Mantle, more Mickey Mantle and still more Mickey Mantle.

Never in the history of baseball has the game known the wonder to equal this Yankee rookie. Every day there's some other glorious phrase as the baseball writers outdo themselves in attempts to describe the antics of this wonder: "He's faster than Cobb . . . he hits with power from both sides of the plate the way Frankie Frisch used to . . . he takes all the publicity in stride, an unspoiled kid . . . sure to go down as one of the real greats of baseball."

Mantle wasn't in the starting lineup when the Yankees arrived in Los Angeles on Friday, March 16, to play the Hollywood Stars. The game was a sellout; across town the St. Louis Browns and the Chicago White Sox played in front of 235 fans, including their unhappy owners. (The Yankees would draw nearly 140,000 during their ten days in California, acccording to the *Los Angeles Times.*)

The next morning, Nick Ferguson arrived bright and early to take his pal out to breakfast at a greasy spoon on Wilshire Boulevard. As Mantle inhaled box after tiny box of cornflakes, Ferguson thought back to the mornings he had spent at the Mantles' home watching Mickey and his twin brothers eat big soup bowls full of cereal. That was one reason the family moved out of Commerce to Dr. Wormington's farm east of town, where they could have some cows and enough milk for all those cornflakes. Mickey milked all nineteen of them before heading off to school.

After breakfast, Mantle and Ferguson drove to Wrigley Field, the minor league park where Gary Cooper had once stood at home plate to deliver Lou Gehrig's farewell speech in *The Pride of the Yankees.* Mantle hit a mammoth home run—"cannonading the pellet over the center field bleacher fence 412 feet from home plate and another wall beyond," according to *The Arizona Republic.*

Gil McDougald, destined to become 1951's Rookie of the Year, saw a scene often repeated by unwary center fielders. "Mickey hit a two-iron shot, and this guy come runnin' over in center field thinkin' he was gonna catch it. He leaps up, and that ball took off like an airplane over the fence. The center fielder was in a state of shock."

The next day, at Gilmore Stadium, Mantle went from first to third with such blinding speed it drew a collective gasp from the crowd of 13,000. After seeing Mantle in Los Angeles, Branch Rickey, the general manager of the Pittsburgh Pirates, wrote Dan Topping, "I hereby agree to pay any price (fill in the blank) for the purchase of Mickey Mantle. *And please be reasonable.*" Topping's facetious reply: Ralph Kiner and a half mill. Frank Lane, the White Sox general manager, fumed at the Yankees' dumb luck: "They got him for nothing. Nothing—do you hear? Why, for a prospect like that I'd bury him in thousand-dollar bills."

New York *Daily News*, March 19: "Mantle very well could be the key to the pennant."

New York Daily Mirror, March 20: "Now the Mickey Mantle madness has spread to the players. Yankees, old and young, openly debate the ability of Mantle, whose speed and power in six exhibition games forced the reporters to all but star him as a daily feature."

"Who is this Mickey Mantle who knocked my Yogi off the front page?" wondered Carmen Berra.

The Bay Area was home to Jerry Coleman, Billy Martin, Jackie Jensen, Frank Crosetti, Charlie Silvera, and Gil McDougald. But it belonged to DiMaggio. At Seals Stadium, where the wind blew in from right field and the fog rolled in nightly off the bay, Mantle hit a 400-foot home run that cleared the right field wall. "You had nineteen or twenty homers in twenty years hit over the right field fence," McDougald said. "Bounced up in the park across the street not far from where I lived," Silvera said.

On Saturday night, March 24, DiMaggio hosted a party at the family restaurant on Fisherman's Wharf for teammates and writers. One impertinent diner inquired whether Joltin' Joe would consider moving to left field to make room for Mantle in center. "There's nobody taking center from me until *I* give it up," DiMaggio replied.

On March 26, the Yankees were back in Los Angeles to play the Trojans at USC—their last West Coast game. USC's new coach, Rod Dedeaux, had played two games for Stengel when he managed the Brooklyn Dodgers. Dedeaux got a bigger bonus in 1935 than Mantle got from the Yankees. Three of Dedeaux's former players—Hank Workman, Jim Brideweser, and Wally Hood—were Yankee rookies. He would continue to send talent to the majors for another thirty years: Tom Seaver, Mark McGwire, Randy Johnson, Fred Lynn, Dave Kingman, Bill Lee, and Ron Fairly, among others.

The Yankees arrived on campus in time for an 11:30 A.M. luncheon at the University Commons, where, according to pitcher Dave Rankin, "sorority girls played bridge all day and hoped for the best." By then snug Bovard Field was SRO. "Additional stands had been erected and the outfield roped off to accommodate any spillage of customers," the *Los Angeles Times* reported—the crowd was later estimated at 3,000. Those unable to find seats could listen to a special broadcast on radio station KWKW.

Cozy, palm-draped Bovard Field (318 feet down the right field line, 307 feet down the line in left) was tucked into a corner of the campus near the Physical Education building, which sat along the third base line. Beyond the right field fence lay a practice field where USC footballers were running spring drills. The impressionable Mantle importuned USC's senior team manager to point out the gridiron stars. Wise guy Phil Rizzuto sent the Trojans' eight-year-old batboy, Dedeaux's son, Justin, to keep Mantle company on the bench—"Hey, rook, I got somebody here your age."

The temperature at game time was only 59 degrees, with a wind from the southeast at 6 miles per hour. Conditions were Southern California dry—it hadn't rained in twenty days. The National Weather Service noted "some haze." Smog had not yet entered the vocabulary. Tom Lovrich, the Trojans' ace, had already beaten the Pittsburgh Pirates and the Hollywood Stars that spring. A sidearming right-hander who threw a heavy, sinking fastball, he would go on to a respectable career in Triple A ball. When Mantle came to bat in the first inning, Lovrich didn't even know that he was a switch-hitter. Dedeaux told him, "When in doubt, keep the ball low."

The count, in Lovrich's memory, went to two balls and two strikes. His intention was to throw the next pitch low and away, trying to entice Mantle to chase something off the plate, which he did. The pitch couldn't have been more than eight inches off the ground. "Our catcher, John Burkhead, kind of dove or fell to his side to block a wild pitch," Lovrich said. "Mantle actually stepped out of the box and reached across the plate. How he reached it, we never knew. You knew the ball was hit. It had *that* sound. A pitcher's unfavorite sound."

Dedeaux stood, mouth agape. "You heard the swish before you heard the sound of the bat as the ball disappeared into the day."

In a 1986 letter to baseball researcher Paul E. Susman, thanking him for his "unrelenting interest" in the matter, the Trojans' center fielder Tom Riach described the play this way: "Riach ran just to the right of the 439 foot sign at the fence. I jumped up on the fence (approximately 8 feet) and watched the ball cross the practice football field and short-hop the fence on the north side of the football field."

Among the football players preparing for the coming season on the adjacent field was Frank Gifford, who was also recruited by Dedeaux as a catcher. He watched the ball bisect the sky. "It went over the fence and into the middle of the football field where we were playing, which was probably another forty-five, fifty yards," he said. "The ball came banging into the huddle. It bounced and hit my foot. I said, 'Who the hell hit that?' Somebody said, 'Some kid named Mickey.' We didn't like baseball players. We thought they were gay. It was like, 'Who are these freaks who would enter our domain?'"

Gifford was the last man on the field to see the ball. "It was never retrieved," Rod Dedeaux said. "We never saw it again."

Mantle was greeted in the dugout with hooting and hollering unseemly for an exhibition game against a collegiate team. "They pounded him," Justin Dedeaux said. "They knew they had seen something."

The batboy regarded Mantle's discarded bludgeon with wonder: "What's in this bat?"

Another towering home run in the sixth landed on the porch of a house beyond the left field fence. In the seventh, a bases-clearing triple flew to the deepest part of center field. In the ninth inning he beat out an infield single on a common ground ball, well played by the shortstop, who, pitcher Dave Cesca said, "would have thrown out any normal human being."

"The greatest show in history," Rod Dedeaux called it later.

Ed Hookstratten, a relief pitcher not then on USC's roster, recalls leading a search party out to the football field, looking for the spot where Mantle's shot fell to the earth. "We walked it off," Hookstratten said. "A shoe is a foot. We got over the fence in the football field and paced it from there. I bet the whole team went out. We were all curious. Six hundred, six-fifty, going toward seven hundred feet, absolutely."

Despite Gifford's eyewitness testimony, reports circulated around campus that the ball had landed in a Methodist church behind the practice football field. Or over it. Or in a dentist's office.

Six decades later, Bovard Field remains sacred ground in Mantleology. Though the field is long gone, grown men equipped with 1951 Sanborn Insurance maps, Google Earth satellite imagery, and lots of free time still try to calculate the precise distance the ball flew when Mickey Mantle announced himself to the world. Estimates range from 551 to 660 feet, depending on whose diagrams, digital readouts, and trajectories you consult. Mantle himself claimed not to remember. Ralph Houk, the Yankees' backup catcher and future manager, said, "I'll say six hundred feet—and I lie a lot."

Years later, Dedeaux told me he doubted that any ball could have traveled 600 feet—science be damned—given the placement of the diamond among the buildings and athletic fields on campus. But to the day he died, Dedeaux swore he saw Mantle hit two 500-foot home runs on March 26, 1951, one left-handed, one right-handed.

In the telling and retelling of the events of that day, memory calcified into fact and a myth was born. In the fine print of history, and the vaults of university film, where fact resides, a different version of Mantle's second home run emerges. According to the box score, Dedeaux used only three pitchers that day; Cesca, the lefty, pitched only the ninth inning, which means that Mantle's sixth-inning home run had to be an opposite field shot hit left-handed. Ben Epstein's game story in the *Mirror* stated: "Mickey obtained all his extra-base shots batting left-handed." The Los Angeles *Herald-Examiner* concurred. None of the profuse dispatches filed by New York writers mentioned a home run from each side of the plate, and none of the Trojans recalled it that way, either.

Proof positive came from the USC School of Cinematic Arts in the form of a two-second film clip in the 1951 *Trojan Review*. There's number six, Mickey Mantle, batting left-handed in the top of the sixth. His bat is a blur as he steps into the pitch. Front toe turned inward, back foot lifted off the ground, he follows the flight of the ball over the left field fence. Everyone is looking that way, the third base coach, ump, runner, and the news photog squatting along the base line; and all the fans clustered behind the chain-link fence in the shade of a eucalyptus tree near the first base dugout. The narrator is solemnly impressed: "Yankee flash, Mickey Mantle, bangs out his second home run to score a pair of teammates."

When the game ended, Mantle was hitting .432; DiMaggio was batting under .200, having left the field after two at-bats, a normal spring training accommodation for an aging, aching star. Fans besieged the press box, wanting to know the answer to Carmen Berra's question: Who *is* this Mickey Mantle? Coeds swarmed the team bus. By the time the Yankees got back to Phoenix the next morning, autograph hounds were offering two of anybody else's signature for one of Mantle's.

The press of expectations was upon the nineteen-year-old Mickey Mantle. Six days earlier, he had been mere background in a Hollywood flack's snapshot, happy just to be in the picture. Now he had moved to stage center, where he would remain for the rest of his life. "It becomes doubtful that Casey Stengel will dare to let him out of his sight," Arch Murray wrote in the *New York Post.*

The future was manifest in the March 26 box score: Mantle, 5 AB, 4 H, 2 HR, 1 3B, 1 1B, 7 RBI. More than a decade would pass before he drove in that many runs again. Houk and Berra looked at each other and said, "My God, whadda we got here?"

Headline writers invoked classical mythology in an effort to convey the epic proportions of the day. "One for the Mantle," the *Los Angeles Times* declared. "Yanks Dismantle Troy."

A half century later, Justin Dedeaux described the wonder of it all more simply: "This was the day the whole world opened up."

2

October 5, 1951

WHEN FATES CONVERGE

A letter from Mantle's father, Mutt, was waiting for him in Phoenix. His local draft board wanted to reexamine him, and wanted to do so within the next ten days. When the Yankees headed east after another week of exhibition games in Arizona, Mantle wasn't with them. He detoured to Miami, Oklahoma, and then to Tulsa to have his draft status reviewed. Again, he was declared medically unfit to serve; Stengel lobbied hard to put him in a Yankee uniform instead Mantle expected and wanted to be sent to the class AA team in Beaumont, Texas.

His fate was still undecided when he boarded a night flight out of Kansas City for New York on Friday, April

13. The Army's decision generated suspicion distilled with typical sports page asperity. "So Mantle has osteomyelitis. What's the big deal? He doesn't have to *kick* anybody in Korea."

Welcome to the big city, kid.

Landing at 7:30 A.M., Saturday, he headed straight for Ebbets Field, where the Yankees were scheduled to play the Dodgers in the second game of the annual interborough exhibition series. Mantle prevailed upon Stengel to put him in the starting lineup, but not before the manager showed him how to play the tough right field corner that had been his turf when he played for the Dodgers at the dawn of the twentieth century. "First time the kid ever saw concrete," Stengel told his writers; they were always *his* writers.

On Sunday, the kid went 4 for 4, with a home run over the 38-foot scoreboard in right field. General manager George Weiss put him on the major league roster. The next DiMaggio had arrived before the original departed. There was tension in that for both of them. "How'd you like to replace George Washington?" said teammate Jerry Coleman.

Stengel didn't make it easier. "Stengel loved Mantle, and didn't like DiMaggio," said future Hall of Fame manager Whitey Herzog, "So Joe held that against Mickey."

That was apparent when the Yankees opened the season against the Red Sox at the Stadium on April 17. DiMaggio pointed him out to columnist Jimmy Cannon: "This is the next great ballplayer." But when an enterprising photographer posed DiMaggio and Ted Williams on either side of Mantle, Joe D. declined the opportunity to introduce them. The Splendid Splinter handled the niceties himself.

There was a time, back in high school, when Mantle idled through fourth-period study hall with a two-page magazine spread devoted to Joltin' Joe. He bragged to his classmate Joe Barker, "I'm going to take his place in center field at Yankee Stadium."

But Stan "The Man" was Mantle's boyhood idol. He mentioned Musial the day he signed with the Yankees. Weiss immediately corrected that misconception, telling his young star the story line he was expected to follow—Joe D. was his hero. Given DiMaggio's intimidating frostiness, it was little wonder Mantle never asked for help. Gil McDougald said, "When Joe was in the dugout, nobody would say a word. Joe would sit down. He'd like to smoke, so he'd be sittin' down close to the runway. Everybody else would be sittin' down toward the water faucet at the end of the dugout."

DiMaggio gave three years to the war effort, helping to make the world safe by playing baseball for Uncle

Sam. He returned to the Yankees in 1946 with an ulcer, an unwanted divorce, and bone spurs that would plague him for the last six years of his career. By 1951, it was clear to everyone, including Joe's brother, Dom, the Red Sox center fielder, that this season was likely his last. "He staggered through it," Dom said.

Two photographs taken in the Yankees' locker room documented the imminent succession. Pete Sheehy had given Mantle the locker next to DiMaggio's. In the first photo, taken five days before opening day, Sheehy is filling the empty cubicles in preparation for the new season, hanging Mantle's crisp number 6 beside DiMaggio's venerable number 5. A week later, *Life* magazine photographed them in street clothes in posed postgame chat, Mantle sitting on his stool in pants too short to cover his white socks, with DiMaggio towering above him. Joe looks immaculate, the way he always did—white shirt, braces, Countess Mara tie—the whole upwardly mobile bit. "Looked like a senator," Billy Martin said, and Mantle agreed: "Like you needed an invitation to approach him."

Red Smith devoted his opening-day column in the *Herald Tribune* to the rookie who played his first major league game wearing "impoverished baseball spikes" and "soles flapping like a radio announcer's jaw." The Commerce Comet hadn't gotten any sleep the night before. He gave the cabbie who took him to the Stadium the next morning a nickel on a $3.00 fare. The son of the undertipped hack said his father never forgot the slight.

Another Yankee rookie, public address announcer Bob Sheppard, filled the Stadium for the first time with a stentorian voice that sounded older than his years. *In right field, number six, Mickey Mantle. Number six.* A speech teacher who became an adjunct professor at St. John's University, Sheppard immediately appreciated the qualities and construction of Mantle's name: soft, flowing sounds that conveyed his grace alternating with hard, tough consonants that suggested his power—a staccato rhythm that implied speed. It was almost as if his name embodied the traits that defined him. "It said, 'He's an American guy from the Midwest,'" Sheppard said. "If it was Michael, it wouldn't be as good."

Mantle went 1 for 4 in his first big league game with a single and an RBI. But he was monosyllabic with reporters who wanted to know his every thought. How could he explain himself to people who believed Commerce, Oklahoma, was a made-up name?

"I was shy, scared," Mantle told me decades later. Too scared to get off the bus and go to the reception at Whitey Ford's wedding, too scared to face the scrum of sportswriters who dogged his locker room. "They called me aloof. I thought that meant horny."

Hank Bauer and his roommate, Johnny Hopp, took him in and took him on as a project. They worked to purge the rube of his wardrobe and introduced him to the finer things in New York: corned beef and other pleasures of the flesh. They had an apartment above the Stage Deli on Seventh Avenue. Hymie Asnas provided food on the house; Bauer took care of him when World Series time came around. "He weighed 170 pounds," Bauer said. "By the end of the season he weighed 190."

In those days, veterans taught rookies how to look and how to act major league; they enforced the code of clubhouse etiquette and on-field behavior. They took care of one another. Bauer's generous example—as well as DiMaggio's cold shoulder—would inform the way Mantle treated rookies for the rest of his career. "I knew he didn't have the proper attire," Bauer said. "He came in Hush Puppy shoes, white sweat socks, rolled-up pants a little bit short, and a big white tie with a peacock and a tweed sports coat. Next day, I said, 'Come with me, and I'll buy you a couple of sports coats.' Took him to Eisenberg & Eisenberg."

Bauer made him presentable, but he couldn't make him savvy. Mantle spent $30 apiece for "cashmere" sweaters that turned out to be made of a flammable synthetic. He would later purchase $2,700 worth of stock in a nonexistent insurance company. He was an easy mark. "Very naive about people," Merlyn Mantle told me. "He would trust shady characters."

Older, wiser teammates tried to help. Bobby Brown took him aside in the outfield one day: "You've got the world by the tail. Take good care of yourself, work hard, stick with the straight and narrow. You'll make a lot of money."

Mutt Mantle knew better than anyone that his son was easily led—he made him that way. He approached Red Patterson, the Yankees' publicity man, for help. In a rough draft from an unpublished memoir provided by Patterson's son, Bruce, he wrote: "During a rainout in St. Louis, he came to my room and asked if he could discuss a matter with me which had him disturbed. 'I would like you to take good care of Mickey when he goes all the way up to the Yankees. He is going to get a lot of attention and there will be people making him offers but I wish you would handle him. He can use all of your advice.'

"In effect, he was asking me to act as a sort of agent for Mickey. I explained that as a Public Relations man for the club I would give him as much help as possible but I could not be his agent.

"True to his Dad's concern Mickey was beset by

agents and somehow got tangled up with two at one time. It took legal action by the club to straighten out the mess. I can recall one conversation Tommy Henrich and I had with Mickey in which we asked him if he had obtained a lawyer to represent him in his transactions. 'No, I didn't have to. They had a lawyer up in their room.' "

An opportunistic agent named Alan Savitt waylaid Mantle in the lobby of the Concourse Plaza Hotel his first week in New York, promising $50,000 a year in endorsements, to be split fifty-fifty. Short of cash, Savitt soon sold a 25 percent interest in Mantle futures to a showgirl named Holly Brooke, who introduced the rookie to scotch and the art of picking up a check.

Carl Lombardi, Mantle's minor league teammate and friend, tried to warn him off the deal—and the girl who came with it. Lombardi recalled trying to reason with his obdurate friend during an evening at a Jersey roadhouse. "I said, 'Mick, before you make a commitment, do yourself a favor, go to the front office and talk to them about it.' Like I said, he was stubborn. After he signed it, he said to me, 'Boy, I'm sorry I didn't listen to you.' "

On the field he did well enough to merit a major profile in the June 2 issue of *Collier's* magazine. But as the pressures and distractions of New York mounted, his production decreased and his temper flared. He kicked water coolers and lost his cool—often enough that DiMaggio confided to *New York Times* beat writer Louis Effrat, "He's a rock head."

Pitchers began to figure him out and exploit the holes in his swing, particularly when he batted left-handed. Satchel Paige, then pitching for the St. Louis Browns, made Mantle look so bad he laughed out loud at him. By mid-July, his batting average had dropped to .260. "I was striking out about four out of every five times up," he told me.

Not true. By then Mantle had bought into the revisionist history that exaggerated his futility. He struck out only .9 times per five at-bats during the first two months of the season, according to Dave Smith of Retrosheet. Between May 30 and July 14, he struck out 25 times in 97 at-bats, 1.56 strikeouts per five at-bats. (For the season, he averaged 1.09, just a little above his lifetime average of 1.06 strikeouts per five at-bats.) The fawning newspaper hacks turned into jackals: "The next DiMaggio struck out on three pitches."

On July 14, in Cleveland, Mantle broke up Bob Feller's attempt for his second no-hitter in two weeks with a sixth-inning double. Asked what he had thrown the rookie, Feller said tartly, "A baseball, I presume."

After the game, the Yankees announced the purchase of pitcher Art Schallock from Brooklyn, meaning that someone had to be dropped from the roster. Four days later, Mantle, the team leader in RBI, cried when Stengel told him that he was being sent down to the Triple A Kansas City Blues. For years afterward, Mantle would recount the mutually tearful conversation. "This is gonna hurt me more than you," the manager insisted.

Joe Gallagher, a young gofer doing stats for Mel Allen, saw Mantle in the lobby of the Cadillac Hotel as he was getting ready to leave. He was still crying. "Kind of like it was the end of his career," Gallagher said. "All I could say to him was 'You'll be back.'"

He played his first game for the Blues in Milwaukee in a fog so thick he said he might wear a catcher's mask in center field. He dragged a bunt down the first base line for his only hit, showcasing his speed. Manager George Selkirk minced no words letting him know he was there to regain his swing, which promptly went south. He didn't hit a home run for twelve days. Johnny Blanchard spent four days with him before being sent to Double A. "He said, 'Make room for me, Blanch, I'm coming down,'" Blanchard recalled.

When his minor league buddy Keith Speck came to visit, Mantle told him, "I ain't never getting back up there."

Folks back home and some in the Yankees organization thought he had already acquired too many big-league habits. Hank Bauer later told Tony Kubek that when Mantle moved into the apartment above the Stage Deli, a bottle of Jack Daniel's arrived with him. "A lot of people thought that he'd just got to drinkin' and carousin' so much with Billy [Martin] and some of these other guys, he got completely out of whack," said Frank Wood, whose father played American Legion ball against Mantle in Picher, Oklahoma.

One night, Mantle treated his boyhood friend Bill Mosely and his wife to a night on the town in Kansas City. Mosely was in the Army, stationed at Fort Scott, fifty miles away. The evening was a revelation—not just because of the style to which Mantle had become accustomed but because of the bravado with which he indulged and the conviction that it would not interfere with his ability to play the next day.

"We're drinkin' there and eatin' and everything, things I never heard of before. Caviar. Is that right? And a guy by name of Harold Youngman, he's Mickey's sugar daddy, so he's footin' the bill. I'm thinkin' all we had there comes to around a thousand dollars. In my time that was a lot of money. Still a lot of money. But anyway, we had a ball and I got to talkin' to Mickey and come to find out he's got a doubleheader to play the next day! I don't know how in the hell he did it, but he did."

In fact, he wasn't doing much. Welcomed to town as a "tonic" by the *Kansas City Star* on July 16, he became a "dejected and harassed young man" by August 4. After that bunt single in his first at-bat, he told me, "the next twenty-two times up I didn't even hit the ball at all." His statistical recall was slightly off—he had had three hits in his first eighteen at-bats for the Blues—but the memory of his futility was indelible. "I was pretty scared. Probably I was more disappointed than scared that I wasn't doing better because of my dad, y' know. He lived and died for me to be a baseball player, and it looked like I wasn't going to do it."

He called his father and said he wanted to come home. "I was down, really down."

"You wait right there," Mutt said. "I'll be up there."

The day before he left for Kansas City, Mutt called Ed VonMoss at the Blue Goose Mine to say he wouldn't be at work the next morning. "I gotta go get that lazy kid of mine," Mutt told him.

According to VonMoss's son Jerry, his father reassured Mutt that there was still a place for Mickey in Commerce: "I'll have a job for him."

It is unclear exactly when Mutt delivered his ultimatum. The June 1997 edition of the Ottawa County

Emporium, a historical newsletter featuring reprints of newspaper stories from Miami, Oklahoma, included this report: "Mr. and Mrs. E. C. Mantle and their son, Larry, as well as their daughter, Barbara, and Miss Merlyn Johnson, visited Mickey after his home debut for the Blues on Sunday, July 22nd."

In his many tellings of the woodshedding session, Mantle edited everyone else out of the meeting at the Aladdin Hotel. But Mutt took the whole family along. Merlyn remembered the long, quiet drive to Kansas City. So much was at stake. When they got to the hotel, she recalled, "Everybody was in the room. Then we went outside, but you could hear. I heard him say, 'If that's all the man you are, then get your clothes and let's go home.' Mutt did not yell. He spoke with authority. Mick was crying, of course. He was embarrassed because he wasn't cutting the pie."

Mantle had expected solace and support, not paternal fury. "I thought he was coming up to give me a pep talk," Mantle told me. "He comes up, walks in the hotel room, and starts throwing all my shit in a bag. I said, 'What's the matter?' He said, 'I thought I raised a man. You ain't nothing but a goddamn coward.'

"I said, 'Wait a minute.'

"He said, 'Ah, bullshit, you come and work with me in the mines. I didn't raise a man. I raised a baby.'

"He was crying, and I was crying. I said, 'Well, let me try again.'

"He said, 'Bullshit. Come on. I came on all the way up here. You're going back with me. You ain't got a gut in your body.'

"He made me feel I was about that tall. Finally he says, 'I'm gone. If you can't play, get a bus and come home.'"

The message was delivered with vehemence that the younger Mantle couldn't possibly understand. True, he had noticed that his father's khakis were hanging loose on his frame. True, he had always worried about the old man's smoking. But most nineteen-year-olds don't spend their days contemplating parental mortality. Mantle was too young, too immature, too caught up in the thrall of his new life to see the signs of fatal illness and desperation in his father's gaunt, angry face.

There was no opportunity for Merlyn to be alone with him, much less comfort him. "I wouldn't have," she told me. "Nobody did."

Truth to tell, she wouldn't have been disappointed if he had come on home.

After they all went out to dinner, the group headed back to Commerce at Mutt's cautious 35 miles per hour, leaving Mantle to decide whether to pack for a three-week road trip or buy a ticket for Commerce. "I thought about it a long time that night," Mantle told me.

He opted for the future Mutt wanted for him. And he began to hit—four hits in one game in Milwaukee, a home run in Louisville, two more the next day in Indianapolis (one left-handed, one right-handed), then another one the following day. Two days after that, in Toledo, he hit for the super cycle, banging a single, double, triple, and two home runs. Within the month, he had 11 home runs, 52 runs scored, and 60 RBI.

By late August he was on his way back to the Bronx, but not without another Army-mandated detour—a third reconsideration of his draft status prompted by angry letters to the White House and front-office concern about negative PR. When he finally arrived in the Yankee clubhouse, he found lucky number 7 hanging in his cubicle. Pete Sheehy had given away number 6 in his absence.

The Yankees clinched the pennant in Philadelphia on September 28, the same day the Giants tied the Dodgers for first place in the National League. Mutt, his brother Emmett, and his pals Turk Miller and Trucky Compton drove east for the World Series. The kid showed them the town. In *The Mick*, Mantle described his father's parochial confusion upon seeing the statue of Atlas in front of Rockefeller Center: "Shoot, the Statue of Liberty's smaller than I thought."

The Oklahoma boys didn't know how much money it cost to go to the movies; they didn't know where to get off the subway for the ballpark (and ended up walking three miles). They sure didn't know how to hold their big-city liquor; riding the train, pressed between New York City straphangers, Compton threw up in the hat of an unlucky passenger.

But Mutt knew trouble when he saw it.

Her name was Holly Brooke. Mantle introduced her to his father as his "very good friend." He recounted the conversation in *The Mick*:

Maybe she winked at me. I don't know. But Dad knew something was up—and he didn't like it a bit. Later, he took me aside.

"Mickey, you do the right thing and marry your own kind."

"It's not what you think, Dad."

"Maybe not, but Merlyn is a sweet gal and you know she loves you."

"Yeah, I know."

"The point is, she's good. Just what you need to keep your head straight."

"I know."

"Well, then, after the Series you better get on home and marry her."

I half turned from him, nodding silently. There was nothing more to discuss.

"She was older," Merlyn told me. "She had a kid almost as old as Mick. She more or less got in with this attorney. Mutt saw the situation. He knew it was trouble. Mick could be very easily swayed."

While Brooke trysted with him in major and minor league cities, Merlyn was back in Oklahoma, wearing his engagement ring and receiving love letters penned on Yankee letterhead. In one letter, written early on a sleepless road-trip morning, he pleaded with her to write to him the way the other wives did. Another letter, written in the clubhouse, began:

Honey I sure will be glad to see you—I'm going to make up for all the loving I have missed from you when I get home—The only thing is I will just want to stay there and hate it all the more when I have to leave you again. We haven't been together very much since we have been engaged have we? When we get married we'll make up for it.

It was signed, "All of my love, Mickey M—"

"He wrote like he loved me," Merlyn told me six decades later.

Mantle's dalliance with Brooke set a precedent for a double life that persisted long after the relationship ended and would continue throughout his married life. Nor was Brooke that summer's only leggy temptation. Among them was a Copa girl named Peaches, a close personal friend of the mob boss Joe Bonanno. Mantle was too eager and too innocent to understand his dangerous indiscretion.

"He was gonna have Mickey rubbed out," said Mike Klepfer, a friend in later life whose longshoreman father heard the waterfront scuttlebutt about a contract on the amorous ballplayer. Decades after the fact, Klepfer's father told Mantle, "I remember when they were going to kill you." "Mickey looked like he'd seen a ghost," Mike said.

On October 3, Yogi Berra was making his way home from the Polo Grounds, trying to beat the traffic on the clogged streets of upper Manhattan, when "whatchamacallit" came to the plate in the bottom of the ninth inning of the deciding play-off game between the Brooklyn Dodgers and the New York Giants. With the Giants trailing 4–1, Berra thought the outcome was a foregone conclusion. Like everyone else in New York, Berra was sure the Yankees would face the Dodgers in game 1 of the World Series the next day.

Bobby Brown also missed Bobby Thomson's historic at-bat, which was seen across the country on the first coast-to-coast baseball telecast. Brown was waiting for his father behind the wheel of his new Chevrolet outside the press gate at the Polo Grounds. He had given his dad his ticket to the game. They learned the outcome at a red light on Amsterdam Avenue from the driver of a car in the next lane—Brown couldn't afford a radio in his new sedan.

Mickey and Mutt were still in the ballpark when Thomson stepped to the plate. Like most everyone else in the Polo Grounds, the Yankees were rooting for the Giants. "Bigger ballpark, bigger World Series money," Gil McDougald said.

They saw Ralph Branca lumber to the mound, summoned by Dodger manager Charlie Dressen to relieve the exhausted Don Newcombe. Probably they didn't notice, as Dodger center fielder Duke Snider did, the ominous change in Dressen's demeanor. "Usually Dressen liked to bring the relief pitcher up to date, give him all sorts of instruction," Snider said. This time, Dressen was mum. "I said, 'Charlie's worried,'" Snider recalled. "So I became worried."

They saw Willie Mays, New York's other rookie center fielder, kneeling in the on-deck circle. "Willie, he was scared to death," Snider said.

Mays was still kneeling in the on-deck circle when Thomson rounded the bases at 3:58 P.M. Snider had a better view than anyone else of the ball that broke Brooklyn's heart, a line drive that sent Andy Pafko to the left field wall. "I ran over," Snider said. "It was a low line drive. I was there to receive the carom. I thought I was going to hold him to a double."

Thomson's home run would soon be known as the Shot Heard Round the World and the Miracle at Coogan's Bluff. When Snider saw it dip over the fence, he said, "I took a right turn and went into the club-house in center field and didn't break stride."

Snider left the Polo Grounds with his parents, count-ing effigies of the luckless Ralph Branca hanging from Brooklyn light poles. The Yankees headed to a pre-Series bash at the Press Club; Mantle went to the hotel room of Tom Greenwade, the scout who had landed him there. Greenwade's wife, Florence, answered the door. "He said, 'Would you mind if I came into your room and just stayed here awhile 'til he gets back?'" their son Bunch recalled.

He didn't offer an explanation for his appearance at her door or for his glum mood. Florence Greenwade assumed homesickness was the cause. It was a Mantle story she often told, and her daughter Angie remem-bers it well: "So he spent his evening sitting with my mother. She said he was so miserably unhappy; wondered what he'd gotten himself into. I think he was scared and nervous. He knew that so much was expected of him."

Mrs. Greenwade didn't ask any questions. She didn't ask anything of him at all. Angie said, "He basically sat there quietly. I'm sure he knew nobody would be knockin' at the door looking for him. He didn't have to do anything but *be*."

On October 4, 1951, parking meters were installed in downtown Brooklyn, adding insult to the injury of the heartbroken borough. *An American in Paris* opened at Radio City Music Hall in Manhattan. And in the Bronx, Mantle played in his first World Series game.

Baseball was on the cusp of radical change. Babe Ruth was three years dead; DiMaggio was taking his curtain call. His successor, Mickey Mantle, the first telegenic star of the new broadcast age, was installed in right field. Mantle's charismatic foil, Willie Mays, was playing center field for baseball's first all-black outfield.

Unlike Mantle, Mays arrived in New York without tabloid fanfare. Unlike Mantle, Mays pleaded to be sent to the minors when he struggled during his first days with the Giants. But they had more in common than it appeared, more than a shared future on similar real estate.

Born the same year to fathers who rolled baseballs across the floor to baby boys who could not yet walk, they were in their major league infancy. What the 65,000 paying customers at Yankee Stadium saw that afternoon were two works in progress whose unlimited potential would fuel unending debate. They would improve each other and everyone who played with them and against them.

Arriving at the Stadium that morning, Mantle was startled to see his name in the starting lineup—batting leadoff. The Giants had had no time to catch their breath or sit still for the usual briefings. "We just went out and played," outfielder Monte Irvin said. "We didn't know anything about anybody."

Their scouts had alerted them to Mantle's uncompromised speed; nonetheless, when they saw him on the base paths in game 1, they were stunned by the fact of it. "Fastest white guy we've ever seen to first base," shortstop Alvin Dark said.

Adrenaline carried the Giants to a 5–1 victory in the first of the sixty-five World Series games Mantle would play. The second of them would be the most pivotal game of his career.

Fifth inning, game 2. DiMaggio is in center; Mantle

is in right. Mays steps to the plate. The collision of fates is almost operatic, triangulating the future of the game. On the mound, Eddie Lopat goes into his windup. Mays gets wood on the ball but not a lot. The result is a tepid opposite-field fly ball, not deep, not well hit, not difficult to catch except that it's what ballplayers call a tweener, splitting the difference between DiMaggio in center and Mantle in right.

Here's DiMaggio, shaded over toward left center, asserting his proud prerogative—*This is my turf! Mine!* And here's Mantle, chasing the future across America's most famous lawn. Isn't that what Stengel had told him to do? *The dago's heel is hurtin'. Go for everything.*

He was new at this outfield play. Hell, everything was new for him. Maybe he didn't understand the etiquette—if the center fielder can get there, it's *his* ball. Especially if that center fielder is DiMaggio. Hank Bauer learned that the first time he made the mistake of taking a ball hit between them. Jogging back to the dugout, DiMaggio gave him a lethal stare. "I said, 'Joe, did I do something wrong?'

"He said, 'No, but you're the first sonofabitch who ever invaded my territory.' Center fielders don't call for nothin'. When I heard the grunt, I got the hell out of the road."

The past and the future converged on a routine

fly ball in Mantle's ninety-eighth major league game. Imagine Mutt watching. He sees the geometry of disaster. The ball is dropping. Joe's coming. Mickey's charging. "I was running as hard as I could," Mantle told me. "At that time, I could outrun anybody. I ran over to catch it. Just as I was getting ready to put my glove up, I heard him say, 'I got it.' Well, shit, you don't want to run into Joe DiMaggio in center field in Yankee Stadium. I slammed on my brakes like that."

Embedded in the outfield sod that sloped downhill at perhaps a 10-degree angle from the right field fence was a six-inch round depression. "Actually it was a sewer drain, maybe four by four inches," said former batboy Frank Prudenti. "There was, like, a piece of metal in the center. You could pull it up, and you could push it down. Like a cork on a bottle."

The cover was made of thin plywood with a rubber coating, Prudenti says, maybe three-fourths of an inch thick. "It was wedged in there, belowground. You had to hit it with your heel, wedge it down real tight. If it wasn't, somebody could definitely trip on it."

Generations of Yankee outfielders and their opponents were well acquainted with this ancient piece of Stadium infrastructure. "Been in it, been on it, been around it, and fell on it," said Bobby Murcer, another of Greenwade's Oklahoma finds.

Bauer used it as his anchor. Berra was taught to play off it. "Never stand on the drain," Tommy Henrich told him.

Gil McDougald, the second baseman, had retreated into right field, following the flight of the ball. "You could see the whole thing coming in your mind. I knew that it looked like trouble. Mickey, you gotta understand, was playin' pretty deep because he had to come down that hill, or incline, I guess you'd call it, out there. So it wasn't what you'd call a short fly ball. It was like a humpback job. It was Mickey's ball, but DiMag, being the icon he was, and Mick being a rookie, he gave way instead of really taking charge."

From the visiting dugout Al Dark also tracked the flight of the ball. "All of sudden, Mickey throws on the brakes and his legs went out from under him and he slipped as you would slip on an ice thing. Then he couldn't get up and it didn't look like he wanted to get up."

Mantle was motionless. Yankee Stadium was still.

A sequence of news photographs documented the progression of the disaster in right center field.

Click.

There's DiMaggio camped under the ball, his glove open at his side, looking up into the sun. There's Mantle splayed on the grass in front of the 407-foot sign. The shadows of the championship banners ringing the

Stadium point toward his fallen form. His right leg is folded beneath him, the injured knee bent backward at an ugly angle. His left leg extends upward toward the sky. To his left, there is a faint indentation in the grass.

Click.

Now DiMaggio cradles the ball, his glove pressed against his stomach, and turns toward Mantle. He lowers his uninjured leg like a drawbridge, shifting his full weight onto his side. He buries his head in his arms on the turf. Behind him, the polite grandstand crowd, some in fedoras, some in coats and ties, begins to rise, Windsor-knotted necks craning to see.

Click.

DiMaggio kneels beside him, whispering words of reassurance, a consoling hand resting on his shoulder. *They're coming with the stretcher, kid.* Mantle said it was their first conversation of the year.

Click.

Now his teammates come running from the bullpen, their spikes churning up an urgent trail in the warning track dirt. The backup catchers, Charlie Silvera and Ralph Houk, are first to reach him. "The only time Houk and I got our picture in the paper," Silvera said.

Mantle lies curled in an almost fetal position. "He was going full speed," Houk said. "He was about to get the ball."

"Joe more or less ran him off it," Silvera said.

They told him not to move, as if he could. "He was kinda moaning," relief pitcher Bob Kuzava said. "The trainers, they wanted him to stay still because they didn't know what happened. They tried to immobilize him so he isn't gonna injure himself anymore."

Click.

It looked like he'd been shot. He wasn't sure he hadn't been. "I was running so fast, my knee just went right out the front of my leg," he told me, trying and failing to reproduce the sound of rupturing flesh and broken promise.

It was so sudden, so painful, so shocking that he soiled himself. "Shit my pants," he told me, and dared me to write it.

"Must be like giving birth," he told his friend Mike Klepfer years later.

Newsweek reported, some spectators thought he'd had a heart attack. "He lay like he's dead," Jerry Coleman said. "Seemed like he was there twenty minutes before they finally got around to getting him out of there."

Five Yankees carried him off on a stretcher, three on one side, two on the other, like pallbearers. Mutt was waiting in the dugout.

A pool photographer was allowed in the trainer's

room, an exception to the "off-limits" norm. Still in uniform, his sanitary socks dirty with exertion, a towel demurely draped across his waist, he props himself up on his elbow, looking in blank disbelief at what used to be his right leg; surely it no longer felt like his own.

Sidney Gaynor, the team physician, stands impassively at his side, checking the ice pack fixed to his leg. Gaynor initially diagnosed the injury as a torn muscle on the inside of his knee. A day later, he called it a torn ligament. Over time, it would be variously described (by Mantle and a legion of reporters) as torn cartilage, torn ligaments, torn tendons, and a combination of all of the above.

Later, in the locker room, Mutt squatted by his son's side as he struggled to put on his argyle socks and his wing-tip shoes, glancing up at an inquiring photographer from beneath an errant forelock. Mantle never looked that young again.

The Yankees sent him back to his father's hotel, his leg splinted and tightly wrapped. "Come all the way up here, and you bung your knee," chided one of Mutt's Okie pals.

"Thought you fainted," Mutt said. Mantle wasn't sure he hadn't.

"Naah," he replied with youthful bravado. "I felt like fainting my first game in the Yankee Stadium."

"Yanks' Joy over Triumph Is Tempered by Loss of Mantle for Remaining Games," the *New York Times* declared the next morning. It was the first time he appeared on the front page of the paper of record.

On October 5, 1951, a game was won and a fate was sealed. The drain in right center field became a baseball landmark. On opening day of the 1952 season, Mantle would make a pilgrimage to the spot "where he had come to grief," as Arthur Daley put it in the *Times*. "I couldn't find it," he told the columnist, grinning and shuddering at a memory. Daley wrote, "He still could not fully comprehend or remember."

More than fate was at play. When Howard Berk, the Yankees' vice president for administration from 1967 to 1973, reviewed plans for the Stadium's renovation in the early 1970s, architects told him that a groundskeeper had forgotten to put the rubber cover on the right field drain. "Not the first time a groundskeeper forgot to put something on the field that endangered a player," Berk said.

DiMaggio chose Willie Mays Night at Shea Stadium in 1973 to offer his account: "I said, 'Go ahead, Mickey. You take it.' I called out to him as we converged . . . Luckily, I was close enough to make the catch."

Mantle never blamed DiMaggio publicly. "He had his own opinion, but he never said it," Merlyn told me. "He ruined his career."

The morning after, his knee was so swollen he couldn't walk. Mutt took him to Lenox Hill Hospital for X-rays. "I couldn't put any weight on my leg," Mantle told me. "So I put my arm around his shoulder. Now, this guy's as big as me, maybe a little bigger. When I jumped out, I put all my weight on him and he just crumpled over on the sidewalk. His whole back was eaten up. I didn't know it. But my mom told me later he hadn't slept in a bed because he couldn't lie down for, like, six months. And no one had ever told me about it. They never did call me.

"So when he crumpled over, we went to the hospital and we watched the rest of the Series together. That's when they told me when I got home I'd better take him and have him looked at because he's sicker than I think he is."

"Hodgkin's disease," was the diagnosis.

Mutt's illness was not disclosed. His distress was said to be profound. The *Times* reported: "Mantle's father became so upset when his son slipped that he too required hospitalization."

They watched the last four games of the Series on a small black-and-white TV with rabbit ears. Mutt seized the opportunity to point out things Mickey might have done better. Pain became a teachable moment. The Yankees won their eighteenth world championship. Mutt was sent home to die.

Mantle's knee was slow to heal. The front office decided to send him to Johns Hopkins Hospital in Baltimore for a second opinion. The verdict came on October 22: no surgery needed. Go home and rest.

In less than twenty-four hours, all the supporting structures of his life imploded. His father had only months to live; his potential was irrevocably circumscribed; his knee and his heart were never the same. A wire service reporter filed a prescient deadline dispatch: "His mind is already shackled with the thought that the knee might pop out whenever subjected to strain."

That October afternoon was the last time Mantle set foot on a baseball field without pain. He would play the next seventeen years struggling to be as good as he could be, knowing he would never be as good as he might have become.

3

October 23, 1951

UNDERMINED

1.

That the earth would give way beneath his feet was a grim irony for Mickey Mantle. Growing up in Commerce, Oklahoma, in the dead center of the Tri-State Mining District, fatalism was an inheritance. It percolated up from the tainted, unstable earth. That forgotten corner where Missouri, Oklahoma, and Kansas meet was hardly the Oklahoma of Rodgers and Hammerstein. A century of mining lead and zinc from the ancient bedrock had left the ground as hollowed out as the faces of the men who worked it.

The lead went into munitions used to fight the Hun

in World Wars I and II, into lead-based paints and pigments, into sinkers for fishing rods and weights for balancing tires. It was also crucial to the manufacture of lead-acid storage batteries. Zinc was needed to galvanize steel, to cure rubber, and to line sinks and washstands. It was an essential ingredient in pharmaceuticals and cosmetics.

Two blocks west of the front door of Mantle's boyhood home was a hulking, ashen heap of mineral detritus disgorged from the abandoned Turkey Fat Mine, where the first shaft was sunk in Commerce. Three blocks from the house he purchased for his parents with his first World Series check was a crater twenty to thirty feet deep, an insidious reminder of how easily life could give way. That house at 317 South River Street, with the family's first telephone, was where he went to recuperate when doctors at Johns Hopkins told him he could go home. By the time he arrived, Mutt had gone back to work as the ground boss at the Blue Goose Mine.

Even before Mutt got sick, before Mantle ripped up his right knee trying not to run into Joe "Fuckin'" DiMaggio, he had reason to doubt his own longevity. Everything about the world that produced him undermined confidence in long life. When a reporter came to call, Mantle told him that three hundred feet below his chair, men like Mutt were trying to claw out a living from exhausted mines. It was only a slight exaggeration.

Mining had created a tenth circle of Hell, turning a verdant swath of the Great Plains into alien terrain, flat except for the mammoth piles of mineral waste known as chat. Locals call this range of bleached man-made dunes the "Chatanagey" Mountains. Cruel Billy Martin called Merlyn "Chat Pile Annie."

The highest of them, at the Eagle-Picher Central Mill, a mile and a half northeast of the Mantles' home, was a twenty-story behemoth built from over 13 million tons of chat. Long after the ore played out, the metastatic landscape remained disfigured by 5,000 acres of tailing piles and sludge ponds so toxically opaque that no shadows were cast upon them; 1,200 open or collapsing mine shafts lurked beneath the overgrown, contaminated grass; 40,000 drill holes and hundreds of water wells reached deep into the Roubidoux Aquifer.

The worst desecration was centered in Ottawa County, Oklahoma, where 2,500 acres were left undermined, 50 of them punctured by cave-ins. Also left behind when the ore played out were 300 miles of tunnels that wound their way through parts of three states and underground caverns, one as big as the Houston Astrodome. Some folks swear you can walk the twenty-eight miles from Commerce, Oklahoma, to Joplin, Missouri, without ever seeing the sky.

Only 6 percent of what miners like Mutt hauled out of the ground was ore-grade—thus the aptly named Discard Mine on the Kansas-Oklahoma border. Waste laced with cadmium, magnesium, copper, and gallium was strewn over 41 square miles.

When the Mantles arrived in the area in 1935, a time of unprecedented and violent labor unrest in the Tri-State region, these towering buttes were a source of pride for a workforce known for its fierce independence and anti-union ways. Adults regarded them as protectors against the tornados that spiraled across the land. Children rode their bikes up the dusty slopes in summer and slid down them on rusting car hoods in winter. In the shadow of these bleak mounds, they roasted wieners, ate cake, and sang "Happy Birthday."

Boys learned to play baseball on dried-out sludge ponds of chemical residue, alkali flats as smooth as the most manicured major league infield but not as forgiving. There was one a block from Mantle's home and the ball would roll forever, which is why, he confessed later, he preferred to play the infield. "They was full of lead and zinc—it's a wonder we all didn't end up with lead poisoning in our blood," his boyhood friend LeRoy Bennett said. "There was no grass that growed on 'em because it was so heavily slanted one way or the other on the chemical chart."

Everyone but Mantle learned to swim in the quarries created by cave-ins, leaving them with rat-red eyes; his mother would haul him home in a fury when she caught him so much as wading. He could barely manage the dog paddle. Cave-ins were routine yet shocking. One night driving home from work, Merlyn's uncle felt the pavement give way beneath his wheels. "Went over and it crashed in," Merlyn told me.

Route 66, the Mother Road connecting Chicago and Santa Monica, California—and Mickey and Merlyn's hometowns—was not immune. "I never will forget, one time, the highway splittin' wide open," Bennett said. "Miners were always gettin' killed and that kind of thing, but it was kind of expected. Eventually it was gonna happen and you couldn't do anything about it anyway, so most people just accepted it."

Everyone knew the air the miners breathed wasn't good, that the work was lethal, the earth's crust precarious, but Paul Thomas, the undertaker who buried Mickey's father and Merlyn's mother, never thought he would have to bury the entire place. No one could have imagined that one day the government would pay citizens to leave their toxic homes. Hanging from the beams in Thomas's Picher garage above three shiny hearses were rusty relics of the miners he interred: helmets with carbide lamps, lanterns, and kettles. The walls were covered by wide-angle portraits of proud mining crews, including Mutt Mantle's at the Hum-bah-wah-tah Mine. Posed in front of the doghouse where they changed their clothes, lunch pails at their feet, the roof trimmers, hookers, bumpers, and rope riders peer at the camera through masks of exhaustion and soot. An Eagle-Picher sign declares: WE USE SAFETY HERE. The photo is dated June 8, 1941.

In the language of the mines, men worked *on top* or *in* the ground, never above or below. In 1935, when Mutt went to work for Eagle-Picher, a common underground worker earned $2.80 a day, according to *Union Busting in the Tri-State*, a definitive history of the industry written by George Suggs, Jr. In an eight-hour shift, miners filled as many as forty-five to sixty 1,250-pound cans, all by hand, lifting up to 75,000 pounds every day.

They rode to work in the buckets they loaded, falling into the darkness at the force of gravity. Everything needed below went down the same five-by-seven-foot shaft, including the air they breathed—there was no other ventilation. The mules that pulled the cans to the shaft lived out their lives in the ground inhaling the smell of mother earth in a climate-controlled 65-degree tomb. The working conditions for man and beast were appalling. Miners carved the rock face, inhaling the dust generated by their labor. A roof trimmer standing atop an 80-foot ladder chipped ore from the ceiling while four rope riders steadied the precarious perch. Their backbreaking labor required teamwork and bred a mordant camaraderie not unlike that of baseball teams. It's no accident that so many of them, including Mutt and his brother Eugene —known as Tunney— spent their off days playing baseball in the sunshine and arguing over pitch selection.

The mine whistle summoned Mutt and his crew at 7 A.M. and sent them home at 4 P.M., fifteen minutes after the dynamite charges were lit in preparation for the next day's dig. The ground shook; wives and children went about their business and hoped for the best. "Sirens were a dreaded, scary thing, because that did mean there had been some kind of a cave-in and somebody was hurt," said Ben Craig, a banker in Kansas City who played sandlot ball with Mantle. "It was just, always, hold your breath."

"It would be rare to have two in one day, but you didn't go very many days at a time without hearing one," Craig said. "It was not unlike the tornado warning sirens we have out in this part of the country. I don't think it was as loud as these are now, though. Nobody wanted to spend that much money."

Mantle's best friend, Bill Mosely, lost his father one

day when he set the charges and didn't get out in time.

Between 1924 and 1931, there were 24,464 reported accidents in the Picher field, according to statistics in documents kept by the Tri-State Zinc and Lead Ore Producers Association, to which only half the mine operators belonged. During that same period, 173 miners were killed, many as a result of falling rock. Archives from the Tri-State Mineral Museum in Joplin document the grim ordinariness of death in terse, numbing language: "Machine man. Killed by a falling slab." "A shoveler. Killed instantly by a falling rock." "Bumper. Killed by a fall of rock." "A miner. One of 4 miners killed by falling slab."

Some slabs were larger than a city block.

Little wonder Mutt didn't want this for his son. Mosely says a school field trip to the bottom of the earth was as close as they ever got to going in the ground. In the summer, when he wasn't playing baseball, Mantle worked for Eagle-Picher, hacking away combustible blue stem grass that grew up around the poles that carried electrical wires to the mines. "They'd take him ten miles out of town and have him dig ten-foot circles, one foot deep, around every telephone pole," Merlyn told me. "He had to dig his way back in."

In the winter of 1950–1951 Mantle worked as a roustabout earning $33 a week. "I know every job Mickey had with Eagle-Picher," said Frank Wood, a metallurgical engineer, whose father owned a store that supplied equipment to miners, and who himself worked on the subsidence report to the governor of Oklahoma. "He worked with the Eagle-Picher pump crew, maintaining all the pumps that were above ground, pumping the water out of the mines. Now, this is a rascally bunch of hard-drinkin', hell-raisin' fellas that did this work, but they were very diligent and did very professional work. Mostly what Mick did for 'em was a gofer. They'd send Mick over to get bearings and repair parts for the pumps, and he'd set there at the engineer's desk and practice his signature on the Yellow Transit notepads. I got three notepads of Mickey Mantle signatures."

"The bull gang, they called it," his cousin Max Mantle said. "They'd go from one mine to another working on the equipment. They had to go down in. As far as being a miner, he wasn't."

2.

The best days of the Tri-State Mining District were ten years gone when Mutt moved his family to the region. The land's lucre was first discovered in 1848, the year Mantle's great-grandfather, an English coal miner, immigrated to America. The Twenties were the glory days. Between 1908 and 1930, the ore that came out of the mines was worth more than $300 million. The human cost of extracting the wealth was clear as early as 1915, when doctors noted pulmonary disease in almost two out of three miners. Laws passed in 1923 by the Oklahoma Department of Mines required operators to wet the muck piles prior to shoveling the ore into buckets. But miners paid by the bucket were reluctant to waste precious time wetting down the ore, and the laws were loosely enforced. They choked on the air they breathed, and when they tried to cough up the fragments of chiseled rock caught in their lungs, they choked on their own blood.

Silicosis was more feared and far more common than the random but inevitable collapse of rock. A clinic opened in Picher in 1927, but it was for the benefit of the mine operators, who were anxious to cull the sick from the workforce. Doctors provided advice but no treatment. Annual X-ray examinations were compulsory. Miners were required to carry a wallet-sized health card certifying that they were free of disease. Those whose X-rays came back positive were fired the same day and could never be hired by another mine. An attorney for Eagle-Picher explained the company's methodology for ridding the area of silicosis and the rampant tuberculosis that ensued: "When they get sick and can't work, we throw them on the dump heap."

That explains why Mutt never went to the doctor.

Between 1927 and 1932, almost 30,000 miners were examined; more than 5,000 of them had both silicosis and tuberculosis, which spread throughout the mining towns as quickly as they were built. Picher was the corporate, civic, and cultural center of the area, a town of 10,000 people that grew to have 5 movie theaters, 43 grocery stories, 28 boardinghouses, 2 hospitals, and no place to park. "About every other business was a bar," Paul Thomas said.

In the beginning, the mining camps were little more than shantytowns with flimsy houses built one on top of another. Later, mining companies built "shacks" for their workers like the one the Mantles lived in on Quincy Street. "Everybody had a little wood frame house with a porch on the front," said J. Mark Osborn, a physician from Miami, Oklahoma, who played a central role in bringing the government's attention to health problems in the blighted region. "One day you'd walk along and see a guy, and he'd be coughing up blood into a spittoon. In a couple of weeks he'd be gone and you'd go three or four houses down and there'd be another guy coughing up blood and dying in a couple of weeks."

His grandfather was one of those men.

Whatever fears and prejudices Mantle took with him when he left Commerce were the residue of growing up in an insular, homogeneous world fraught with unifying peril. The indigenous population was Native American; the eviscerated land was a Quapaw reservation until 1897. Blacks were not welcome after dark. "It was an informal thing, and the police departments and the county sheriffs and 'the country club set' set the rules," said Bennett, who left town for the Naval Academy and graduate school at MIT. "As far as I can recollect, there was not one black in Commerce, not one. I didn't know what Jews were."

One night years later while visiting Mantle in New York, they went to hear Les Paul and Mary Ford perform in New Jersey. On the way back in the car, Bennett recalled, Mantle saw "a black person, and Mickey purposely rolled down the window and yelled to this guy, 'Hey, you black bastard, go home and take a bath.'"

Parochialism and prejudice were offset by public schools that were the hub of the community and teachers who saw the best in their students. Among them were Ed Keheley, a nuclear engineer who returned to Picher after retiring as site manager of the Lawrence Livermore National Laboratory in California, and Kim Pace, a learning specialist who became principal of the elementary school she had attended as a girl. Keheley came home to raise cattle but found the pastureland toxic, and his hometown despoiled—mired in controversy about the impact of long-term exposure to lead on Picher's children. It wasn't so much a question of the level of lead in the blood as the length of exposure that caused the problem—damage suffered before the age of six is irreversible. "These kids were known to the schools," said Keheley, who has conducted historical research on the Picher Mining Field for the United States Department of Justice and private organizations for more than a decade. "They passed them from class to class, gave them diplomas. It was certainly better than ten percent."

Pace says her students needed seventy-five repetitions to master reading skills that average students retained after fourteen to twenty-five repetitions. What should have taken two or three weeks to learn took six to eight weeks for her children. Once they fell behind, they stayed behind.

No one wanted to believe it, least of all the parents of the children ridiculed as "chat rats" and "lead heads." Suspicion redoubled when, in May 2006, workers involved in a University of Oklahoma study confessed to having submitted fraudulent blood samples. But everyone knew somebody who struggled to learn to read.

Merlyn Mantle told me she was among them. Two of her sons were later diagnosed with dyslexia, a learning disability in which letters and symbols are reversed. "I think it came from my side of the family," Merlyn told me. "My aunt had a child with problems. It was all on my side."

She never saw signs of learning disabilities in her husband, but Mantle's friend Pat Summerall did. He too has a son who is dyslexic. He says he and Mantle talked about their shared experience on several occasions. "He couldn't pronounce the word, but he knew what he was saying," Summerall said. "He thought he had that disease. He'd see different things in road signs. He'd see it going one way when it was going the other."

At the end of Mantle's life, doctors tested his blood as well as that of his sons for elevated levels of lead. "Some scientist out of Tulsa was saying 'What's wrong with his liver was the lead,'" Danny Mantle told me. "They made all of us kids, me and David and Mick, do all this blood work. It wasn't in our blood. They did find it in Dad's."

3.

"You haven't got a problem that God cannot solve."

That optimistic promise greets worshipers at the entrance to Jerry VonMoss's Exciting Southeast Baptist

Church, just off Mickey Mantle Boulevard—old Route 66—in downtown Commerce.

His father was a supervisor at the Blue Goose Mine. His parents and Mickey's parents were friends. On the desk in his office he keeps a reminder of a dead way of life: a ten-pound ingot of lead ore his father made into an ashtray. Zinc was far more abundant than lead in the Tri-State area, by a ratio of six to one. Miners called it Jack. Until Mickey Mantle came along, it was the only name in town.

On the shelf behind his desk, VonMoss keeps a Bible, a photocopy of a *Sporting News* questionnaire filled out by Mantle when his nickname was still "Muscles," and his father's map of the mining district. Mine operators named their stakes after their children, their wives, their lovers, lives and luck lost and found: Cactus, Emma Gordon, Lead Boy, Nancy Jane, Dew Drop, Prairie Chicken, Bull Frog, Skeleton, Lawyers (because it was full of snakes), Darling, and No Dinero.

After the ore played out, after B. F. Goodrich closed its Miami plant in 1987, there wasn't much commerce left in Commerce or the rest of the Tri-State region. Picher became a ghost town. Rusted industrial skeletons stood sentinel on city street corners behind government-mandated chain-link fence—derricks, pinions, and those massive 1,250-pound cans. Colored banners adorning public buildings warned STAY OFF

THE CHAT! and offered the only respite from the bleak palette.

After nearly three decades as the most toxic waste site on the EPA's Superfund list and, Keheley says, after the infusion of more than $240 million of taxpayers' money, government officials accepted the findings of the 2006 subsidence report his committee prepared for Oklahoma senator James Inhofe. The money had been spent on a well-intentioned but misbegotten attempt to rid yards of the mineral residue that arrived on every breeze; to plug open mine shafts; and to rid Tar Creek of the acidic water that reached the surface and municipal water supplies a decade after the last mine closed in 1970. But it was all to no avail, Keheley's committee found, because Picher's underpinnings were as unstable as Mantle's.

Merlyn's hometown was declared unsalvageable. The government began offering buyouts to residents who never wanted to leave and often couldn't afford to do so, a protracted process that generated enmity and litigation.

Finally, on May 10, 2008, Ed's wife said, "The Lord looked down and said, 'Enough.'"

An EF4 tornado roared across the Kenoyer chat pile, killing seven of Picher's citizens and destroying 114 homes, including one belonging to Merlyn's late mother, one to her sister, and one built by Picher high school students for Sue Sigle, an elementary school teacher who grew up across the street from Merlyn. The only thing Sigle cared about was the unlocked safe holding her late husband's collection of Mantle memorabilia, including a ball he had autographed for her son. He had planned to open a baseball card shop when he retired.

She was in Branson, Missouri, when the tornado struck. When she got home, all that remained of the two-story brick home was the fireplace. Highway patrolmen escorted her through the rubble, past the camera crews stationed on the front lawn. There, among the ruins, she found the open safe, with the precious family heirloom. Neighbors had retrieved it for her. She felt blessed.

After decades of despoliation, Mickey Mantle's memory was the only resource left to mine. Everyone became a prospector. And that ore was pretty much played out, too. Memorabilia hunters had excavated every attic for anything he might have touched, used, or signed. Jerry VonMoss provided a guided tour of Mantle-area landmarks, with his wife, Corrine, and friend Jim McCorkell in the backseat. She pointed out an empty pasture west of town where Bonnie and Clyde spent the night before robbing the bank, killing the constable, and abducting the chief of police. The scandalous thing was . . . Bonnie wasn't wearing underwear.

Across the way lay a field where the charred remains of a house once occupied by the Mantles had once been hidden by high grass. "This guy stomps around and finds some boards that were on the house," VonMoss said. "He takes those boards and sells 'em to an old boy in Florida, who takes the boards and cuts them up into little pieces and sells them. Now, who made the most money?"

VonMoss eyed McCorkell in the rearview mirror. "Jim's the guy who sold 'em. The deacon of the Southeast Baptist Church selling boards off of Mantle's house!"

McCorkell sold each ten-inch board for $200 to a Florida merchant, who cut them down further and resold them at a generous markup—marketing them as relics of the house at 319 South Quincy Street, where Mantle had taken his first swings as a switch-hitter.

Two local entrepreneurs purchased the house, hoping to develop it into a tourist attraction. They were optimistic about their prospects. After all, Branson, Missouri, the entertainment mecca, was only two and a half hours away and, said Miami mayor Brent Brassfield, "nine to fifteen thousand cars pass by Miami on the Interstate every day." His brother owns half interest in the house.

They were surprised by the modern amenities and hung a plaque by the bathroom door: MICKEY MANTLE IN-DOOR PLUMBING. They planned to straighten the humpbacked shed the Mantles used as a backstop. But when Mantle visited, he told them, "It leaned when I was a kid." So they braced it to lean forever. A sign advertising RESTORATION IN PROGRESS, MICKEY MANTLE COMMERCE COMET BOYHOOD HOME was stolen off the front porch.

For a time, hopes rested with plans for a Mickey Mantle Museum. Board members of the Mickey Mantle Memorial Trust, many of them childhood friends and classmates, envisioned a 33,000-square-foot educational facility that would celebrate the history of the region and its contributions to America—Mick and Jack. But the plan died for lack of funds, support, and potential visitors. The Mantle family informed the trust that they preferred to see a statue of The Mick erected at Mickey Mantle Field, according to Brian Waybright, chairman of the trust and director of the annual Mickey Mantle Wooden Bat tournament. A nine-foot, nine-hundred-pound bronze likeness—stationed behind center field—was unveiled at 6:07 P.M. on June 12, 2010.

Nonetheless, townsfolk like Ivan Shouse, Mantle's high school classmate, were disappointed and perplexed by the fate of the museum and of the town he left behind. "Why *did* Mick move out, anyway?" he asked.

4

May 27, 1949

PATRIMONY

1.

The Mantles of Brierley Hill, a soot-draped coal-mining town in England's West Midlands, fled the "Black Country" fifty years before the ore played out. Elihu Burritt, the American consul to Birmingham, described the landscape, pitted by collieries and iron-works, in 1862 as "black by day and red by night."

Fourteen years earlier, Mutt Mantle's great-grand-father George brought his family to America, seeking light and air and a new way of life. They arrived in New Orleans after a months-long trial at sea. When the wind died and the *Sailor Prince* was becalmed, the women set about washing clothes on deck, only to see the wash barrels swept overboard when the breeze returned. It was a harbinger of the life of privation that lay ahead.

A riverboat ferried them up the Mississippi to St. Louis, where George and his sons found initial employment in nearby coal mines, according to a family history shared by Max Mantle. Three years later, they headed west for Missouri's Osage country to try to eke out a living aboveground as farmers and grocers. But within two generations, Mantle men would be working in the ground again.

Elven "Mutt" Mantle was eight years old when his mother, Mae, died of pneumonia a month after giving birth to her fifth child, Emmett. His father, Charles, never remarried, and struggled to raise the children alone. An aunt and uncle reared the new baby as their own thirty miles away in Pryor, Oklahoma. Mutt was eighteen when he met and soon after married Lovell Thelma Richardson Davis, a divorcée—a rarity in that time and in that place—who was eight years his senior. Family lore has it that when Mutt arrived to call on Lovell's younger sister, she stepped forward and declared she would have him for herself. "Mutt married himself a mother," relatives said.

That was true enough; Lovell already had a daughter and a son from her marriage to Bill Davis. Still a teenager, Mutt took on the responsibilities of a much older man.

It was a union of opposites. "Daddy was a very passive individual," his youngest son, Larry Mantle, said. "My mom was a hellcat. He did whatever she said."

The first of their five children, Mickey Charles, was born on October 20, 1931, in Spavinaw, Oklahoma, in the depths of the Great Depression. Mutt picked the name before he knew the child was a boy in honor of his hero, Hall of Fame catcher Mickey Cochrane, and his father, Charlie, a semi-pro southpaw pitcher. He ordered Mickey's first baseball cap six months before he was born. He placed a baseball in the newborn's crib and seemed surprised when he showed equal interest in his bottle.

Before the boy could walk, Mutt and Charlie propped him up in a corner and rolled balls across the floor to him. Lovell fashioned his first sliding pads from Mutt's old wool uniforms and had a cobbler fix spikes to an old pair of shoes to fabricate his first pair of cleats.

When Mutt lost his job grading roads in Spavinaw, he took up tenant farming, tending 80 parched acres for four futile years until drought chased him from the Dust Bowl for good. He quit the land for the promise of employment in the mining towns forty-five miles to the northeast. Within a decade, he had ten mouths to feed: his father; Lovell's children, Ted and Anna Bea Davis; Mickey, his twin brothers, Ray and Roy, Larry, the baby, known as Butch, and the only girl, Barbara, who was called Bob. She never knew why.

Mutt moved his family first to the small mining town of Cardin, then to the Quincy Street house in Commerce, where they slept four to a bed for ten years. The modest one-story structure, measuring twenty-five by thirty feet, had four rooms, including the kitchen, which had a wood-burning stove, and tin can lids pressed into knotholes in the plain pine floors. "I cannot believe that many of us lived in that little bitty house," Barbara DeLise said.

In 1944, much to Lovell's consternation, Mutt traded it for an old farmhouse with a calf on the outskirts of town. He wanted better air for his ailing father and to live off the ground, not in it. "She was so mad," DeLise said. "The house was a two-story, and it was a wreck. The cracks in the floor were so bad. And when the wind blew, the linoleum would be standing this high off the floor. So we didn't stay there too long. It had no bathroom. Had to take a bath in the washtub. We didn't have nothin'. You just went outside and went to the bathroom. That was pretty bad."

Grandpa Charlie died soon after, just as his oldest grandson was entering eighth grade. He was laid out in the front parlor, such as it was. "Say goodbye to Grandpa," Mutt said, escorting the oldest boys to the open coffin.

"From there we moved to Dr. Wormington's place," DeLise said. "He lived in town, and Dad took care of his farm. We weren't growing anything that I know of. Dad just took care of the animals. We had cows and chickens and four or five horses. We had one rooster. This is probably why Mickey got so fast. We had a rooster that was meaner than any dog you ever, ever saw. Every time you stepped outta that house, that dang rooster was right there. And, man, it would jump on you. He would take a ball bat and run to that bathroom, just trying to beat that rooster."

There was yet another move, to Whitebird, before the family settled back in Commerce. There was always enough to eat—especially biscuits and beans. Enough, Larry recalled, to feed Mutt and Lovell's friends Jay and Eunice Hemphill, who often showed up just in time for dinner and left as soon as it was done.

Pauline Klineline, a cousin on Lovell's side of the family, said her mother always laughed when she saw childhood pictures of Mickey in a clean white shirt because he never owned one. Lovell took in ironing to supplement the family income. "They could just barely eke out a living," said LeRoy Bennett, Mantle's first childhood friend.

Though Lovell's father was a church deacon and Mutt's English forefathers were known as "dissenters" because of their fidelity to the Primitive Methodist Church, religion was not stressed in the Mantle household. Nor was education—Mantle later said he never saw his father read anything but the sports page.

In the Mantle canon, Mutt is portrayed as a tough man in a tough world who was tough on his oldest son. Kind? "That's an interesting question," Bennett said. "Yeah, I'd say so. But he probably didn't realize that himself. He was just an ordinary, hard-rock miner."

Mutt was a surrogate father to two of Mickey's pals, Nick Ferguson and Bill Mosely, and to his nephew Max. "He'd just grin at you," Mosely said. "You could tell when you were doin' somethin' wrong and everything, but he was pretty quiet. Everybody wanted to please Mutt, seems like. He's the type of guy, he didn't have to really tell you a lot what to do, but you just felt that you wanted to do what he wanted."

In the family he was known for his prowess in cards and dominoes and his caution behind the wheel. "Mutt had a team of horses and a wagon," Max Mantle recalled. "He drove the horse and wagon down to Afton, traded it for a car. Ray said, 'We made just about as good time gettin' down there with the horse and wagon as we did in the car on the way back.'"

When Mickey got old enough to drive, Mutt's speed limit was strictly enforced. "Mick'd kick it up to forty-five or fifty miles per hour," said Jimmy Richardson, Mantle's first cousin on his mother's side. "Mutt, he'd be squirmin' around, and he'd say, 'Slow down, son! You're airplanin' it!'"

Mutt's two youngest children remember a gentle man, worn out and worn down by the mines, who came home every day, lay down on the divan in the parlor, and had them brush his hair. He was very particular about his dark, reddish black hair, which he wore combed straight back from his brow. "He'd say, 'Okay, Bob, get your brush and your comb,'" DeLise said. "And I'd set down on the divan and comb his hair for hours and he'd take a little nap. I'd get a nickel for every hour."

Larry didn't get paid.

Like her husband, Lovell came into a world fraught with uncertainty and peril. A tornado demolished the family home when she was an infant, injuring an aunt who also had a newborn, Pauline Klineline's mother. "Mickey's grandmother took my mother and Mickey's mother and nursed 'em both," she said.

Lovell grew to be "a fair-sized woman," Max Mantle said, who was also stout of opinion. She was patient with Mickey's crew on winter afternoons, when they ran wild in small quarters. But she had no tolerance for anyone who messed with her boys. Mutt refused to sit with her at Mickey's ball games. Her bellowed motherly support would have made her deacon father wince.

Nor did she shy away from occasional fisticuffs. At one Friday-night barn dance, Commerce men took umbrage at the attention lavished on their women by some out-of-town dandies. "Daddy stepped up, said, 'Ain't no women for you to pick up. Ya'll need to leave,'" Larry said. "Sure enough, Daddy and this guy start out in a fight. Then here comes Lovell, getting in the road."

Mutt tried to move Lovell out of the way. But, Larry sighed, "he couldn't keep Mom out of it."

When one of the twins got banged up on the final play of a football game, the enraged Lovell lit out for vengeance. "She grabbed ahold of my hand, and off we went across this football field," Larry said. "'Who's the Afton coach?' Guy says, 'I am.' She hit him, *wham*, hit him over the bench."

Ted, Lovell's son from her first marriage, spent much of his childhood at his grandmother's house, his widow, Faye Davis, recalled. He suffered from osteomyelitis, the bone disease his half brother Mickey would contract. Ted told his wife that his mother had little patience with his infirmity. "When he was seven or eight years old he used to cry because it hurt," Davis said. "But Lovell told him, 'Shut up, people have to go to work in the morning.'"

Expressions of tenderness were few. Merlyn figured that was the reason her husband didn't know how to show his feelings. "Mick's family was cold," she told me. "His mom was cold. I never heard her call her children 'honey.'"

"She used to whip him, too, something he didn't like to admit," David Mantle wrote in the family memoir, *A Hero All His Life.*

Young as they were, Larry and Barbara don't remember much about their parents' marriage except, he says, that she ran everything. Were they affectionate? "No, not that much," Barbara said. "I don't remember them ever bein' smoochie smoochie."

Larry Mantle's warmth was the exception to a familial reserve handed down through generations of Mantle men. When he tried to hug his nephew Mickey, Jr., at a family gathering, "he almost jumped straight back." Leaving a holiday party with Mickey, Sr., one year, Larry paused to embrace their mother. "We get outside, and Mick said, 'I wish I could do that.'

"And I said, 'What?'"

"He said, 'Kiss Mom on the cheek like that and hug her.'"

"I said, 'Just walk up and do it.'"

"I really felt sorry for him that he couldn't. Because, my goodness, that must be terrible."

2.

Mutt and Lovell's oldest son was as quiet as his father and as pugnacious as his mother. An "ornery little varmint," Cousin Max called him.

Everyone else called him Little Mickey. He didn't weigh but ninety pounds when he was a freshman in high school, qualifying him to play on the Midget basketball team. What there was of him was all boy. He set fire to the trash and raced the flames to the outhouse, trying to douse them with a bucket that had a hole in the bottom. He was five then. He tied himself to the hind quarters of a calf, pretending he was a rodeo rider. "That old calf bolted out the side door of the barn," DeLise said. "We thought it killed him."

He was the big brother who organized the games, made the rules, and played the pranks. "Usually it was fun for him at other people's expense, like mine and Barbara's," Larry said. "We had this old barn with a big wasp nest. He'd do a deal where he'd go in and tear down the wasp's nest. No one could run 'til it started falling."

At Whitebird, he turned the porch into a fort with dynamite crates from the mines. "We used to build rubber guns from tires that had inner tubes," Larry said. "I was on his side a lot. Cannon fodder."

But he was scared of heights, particularly the roller coaster the twins rode to death in St. Louis, and as a little boy scared of bugs. Mike Meier's grandma babysat for the Mantles. "Granddad used to laugh," Meier said. "He was scared to death of everything. Granddad said, 'I never dreamed he would grow up to be what he was because he was such a sissy.'"

He loved country music—especially Bob Wills and the Texas Playboys. Every time they came through Commerce, they saw the same red-haired, freckle-faced boy waiting by Route 66. One day, Mantle wrote, they stopped to introduce themselves. Their young admirer asked to tag along to their performance in Joplin. Wills told Mantle, "You get your parents' permission." Next time, the Playboys took him along.

He hid his own musical ambitions and his guitar in the culvert in the front yard, his artistic impulse trumped by competitive zeal. He used the money set aside for guitar lessons to play pool instead. "He didn't worry a lot about world news or wars or things," said Bennett. "He was sort of loose as a goose. He just wanted to play baseball. He was pretty simple, really—his main worry was hittin' a big curve ball."

In the Commerce High School yearbook—he was sports editor—the caption under his senior picture read: "They're great pals, he and his baseball jacket." He was also listed as Most Popular on the Who's Who page, assistant editor of *Tiger Chat*, the school newspaper, a member of the Engineers Club, and a cast member in the senior play, *Starring the Stars*.

His siblings remember him as an enthusiastic babysitter and compelling storyteller who drove the young 'uns under the bed with tales of the Headless Horseman lurking outside the window. "And Mickey sat in the living room just dying laughin'," Barbara said.

Max Mantle recalls a different ending to the tale: "He was under the bed. He was the furthest one under the bed."

But Cousin Mickey would never spend the night at Max's house; he was always heading home at bedtime. He wet his bed until he left home for his first year in minor league baseball. It had to be embarrassing in a house where everyone lived and slept in such proximity. Perhaps that's why Lovell was so diligent about starching and ironing his boxer shorts every morning.

Dick Cavett broached the delicate subject when Mantle appeared on his late-night talk show in the spring of 1969 with Whitey Ford and Paul Simon. Cavett was asking about the psychological impact of being made to switch-hit, and noted that when "parents teach a kid who's right-handed to become left-handed," it can lead to emotional trauma.

Then he threw Mantle a spitter. "I wondered if any troubles showed up in your personality because of that? Maybe I can just ask Whitey, 'Was he a bed wetter?'"

"It's true," Mantle replied with an easy grin. "'Til I was about sixteen years old. You think *that's* what went wrong?"

The camera cut to a reaction shot of the shaken singer. "*Mickey Mantle* wet his bed?" Simon gasped, as Ford steadied his arm.

Shortly after the show aired, Daniel Zwerdling, a *Washington Post* reporter working on a story about new treatments for bed wetting, decided to call his childhood hero. "I thought he'd hang up on me," Zwerdling said. "I was more embarrassed than he was. He said something like 'ShitIdunno, all I know is, I was pissin' in my bed.'

"I asked, 'Did you have any scars? Did you go to therapy?'

"He said, 'Hell no, my daddy was a lead miner.'"

3.

In a family where doing without and making do were the norm, Mutt and Lovell always made room in the meager budget for baseball. "Mickey came from a *baseball* family," Mosely said. "They'd give up anything, but not baseball."

Mickey was often the beneficiary of their largesse. At Christmas, when all the other children got a pair of socks, there was always enough money to buy him a new baseball glove. He would cry, he later told a friend, because he didn't get any toys. When he was fourteen, Mutt took him to St. Louis to see his hero, Stan Musial, and the Cardinals. Providence offered a chance meeting in the hotel elevator, but Mutt wouldn't allow Mickey to ask for an autograph. A glimpse of a hero was enough.

Lovell was equally devout about baseball. Mosely recalled: "During the day, when the kids were in school and her husband was workin' in the mines, she had the St. Louis Cardinals game on the radio, and when she was ironing or doing her housework, she was keepin' score of what every one of those guys did!"

At the dinner table, she would re-create all nine innings. Lovell knew as much about baseball as any woman, but she was uncharacteristically low key in offering her opinions. "She could critique Mickey, but she would do it real quiet so Mutt wouldn't hear anything," his pal Nick Ferguson said.

Mantle got his fierce, competitive intensity from her. "She is the one that instilled all this fire that made him not the ballplayer but the person that he was," said Larry, the baby brother who was on the receiving end of bullet backyard passes and lethal glares when the football wasn't caught. "Mickey didn't tolerate people not giving their best. Those passes were ninety miles an hour. If you didn't catch it, you'd get this terrible look."

The look Bil Gilbert of *Sports Illustrated* later likened to "a nictitating membrane in the eye of a bird."

"It wasn't that much of a fun game," Larry said. "I quit all the time."

But his threats were meaningless; his big brother would never let him leave the field.

On Sundays, instead of church, the Mantles attended Mutt's semi-pro games in Spavinaw and Whitebird. No one was allowed to get out of the car if the sports report was on the radio. "Mutt was a catcher/pitcher," said Jerry VonMoss, whose father, Ed, managed the Whitebird Bluebirds. "His brother Tunney was also on the team. One day, Mutt was pitching and Tunney was catching. Tunney called for something, Mutt threw something else. They threw down their gloves, met midway between the mound and the plate, and had a fight."

Mickey always bragged on his father. "Best semi-pro ballplayer in Oklahoma," he told me. But others remember his talents more objectively. "He was a very mediocre ballplayer," Ferguson said. "He was not as good as a lot of other players on the teams there. Nowhere near."

"He was good until he broke his leg sliding into second," Barbara said.

If Mickey Mantle was the product of Mutt's thwarted ambition, he was also the beneficiary of his undivided attention during the best, healthiest years of his life. Mutt would not have the time, energy, or drive to invest in his younger sons. His health was failing by the time Ray and Roy and Larry came of age for team sports. Mickey was his one chance to get it right. Maybe that's why, VonMoss says, Mutt "drove him like a nail."

Playtime was over when Mutt got home from work. Every afternoon was punctuated by the rhythmic bang of the ball against the corrugated metal siding of the ramshackle shed. "Every day at 4 P.M., Mickey had to be home, no matter where he was or what he was doing, to do batting practice," Max said. "They'd throw a tennis ball. They'd stand him up against that leaning shed. He'd hit it up against the house. If it hit the ground, it was an out; below the window, a double; above the window, a triple; over the house, a home run. Every day."

Everyone in town knew about the day Mutt came home early from work and caught Mickey batting right-handed against a right-handed pitcher in a Gabby Street League game at the Paul Douthat field outside Picher. Climbed his ass. Raised holy hell. "Boy, the crap hit the fan over that," Max said. "That was a no-no. Mickey never done it again."

Of Mutt's hard schooling, Larry Mantle said, "I don't know how good friends they were, Mickey and Daddy. Daddy was Daddy, and Daddy was the boss."

Mutt was a baseball savant far ahead of his time in envisioning the future of baseball specialization. He was also a realist who recognized his son's personality and talent and the discipline needed to harvest it. "He wanted him to be good and knew what he had to do, and that was it," said Mosely, who made his living as a high school phys ed teacher and football coach.

Like any father, Mutt wanted better for his sons. By the early Fifties there was precious little ore to gouge out of the earth—except for the supporting pillars, in the mines the ore in the Tri-State area had pretty much played out. A job at the B. F. Goodrich plant in Miami represented the highest ambition for most boys Mantle's age. "Get out of school, get rich, marry your high school sweetheart, and buy a new car on time—that was the hope and the prayer," his cousin Jim Richardson said.

If Mickey got out, they all got out. It was a huge—if unarticulated—burden to place on one boy's shoulders. "He was their summer wishes and their winter dreams," his oldest son, Mickey, Jr., wrote later.

4.

Will Rogers, Oklahoma's most famous export before Mickey Mantle, once said, "Oklahomans vote dry as long as they can stagger to the polls." They continued to do so until April 1959. Prohibition was written into the state constitution in 1907, when the "wets" lost the battle for the soul of the Oklahoma and Indian territories. Although near beer was legalized nationally in 1933, bootlegging and home brew were as much a part of Oklahoma culture as going to church. Ted Davis was driving for a bootlegger, bringing booze in over the state line, when he was seventeen. Uncle Luke, one of Lovell's brothers, made home brew, Barbara said. "If he didn't sell it, he gave it away."

In Commerce, dreams and diversions were few:

beer, baseball, and brawling. "That's what we done, drink and fight," said Herman Combs, who worked for Mutt at the Blue Goose Mine. Mutt's brother Tunney (after boxer Gene) got his nickname after knocking a guy's eye out in a barfight while trying to rescue one of Lovell's brothers.

Alcoholism wasn't recognized as a disease, much less a hereditary one. But it ran deep in Lovell's family. "The alcoholism came from Mickey's mother's side," Merlyn told me. "Her brothers all had a problem. Two or three of them were alcoholics. I don't know if they died of it, but they were real alcoholics."

Lovell's sister Blanche did. Aunt Blanche was "a sneaky alcoholic," Barbara said. "You never saw her drink, but she did every day. She lived with us for a while, and you knew she was drinking."

How? "Because she was drunk."

Blanche's body was found in her apartment a week after her death. "Ted was one of them who found her," Faye Davis said. He had been sober for five years when she married him, but the damage was done. "He didn't drink anymore, but he lost his mind. He just went nutty as a fruitcake. He started carrying a pistol and seeing things that weren't there, and I thought, 'Better take him to the doctor.' And he had to go to a nurse's home."

His father, Bill Davis, was an alcoholic, she says;

Lovell didn't drink. Mutt kept a bottle of liquor in the icebox. Larry remembers the look on his father's face when he downed a shot. "He'd take a couple swallows of whiskey and then drink the Coke right quick and then give this terrible sound—*ps'shooow*—and shake all over. If somethin' affected me that bad, I don't think I would drink it."

None of the family members I spoke with thought Mutt was an alcoholic. Mantle offered contradictory accounts of his father's drinking. In his 1994 *Sports Illustrated* confessional, he said, "Dad would get drunk once in a while, like when he went to a barn dance and might have five or six drinks. Hell, for me five or six drinks wouldn't have been a full cocktail party!"

Merlyn confided in the wife of Mantle's friend Larry Meli at a dinner in the Eighties. "Mickey can't help the carousing. Mutt was like that."

He described Mutt as a habitual drinker in conversations with Herb Gluck, the ghostwriter of *The Mick*, with his friend Pat Summerall, and in a family history taken by his Georgia physician, Dave Ringer. "I think he thought his dad was an alcoholic," Ringer said. "Didn't talk a lot about it. Didn't talk much about his dad at all. But I do recall that."

Mantle told Greer Johnson, his companion during the last decade of his life, that when he was a boy of nine or ten "his father would take him to the bars, sit him up on the stool while he drank." He wouldn't have been the first guy in town to do that. "He always led me to believe that his dad was an alcoholic," Johnson said.

Nick Ferguson recalls outings to local watering holes with Mutt when he and Mickey were teenagers. "I guess he was about sixteen and I was two years older," he said. "Mutt took us both to a local drive-in-like thing and got beer for Mickey when he was underage."

"Wasn't no such thing as underage," Max Mantle said.

5.

Mutt did not want his son to go out for football. Mickey did it anyway, which may be the only documented act of rebellion in his young life. Mosely was the starting quarterback and star athlete at Commerce High. "I think he just played because all the rest of us did," Mosely said.

He was good and he was fast, scoring ten touchdowns in seven games as a fullback in a single-wing formation during the one season he played. Showed the opposition "a good, clean white ass," he bragged later to minor league teammates. Ralph Terry, a future Yankee whose hometown played in the Lucky Seven

Conference against Commerce, said, "He'd run sixty yards for a touchdown. Two or three plays later, he'd limp off." Max, the team manager, would rub his legs on the bus ride home.

Mutt's worst fears were realized one October afternoon when Mickey was kicked in the left shin during practice just twelve days shy of his fifteenth birthday. He was the second-string quarterback behind Mosely. While the ball was being handed off, there was a mix-up in the timing of the play. He was helped off the field by his teammates and taken to Max's house, which was close to school. His leg turned "black and blue and red and hot," Max said.

The Commerce football coach, Allan Woolard, didn't think it was anything serious until Mantle failed to show up for school the next morning. "I went over to his house to check with Mick and he was on the divan with his ankle propped up," he told a reporter in 1951. "It was swollen terrifically and was as red as watermelon. He also had a temperature of 103.5 degrees. We immediately took him over to the Picher hospital, and they started treating him at once."

In telling the story of his infamous childhood infirmity, Mantle always minimized its impact as well as its duration: he got kicked in the shin, went to the hospital, doctors threatened to cut off his leg, and he and his limb were saved by the heroic intervention of his mother, who told the sawbones, "Like hell you are." A new wonder drug called penicillin also helped.

In fact, he was hospitalized five times over a period of thirteen months. During those forty days in the hospital, he was exposed to more than a whiff of mortality. "It was a wonder he didn't die," said Bennett. "He was yellow in color and pale."

The sequence of events was recounted in a local newspaper in 1951, and reprinted in a 1997 Ottawa County historical brochure:

> Medical records at the hospital show that Mantle was first admitted on Oct. 10, 1946 for treatment of an "infection at the lower end of the tibia on the left leg." Penicillin treatments of approximately 50,000 units every three hours were given. He received approximately 300,000 units a day. He was first dismissed on Oct. 22. At the time osteomyelitis was suspected but there were no definite indications.
>
> A short time later—on November 15, 1946, Mickey again was admitted, received treatment and was dismissed Nov. 18.

Osteomyelitis is a bacterial infection of the bone, usually caused by trauma. Recognized by physicians since antiquity, it was little understood in Commerce, where it was called "TB of the bone" or "cancer of the bone." Neither was a trivial diagnosis in an area ravaged by tuberculosis and in a family that had just lost a grandfather and an uncle to cancer. Prior to the advent of antibiotics, osteomyelitis was treated with maggots, which ate away the diseased flesh, or by amputation. Mantle's half brother, Ted, was treated with maggots during a childhood bout with the disease and suffered a recurrence due to shrapnel wounds he received in the Korean War. Mantle never mentioned receiving maggot treatment, but a physician in Picher later told Max Mantle's wife that he had, his cousin said.

Osteomyelitis can be chronic or acute; and it can recur without warning. Mantle never spoke of having a prior occurrence or a later one. But friends and family remember him as puny and sickly, with boils on his arms and legs. "Bad blood" was the common diagnosis. A nurse for the Tri-State Zinc and Lead Ore Producers Association, a local Florence Nightingale named Ruth Hulsman, inoculated children against typhoid fever, smallpox, and diphtheria and provided them with shoes and clothing. "One of her patients was Mickey Mantle who had osteomyelitis when he was nine," wrote Velma Nieberding in *The History of Ottawa County.*

After Hulsman's death, her daughter told the *Miami News Record,* "One of her biggest thrills was watching

Mickey Mantle reach success since he was one of the children she took to Oklahoma City when he was just a boy for treatment of osteomyelitis. She would always watch his ball games and tell us, 'That's one of my children.'"

This account suggests the possibility of an earlier infection reactivated by the football injury. Certainly, that was Bill Mosely's suspicion. "I felt like he'd had that all along and that it was holding him back," Mosely said.

In June 1942, there was just enough penicillin in the United States for ten patients; two years later, 2.3 million doses were available to treat the wounded during the invasion of Normandy. The price dropped from $20 per dose in 1943 to 55 cents in 1946. But the patent for mass production of the drug was not granted until May 1948. It's difficult to say which is more miraculous: that the drug was available in Picher, Oklahoma, in the fall of 1946 or that Mantle recovered from the disease after receiving approximately 7 percent of today's standard dose. And Mantle wasn't exactly diligent about taking his medicine. His cousin Jim Richardson remembers "seein' all those pills laid out there" on the ground outside the hospital room window.

According to the Ottawa County newspaper reprint,

It was during his third trip to the hospital that definite evidence of osteomyelitis was found, showing up in x-rays of the bone. This was Mantle's longest stay in the hospital, beginning March 27, 1947 and ending April 10. The swelling of his leg became more prominent and the pain was more noticeable according to a hospital physician. During this stay an operation was performed on the injured leg with an abscess being opened and drainage started.

The procedure didn't have a fancy name. "Scraped the bone," Max Mantle said. It left an indentation in his shin large enough that "you could lay a pickle jar lid over it," said Don Seger, a former assistant trainer for the Yankees. "And it wasn't a pretty scar, either. It had a keloid effect."

When they brought him home from the hospital, his mother had to carry him on her back to and from the outhouse. The infection and the mortification persisted. Lovell applied for public assistance in order to get the funds for him to stay at the Crippled Children's Hospital in Oklahoma City in late July and early August, where he received increased treatments with sulfa drugs and penicillin. "They gave him fifty shots of that stuff every thirty hours," Max said.

He was in Children's Hospital when Max's father, Tunney, died of cancer at age thirty-four. Mantle was hospitalized in Picher for one more week at the end of November.

By the next baseball season, he began to look like a ballplayer. Probably it was just the natural order of things, a boy growing into a man, but the change in him was so immediate and so dramatic, it reinforced belief in a connection between the penicillin and the growth spurt that followed. "When he got that penicillin in him, boy, his body shot out and the muscles in his arms jumped out," Mosely said.

6.

In the sickly summer of 1947, Mantle was invited to join Barney Barnett's Whiz Kids, a prestigious semipro team in the Ban Johnson Baseball League. He was so small that Barnett couldn't find a uniform to fit him. Though he played in only four games (batting .056), Barnett saw something in him. Like Mutt, Barnett was a ground boss for Eagle-Picher. He called his boys "honey."

Later he would talk Mantle up with the local birddogs and build him up with off-season jobs digging graves and hauling gravestones. He also got him a job as a lifeguard, which struck everyone as funny because Mantle couldn't swim. In spite of his best efforts, when Mutt took Mickey to St. Louis for an early workout with the Browns, they took one look at him in uniform and sent him home. He didn't even get on the field.

By the summer of 1948, Mantle had put on nearly forty pounds and four inches and had outgrown everything but his shyness. The Whiz Kids played in a small ballpark tucked into a hollow beside the Spring River. The water's edge was a long poke from home plate—400 feet at least. One night Mantle hit three home runs—two right-handed and one left-handed—that headed straight for the water's edge. The folks in the stands passed the hat in his honor—the $53 they collected briefly caused him to lose his amateur status.

That was the summer Tom Greenwade got his first look at him. Greenwade was a free-range baseball scout, though he didn't dress the part. You wouldn't catch old Tom in chaw-stained socks. Tall, lanky, and resolutely thin—the consequence of a bout with typhoid fever in his twenties—he cut a swath through the ball fields of Missouri, Kansas, Arkansas, and Oklahoma in a three-piece pinstripe suit and a crisp felt hat. Greenwade traveled the back roads in a shiny new Cadillac with a Babe Ruth jersey stowed in the trunk, a useful prop when trying to pry raw, young talent such as Tom Sturdivant, George Kell, Rex Barney, Bill Virdon, Jerry Lumpe, Hank Bauer, Ralph Terry, and Bobby Murcer away from the competition. *How would you like to play baseball in the biggest city in the world, son?*

Greenwade had been a prospect once, with an arm

live enough to stone rabbits for dinner. His pitching arm had gone dead one cold night in the Northeast Arkansas League when the temperature hovered in the thirties. After his playing days ended he went to work for a pipeline company and for the Internal Revenue Service; he studied law, managed in the minor leagues, raised tomatoes, voted Democratic, and befriended Harry Truman, who stopped by on occasion for a piece of pie at the kitchen table. He was hired by the Yankees in 1946 after scouting for the Browns and the Dodgers. He knew everyone in the territory, and everyone knew him. But Greenwade didn't become a legend until he discovered Mickey Mantle.

It may be baseball's most frequently and variously told tale. The way Bunch Greenwade heard it at his father's knee, Tom was heading back home from a scouting trip in 1948 when he saw the bright lights Barney Barnett had installed for the Whiz Kids at their home field in Baxter Springs and stopped to catch a few innings. That first night, Bunch Greenwade said, "Dad went and talked to Mickey. He wanted to know how old he was. Mickey told him he was a junior. Dad told him just point-blank, 'Well, I can't really talk to you right now, you know, because I can't 'til you've graduated from high school.'

"But he said, 'I'm kind of interested in you, and I'll

be back sometime to watch you play. Would you be interested in ever playing ball for the Yankees?'

"Dad always kind of did that, to get their attention you might say."

Mantle remembered the promise, but the way he heard it, Greenwade had come to scout the Whiz Kids' third baseman Billy Johnson. Johnson never made it to the bigs and he never exchanged a word with Greenwade until 1955, when he was playing for an Air Force base team. Greenwade asked why he was pitching instead of playing the infield. "Someone's got to do it," he replied.

In another account sanctioned by local cognoscenti, Johnny Sturm, the manager of the Yankees' Joplin farm team, had to stoke Greenwade's flagging interest in Mantle by threatening to go over his head to the front office. Bunch Greenwade says his father was just playing possum. "Dad went back four different times, and he did it as secretly as possible, telling Mutt on the Q.T., 'I want to watch Mickey play some. He might, possibly, someday, turn into something.'

"And then he would come home and worry."

He had reason to be discreet: rules prevented scouts from talking to underage prospects; also, he didn't want to drive up Mantle's price by eliciting interest from the competition.

The night before Mantle was to graduate, Greenwade couldn't sleep. "He sat up all night—he said he smoked cigarettes and drank coffee all night long," his son said. "And he just knew that when he got to Commerce there were going to be at least three or four other scouts there trying to sign him. He was worried about the Cardinals because he knew that Mickey and his dad were big Cardinals fans."

The next morning Greenwade went to principal A. B. Baker for help. Baker had already furthered the Yankees' cause by telling the Indians' scout Hugh Alexander that the school had no baseball team, that Mantle had been injured playing football and that he had arthritis in his legs. The principal's motivation was unclear, but Alexander tossed the piece of paper with Mantle's name away when he got back to his car. He would remember forever the scrap of paper carried away on the breeze.

The principal directed Mutt and the new Commerce baseball coach, Johnny Lingo, to see the superintendent of schools about getting Mantle excused from that evening's commencement exercises. The biggest night in his brief academic career paled in importance to the Whiz Kids' big game in Coffeyville, Kansas. Whatever the adult petitioners said, it was persuasive. Lingo described Mantle's ad hoc graduation ceremony in a 1953 article by Milton Gross for *SPORT* magazine: "Albert Stewart, our superintendent of schools . . . came Friday and he handed Mickey his diploma in advance and told him he was graduated."

The scholar was given no say in the decision. But he did get a new pair of baseball spikes. Because he had graduated, he had to turn in his school-issued equipment. Lingo told his wife, Charlene, "I ended up buying him a pair of cleats so he could play that night."

7.

That evening in Coffeyville, Mantle went 3 for 4 with two home runs, one from each side of the plate. Greenwade, who later claimed he didn't know Mantle was a switch-hitter until that game, played down his talents when he spoke to Mutt. *Marginal prospect. Might make it, might not. Kind of small. Not a major league shortstop.* Imagine how galling it must have been every time Mantle heard Greenwade later boast, "The first time I saw Mantle, I knew how Paul Krichell felt when he first saw Lou Gehrig."

Greenwade told the Mantles he had an appointment to see Jim Baumer, another highly touted shortstop, the next evening but promised to return to see Mantle in Baxter Springs on Sunday night. Heavy weather was expected and it arrived as promised, along with Greenwade, who pulled his car onto the grass behind home plate. When the heavens opened up, the haggling over Mantle's future began in earnest in Greenwade's Cadillac. Whiz Kid Wylie Pitts swears that lightning struck the light stanchions—bulbs popping and fizzing and showering the field with sparks—the moment Mantle became a Yankee. "Just like *The Natural*," he said.

Negotiations proceeded in the dark. Greenwade offered less than what Mantle could make working in the mines and playing semi-pro ball. Mutt objected. The scout affected some math on the back of an envelope and added a sweetener: "a bonus of $1150 to be paid by the Independence club as follows: $400 upon approval of contract and the remainder $750 payable on June 30th, 1949 if player retained by Independence or any assignee club."

The salary for the remainder of the 1949 season was $140 a month. It was New York's biggest steal since Peter Minuit paid the Indians $24 for the island of Manhattan. Mantle accepted, he later told Leonard Schecter of the *New York Post*, because, "I didn't think anybody else wanted me."

It's not as if bonus money was unavailable; baseball didn't impose limits on signing bonuses until 1955. Kal Segrist, who was signed by the Yankees in 1951 with a $50,000 bonus, played in twenty major league games. Jim Baumer received a $25,000 bonus from the Chicago White Sox and played eight games. But Greenwade spent the Yankees' money carefully. He gave pitcher Ralph Terry a $2,000 bonus in 1953. "What gripes you about those scouts in those days is they sign a guy out of poverty and he'd make the big leagues and then they'd brag about how cheap they got you," Terry said.

8.

The Yankees sent Mantle to Independence, Missouri, a 150-mile round trip from home. Mutt wanted him close by—to eyeball him and to keep an eye on him. In June 1949, Mutt delivered him to a boardinghouse at 405 South Tenth Street in Independence, where he shared a double bed with his roommate, Bob Mallon. Mutt unloaded his luggage, spoke earnestly to his son, telling him to mind manager Harry Craft and to be a team player, and left.

The 1949 roster for the Independence Yankees of the Class D Kansas-Oklahoma-Missouri (KOM) League was a melting pot of rawboned boys and veteran ballast, married men and teenagers, prospects and has-beens. The lights were bad, the pitchers threw hard, and Mickey Mantle was just another ballplayer. He answered to his given name, Mickey Charles, and addressed his elders as "sir" or "mister." Bunny Mick, a Yankee instructor, thought, "He was Jack Armstrong."

To his teammates, he was a fun-loving, prank-playing teenager whose idea of a good time was hanging boogers from the ceiling of a friend's car. Dingleberries featured prominently in his comic patter. He liked to go frogging at night with Joe "Red" Crowder, another country boy with a taste for the local delicacy; one held a flashlight to blind the unsuspecting amphibians while the other grabbed dinner. Mantle also liked to spy on Crowder and his wife, who lived in the next apartment. "Mick would say, 'Bob, come here, they're doin' it,'" Mallon said. "You'd hear the bed squeak. We'd get in the closet and listen."

He threw his knuckleball relentlessly if not well. Teammates quit warming up with him. First basemen dreaded being on the other end of his strong but errant throws. "I'm a married man!" cried Cromer Smotherman in 1950.

Mantle proved a generous shortstop, making more than 100 errors in the 184 games he played in two years in the minors. *Hit it to Mantle and run like hell* was the opposing strategy. In 1949, he didn't show much power either, hitting only seven home runs. Overmatched by pitching and by homesickness, he pleaded with his childhood pal Nick Ferguson to try out for the team. "He probably would have came home right then if Mutt hadn't insisted that he stay," Ferguson said.

Early one morning, he sat on the front porch with Mutt and Mallon, confiding his fears that he would never be good enough. "He was hitting, like, .230," Mallon said. "And his dad said, 'You wanna go back in those damn mines? You haven't even given it a chance yet. Here you wantin' to quit.' He mighta said some cusswords too."

Mantle was afraid the Yankees would send him home before his $750 bonus kicked in on June 30. His insecurity was palpable; teammates found him soft-hearted and unexpectedly tender. Keith Speck recalled the last day of the 1949 season, when Mantle cried on the team bus because the guys "weren't going to be together again."

"I think Mickey was probably more fragile than most folks realize," Bunch Greenwade said. "His feelings ran deep."

They all heard about his uncle Tunney, who had died two years earlier, and his grandfather, who had died three years before that. "Every part of my family's dying," Mantle would say, crying on roommate Carl Lombardi's shoulder. They all remembered it because it was jarring to hear a teenage boy say he didn't think he'd see age forty. He fretted about a recurrence of osteomyelitis and limped on the base paths. "He feared that more than anything," Lombardi said. "He said to me many a time, 'You know, this can kick up anytime.'"

His teammates barely recognized him when he reported for spring training with the Class C Joplin Miners in 1950. "What the hell did you do?" demanded Steve Kraly.

He looked like a blacksmith and sprinted like a cheetah. "The ground shook when he ran by," Jack Hasten said.

His strength and speed were equaled by intensity and temper. Teammate Al Billingsley remembered a game in 1950 when Mantle struck out and flailed out in anger and frustration, hurling his bat and several choice words. "That'll be the last damn time he gets me out."

Then he hit two home runs. "When he got angry the best came out," he said. "I think he fed off of it. There was something special about him and maybe he knew it."

Billingsley also recalled a 3-for-4 game, with a home run, after which Mutt reproved his son: "You would have had four hits if you would have hustled on that groundball."

Failure also made him petulant. When Mutt asked

Lombardi how Mickey was doing, Lombardi replied, "He'll be great if he quits pouting."

Manager Harry Craft asked Smotherman to stick by Mantle's side and see if he could steady the boy. He sat beside Mantle on the bench and on the team bus, sometimes passing him a piece of gum on which to take out his frustrations. "I lived mentally with him as much as anybody could," Smotherman said. "He had mood swings. He expected to get a hit each time he was at bat. He was never, ever satisfied."

When the last game ended and the pennant was won, Mantle was despondent. He hadn't gotten enough hits. "He was the league leader in every department, including strikeouts," Smotherman said.

He needed a way to blow off steam. When the Miners were home in Joplin, Mutt took him back to Commerce after the game, in part, Lombardi said, to save the rent, and in part because he was concerned about how much time his son was spending in the local pool hall drinking beer. It was often enough, Lombardi said, that "sometimes Mickey used to come to the ballpark and say, 'Hey, Carl, I don't feel that good. You'd better cover up for me.' Not that I wanna condemn him for it, because we all drank. We all had beer, but it continued when he went home with the boys and went to the pool hall."

Mantle finished the season with a batting average of .383 with 199 hits, 30 doubles, 12 triples, 26 home runs, 90 strikeouts, 94 walks, 136 RBIs, and 141 runs and was named the Most Valuable Player of the Western League. In his end-of-year report Craft called him "just an average ss" and recommended sending him to the Yankees' Double A team in Beaumont, Texas, in 1951 to learn another position and incubate for another year.

The Yankees rewarded him with a two-week call-up to the major leagues. Smotherman took him to a men's store in St. Joe's to buy a big-league suit. Mantle asked him to pick it out. No one figured he was going up to stay. And no one would have predicted they had seen Mantle as whole as he would ever be in a baseball uniform. "I saw him at his peak," Hasten said, "when he was eighteen years old and could do anything on the ball field."

5

May 20, 1952

IN THE GROUND

1.

At the end of September 1951, when Mutt and his friends headed to New York for the World Series, the local newspaper reported that Miss Merlyn Johnson, an employee at a Commerce bank, had gone west to take a job in Albuquerque, New Mexico. Though she was still wearing her engagement ring, no wedding date had been set. She and Mickey had agreed to date other people. A week after Miss Johnson decamped for the Southwest, the same paper noted that she had returned to Picher. The dispatch gave no explanation for the change in plans, but it was just about the time Mutt told his son, "Go home and marry Merlyn. She's one of our kind."

A wedding was all Mutt wanted for Christmas. "He knew he was dying," Merlyn told me. "He wanted him to be settled. He wanted a redheaded, freckle-faced grandson."

She fell in love with Mickey Mantle the first time she laid eyes on him at the annual Picher-Commerce football game. She was a twirler in the Picher High School band. She and her friend Lavenda Whipkey spotted him in the stands sporting his varsity jacket and a crew cut. "I thought he was the most handsome guy I'd ever seen," Merlyn told me.

He was meticulous about his appearance, immaculate, and he smelled so good. "I thought he had a perfect body, big shoulders, tiny waist, muscles he made hisself," she said. "They wasn't fake-looking."

She barely noticed the residual blemishes of youth that were airbrushed out of publicity photos. She loved him so much, she wrote later, "I wanted to crawl inside him and live underneath his skin."

His high school classmate Ivan Shouse made the introduction at the Coleman Theater in Miami. "We were sitting upstairs in the balcony. The girls were downstairs," Shouse said. "He sent me downstairs to negotiate a date."

That first night the boys took them to the Spook Lights, a local lovers' lane illuminated by mysterious, unexplained lights—an Indian holding a torch while looking for his head, legend has it. Mantle was paired up with Lavenda. "Next morning, Mickey said 'I want to make a change,'" Shouse said.

In *The Mick*, Mantle offered a slightly different account: he said he called Merlyn because Lavenda was busy. Pretty soon, Merlyn and Lavenda were cruising the main drag in Commerce looking for him. His friends at the local pool hall teased: "Merlyn'll be by here in a minute."

They didn't have that much in common except being young and country. He lived for baseball; she thought the seventh inning stretch was the time to go home. She was a gifted soprano who would give up a scholarship to Northeastern Oklahoma A&M College to become Mrs. Mickey Mantle. Decades later, she was wistful about that choice. "It would have given me stability in my life to do something else," she told me. "I do regret it."

They set the wedding date after the doctors at Johns Hopkins cleared him to go home at the end of October 1951. Three weeks before the wedding, the Yankees summoned him to New York for a reexamination of his knee. He had been complaining about pain in his right thigh. "We called him back here for a check to determine whether the cartilage in his leg was affected by the accident," team physician Sidney Gaynor told the *Times*. "Examination reveals the cartilage was not damaged and that the torn ligament on the inner side of his right leg has completely healed."

Gaynor gave him a weighted boot and a set of exercises to strengthen the quadriceps muscle and give support to his knee. Mantle ignored his instructions, preferring, he said later, to sit around, watch TV, and feel sorry for himself.

He didn't have to worry about limping down the aisle. The ceremony took place at the Johnsons' home on December 24. The bride emerged from the bedroom on her daddy's arm. The groom made his entrance from the bathroom, hair slicked back, a boutonniere in his lapel. Mutt's best friend, Turk Miller, was the best man. Miller's brother-in-law, Paul Thomas, the undertaker, was the photographer.

They were married in "a setting of flowers and lighted candles," a local newspaper reported. The bride wore "a faille suit with collar and pockets trimmed with seed pearls and rhinestones, a close-fitting chartreuse feathered hat and a pale pink rosebud corsage accented the delicate champagne color of her ensemble."

"Next to me the groom's father was the happiest person in the room," she wrote later in the family memoir, *A Hero All His Life*. "Mick was somewhere in the top five."

Would he have married her without his father's dying command? Five decades later she wasn't sure. "I do know he wasn't ready to get married," she told me. "He was very immature."

Mantle invited Bill Mosely, home on furlough from the Army, to come along on their honeymoon. He and his wife, Neva, hadn't been able to afford one of their own. "I said, 'You kiddin', Mick?' He said, 'No, no.'

"So come to find out, we was goin' to Hot Springs, Arkansas, and I believe the baseball player Johnny Sain had a motel and a bar there. So he told Mick, when he got married, 'You come down, bring anybody you want to. Everything's gonna be on me.'"

The girls took turns behind the wheel on the 340-mile drive. The boys sat in the back. "Maybe havin' a drink or two," Mosely recalled. "We had some good times there in that place. Johnny told everybody it was Mickey Mantle, put up with whatever he gives you, you know. And they did. They had a bouncer there that Mickey started callin' Hoghead. I said, 'Mick, let up. That guy's bigger 'n both of us.'

"Well, Hoghead wouldn't do nothin' 'cause Johnny'd already told him, 'Whatever Mick does, that's all right.'"

2.

Back home, the newlyweds rented a room in a cheap motel near the bowling alley in Commerce—Dan's Motor Court. "I was named for that motel," their youngest son, Danny, would tell me later. "It was a dump," his mother recalled. "It had an open gas fire. Mutt would come every night to see if we were all right. He was scared to death we'd get gassed."

By then, Mutt wasn't sleeping much. He was in too much pain to lie down. "He had a color," his nephew Jim Richardson said. "Kind of a yellowish brown."

After the wedding, Mickey and Merlyn took him to the Mayo Clinic in Rochester, Minnesota. They stayed five nights in a fleabag hotel while doctors performed exploratory surgery. "One night we were both lying in bed reading and watching TV, and Mick scratched his head and he had crabs," Merlyn told me. "We jumped up. We both started washing our hair. Probably a dollar-a-night place. That was awful."

The medical report was icier than the roads they negotiated at fifteen miles per hour on the long, painful trip back to Oklahoma. "Take him home," the doctors said. "Let him die in peace."

Mantle avoided physicals for years.

Mutt had watched his father, Charlie, and his

brother, Tunney, melt away with cancer. "When my Dad died, he had it in his stomach," his son Max said. "We found out on July 4. On August 6, he died. He went from 225 to 90 pounds in a month. When it was Uncle Mutt's turn, he didn't want the rest of us to see."

Three weeks after Mickey and Merlyn went to Florida for spring training, Mutt and Lovell left for the Spears Chiropractic Sanitarium and Hospital in Denver in Paul Thomas's ambulance.

"He didn't want to die around his kids," Larry said.

In Mantle's retelling, Mutt always dies a hero's death, a lonesome, solo voyage into the hereafter, going it alone in order to spare his family. But Lovell never left his side. An aunt and uncle stayed with Larry, Barbara, and the twins. They never saw their father alive again, and they didn't see their mother for two months.

It was a 700-mile drive to Denver on rutted, rudimentary roads and a misery for all. Thomas and his wife, Wanda, sat up front. Lovell sat in back with Mutt. He was too ill to talk, too weak to sit up. "Too sick to make a trip like that," Thomas said. "That stuff was all over the body. We drove out to Liberty, Kansas. I said, 'We better stop and get a motel and get some rest.' That wind blowed that night. I mean, I thought that wind would blow us away."

The Spears Chiropractic Hospital had been the subject of litigation and controversy since it was founded in 1943. Its controversial proprietor, Dr. Leo Spears, papered the Midwest with advertisements promising cures for everything from cerebral palsy to muscular dystrophy using the "Spears Painless System" of spinal manipulation. A glossy forty-eight-page brochure trumpeted a new "Chiropractic Answer to Cancer . . . Sensational Guarantee . . . Cancer *Relief* or Money *Back*!"

Mutt probably did not see the May 26, 1951, issue of *Collier's* magazine that listed Spears among America's most infamous "Cancer Quacks." Spears sued *Collier's* for $24 million. "At trial he admitted that five out of six persons giving testimonials in the Spears cancer pamphlet were actually dead," according to *At Your Own Risk: The Case Against Chiropractic*, by Ralph Lee Smith. "It also came out that Dr. Leo did not recognize a malignancy in a child that was brought to the hospital; she was treated for rheumatism. He lost the case."

Paul and Wanda Thomas left Mutt and Lovell in Dr. Leo's care and headed home. "When my wife and I got back," Thomas said, "they called and said, 'Come back and get him. There's nothing we can do for him.'"

Merlyn's father, Giles Johnson, and Ted Davis volunteered to make the trip. "Merlyn's dad had epileptic seizures, and he was driving the ambulance," Barbara said. "He went into one of those seizures and Theodore was trying to get the wheel, but he was so strong with that seizure that he had to literally kick his leg off the gas."

When they got to Denver, Thomas said, "Mutt told them, 'Well, boys, I'm not going to go. They said they can cure me.' They turned around and come on back."

3.

Mickey and Merlyn Mantle's first marital address was a swank one. In 1952, the Concourse Plaza Hotel at the corner of 161st Street and the Grand Concourse was still the locus of Bronx society—its gilded ballroom hosted everything from bar mitzvahs to the annual Yankees welcome-home luncheon. Its residential apartments offered temporary refuge to generations of Yankee rookies and their brides. The apartments varied. Suites had white linen and room service; efficiencies had Murphy beds. When Frank Scott, the Yankees' traveling secretary, showed Yogi and Carmen Berra their first apartment, Yogi said, "Whaddya supposed to do, sleep standing up?"

The Mantles' efficiency apartment had no air-conditioning and no television. One of those cost $10 a month to rent. They had four walls, a bed, a chair, a closet, and a telephone. In hot weather, they put on their bathing suits and positioned themselves in front of a fan to watch TV in Billy Martin's apartment.

When the Yankees were home, they ordered lobster downtown at Toots Shor's and prayed that Toots would cover the check. When the Yankees were on the road, their wives depended on the generosity of the bellboys to augment their hot-plate meals and accessorize their bleak accommodations. Joey the bellhop saved the leftovers and opulent floral arrangements from the weddings and bar mitzvahs held in the ballroom. "Desserts, lots of desserts," said Donna Schallock, whose husband, Art, pitched briefly for the Yankees in 1952. "He'd say, 'Go down and take what you want before they throw them out.'"

Since blue jeans were not permitted in the lobby, women who wore them were asked to use the freight elevator that serviced the hotel kitchen. That's how Schallock acquired a complete set of professional copper pots and pans, which she was still using six decades later. When the Schallocks moved out, Joey the bellhop asked, "What have you got, half the hotel?"

When the Yankees left Florida for the opening of the season, Merlyn drove north alone, a daunting journey that foreshadowed years of loneliness as a baseball wife. The transition to New York was even more over-whelming. She didn't know how to put on makeup. She didn't know how to dress and she couldn't afford the right clothes anyway. "Didn't know how to be," she told me.

When Donna Schallock took her shoe shopping on Fifth Avenue, she wrote a check and signed it "Mrs. Mickey Mantle." "Yeah, right," the salesman said. "Wait, she *is* Mrs. Mickey Mantle," Schallock told him. "Oh, yeah, she got the shoes."

As a new Yankee wife, Merlyn was still blissfully oblivious to the more blatant expressions of adoration showered on her handsome young husband. She was clueless about the peroxided and painted exotica that fluttered around celebrities at Manhattan night-spots and hotel lobbies. She had not read the New York gossip columns.

Tom Morgan and Gil McDougald were in their second year with the Yankees, and their wives tried to make Merlyn welcome. Tom's wife, Wanda, invited Merlyn to stay at their home in New Jersey while the Yankees were away on a road trip. When Lucille Mc-Dougald met the two women there, she blurted out to Merlyn, "Thank God you two got married and you're here.

"And she said, 'Why do you say that?'

" 'Oh, well, now we can be done with these nasty headlines.' "

When Lucille left, Merlyn asked Wanda Morgan just what she had been talking about. She told Merlyn about the juicy tabloid items devoted to her tomcatting young husband. "Apparently, when he came back from the road trip, she lit into him like a blue dart," McDougald recalled. "And he would never talk to me again after that."

Merlyn couldn't help but notice the girls who dawdled on 161st Street, waiting for him to make his way up the hill from the ballpark. She loathed their audacity, how they grabbed at him as if he belonged to them, not to her. Their bold sense of entitlement was her first intimation that she had married public property.

It was an uncertain time. He was still limping when he reported to spring training as DiMaggio's heir apparent. The *Times* labeled him "an uncertain factor on the physical side," pointing out that he had not "shown any of the hustle and assurance one would expect of a young man about to move into an important post."

On May 3, the Yankees traded the other Golden Boy, Jackie Jensen, to Washington for Irv Noren. Jensen had been Mantle's rival for the throne but had fallen into disfavor. Noren was an excellent outfielder and contact hitter. The trade gave the Yankees insurance in case Mantle's knees didn't hold up. But it also cleared the way for him to make center field his own, news he ordinarily would have shared with his father. He had not spoken to Mutt since leaving for spring training the third week of February. He had not written to his father, either.

4.

Tuesday, May 6, was a rainy day in New York. Showers would dampen the gate at the game against the Cleveland Indians that evening. The Yankees had been expecting a big early-season crowd. From the apartment window, Mantle watched the cars on the Grand Concourse splatter the pigeons in residence on the island dividing the Concourse. The field would be wet too.

He was getting dressed to go to the ballpark when the phone rang. It was Casey Stengel. Lovell Mantle, who had grown up in a world where baseball was played in afternoon sun, had called the Stadium assuming that that's where she would find her boy.

Stengel's message was succinct: Mutt Mantle had died at 10:30 A.M. He was just forty years old.

Mantle pummeled the wall with rage and resisted Merlyn's attempts to console him. He broke free from her embrace, saying he would make arrangements to fly home in the morning. There was no need for her to come. He remembered slamming the door in her face as he left for the ballpark. She remembered being ordered from the room, banished to the hall, as she had been the previous summer during Mutt's do-or-die lecture in Kansas City. Shut in or shut out: either way, it hurt. This was patrimony, not matrimony.

He played that night because his father would have wanted him to, he wrote in *The Mick*, a narrative of unyielding filial devotion. The box score from May 6, 1952, tells a different story: the Yankees lost 1–0. Noren played center field and grounded out with bases loaded in the third inning. Mantle remained in the dugout in uniform. He had not been able to get in touch with his mother.

Baseball wives are used to being left behind: to make a home, to raise the children, to kill time while trying not to think about how their husbands are filling their empty hours. But when Mantle left for his father's funeral alone, Merlyn was devastated. What would people think back home? Hopefully, they would consider the cost of travel prohibitive. "I never did know why I wasn't invited to the funeral unless he just wanted it to be with his family," she told me fifty years later. It was a harsh way to learn her place in her husband's life.

Mantle found his father laid out in an open casket in the front parlor of the home he had purchased with his 1951 World Series check. "Had him in a shirt and tie," said Larry. "Probably weighed only eighty pounds," said Barbara.

Larry was ten years old. The boy cowered in the parlor corner, hoping no one else would arrive to pay their respects and make him talk about it all over again. But come they did. "Every time somebody new would come, it would seem like they would come and get me and take me over to the casket and tell me how sorry they was," Larry said. "I would no more get over cryin' and get away from there and about that time, here come somebody else. The next thing I'd know, they'd drag me up there to that casket. It seemed to me like it just happened for two days, just continually."

The funeral was held at the First Christian Church in Commerce at 2 P.M. on Friday, May 9. Though Lovell's father had been a Methodist deacon, the Mantles weren't churchgoing folk, which made them unusual in a community where every day's labor was a leap of faith. "When we pulled up to get out to go in, it just seemed to me like everybody in the world was there," Larry said. "Most people I remember seeing gathered at one place in my life."

Mutt was buried in the Grand Army of the Republic Cemetery along Route 66 between Miami and Commerce. When it was time to leave him in the ground for good, Mickey refused to go with the rest of the family. He stayed behind, berating himself for never having told his father he loved him.

5.

The family plot, with its hard stone markers, became the locus of the legend of damned Mantle men claimed before their time by non-Hodgkin's lymphoma. "It just seemed that all my relatives were dying around me," he wrote in *The Mick*. "First my uncle Tunney, the tough one, then my uncle Emmett. Within a few years—before I was thirteen—they had died of the same disease."

This belief became so central to his being that he conflated ages, gravesites, and perhaps even cause of death. In fact, Grandpa Charlie lived until age sixty and is buried some fifty miles away in Adair, Oklahoma. There is no hard evidence of the kind of cancer that took his life. Tunney, the first Mantle man to die too young, is buried near Mutt in the G.A.R. Cemetery. His son, Max, who lived at Mickey's house while his father lay dying at home, said Tunney died of stomach cancer. "I didn't hear about Hodgkin's until Mutt died."

Like Tunney, Emmett Mantle, the youngest of

Charlie's sons, also died at age thirty-four, but not until 1954. He is buried in Tulsa, Oklahoma. By then, the convergence of familial and cultural fatalism and Mantle's own brush with a life-threatening disease had merged into a personal narrative that imposed structure on fear and attempted to keep it at bay.

Mutt's dread diagnosis of non-Hodgkin's lymphoma was first made at Lenox Hill Hospital in October 1951. A year earlier, Mantle told a teammate that Mutt had miner's lung; he told Pat Summerall the same thing thirty years later. Paul Thomas assumed as much and was surprised at the cause of death penned on the death certificate still on file in the garage of his funeral home fifty years later: "carcinoma of the bowel with generalized metastasis."

Whatever the cause or causes of Mutt's death, it was the defining moment of his son's life. Mutt had decided how he would make his living, what position he would play, and what side of the plate he would bat from and when. Mutt decided when he would receive his high school diploma, whom he would marry, and when he would marry her. They weren't bad choices, but they weren't his own. "There was never any talk about what I'd be in life," Mantle once said. "Dad and I knew I was going to be a ballplayer."

Someone else would always decide—father, coach, manager, the American League schedule maker. As his friend Joe Warren said, "When you don't raise your children to make their own decisions, then they grow up and they don't know *how* to make decisions."

Just barely out of his teens, he accepted Mutt's responsibilities (much as Mutt had accepted responsibility for Lovell's two young children) and took on the obligation to live for him. "When he was alive, I was Dad's life," he would say. "Now, making good for Dad is my life."

Without Mutt, there was no one with the moral authority to insist, no one to say no to Mickey Mantle. He would never grant anyone that authority again. And, his brother Larry said, "No one challenged him."

Without Mutt, he was adrift, save for the organizational imperatives imposed by the baseball season. Free to make his own decisions, he made bad ones. "He wasn't under anybody's finger anymore," Merlyn told me. "He could do what he wanted."

A day after the funeral he headed back to New York and to the life and the wife his father had bequeathed him. "He was different," she told me. "For one thing, he was going to have to take care of a wife, mother, and four kids. He was worried. He was making peanuts. The Yankees told him if he was still there in June or July they would raise him to $10,000."

Two weeks later, Mantle took sole possession of

center field—47,000 square feet of prime Bronx real estate. Built atop landfill displaced by the construction of the Grand Concourse, on the site of a former silent movie studio, it was home to a legion of ghosts, Yankee greats and great Yankee fans who importuned the right people to have their ashes scattered by the monuments in center field. "You got to feel the glow of the ghost," former tenant Mickey Rivers said. "Not just the living ones but the dead ones, too."

Mantle was possessed by his father's ghost. Sometimes they had imaginary conversations in the outfield—one-sided as those talks might be. *That home run, the one that went 500 feet? It should have gone 502.* Mutt's ghost would remain the animating force in his son's life for the next forty years.

A last family photograph, taken on December 15, 1951, captured the Mantle men at the dining room table playing canasta. Mickey is the center of familial and photographic attention. Sleeves rolled up, collar undone, he reaches forward, putting his cards on the table. His crisply pressed dress shirt strains against the seams. Mutt sits to his left, his chest sunk inside a flaccid undershirt, his proud thinning hair, brushed away from his brow, the way he liked it, accentuating the hollows of his cheeks and the caverns of his eyes.

The twins fill out the foursome. Butch stands at

Mutt's shoulder. No one at the table is making eye contact. Lovell presides over her brood, Donna Reed style, gazing over Mickey's shoulder at his cards. Her right hand rests on his back as he plays the hand he has been dealt.

6

April 17, 1953

ONE BIG DAY

1.

Standing in the capacious outfield at Washington, D.C.'s, Griffith Stadium during batting practice, Irv Noren glanced at the Mr. Boh sign atop the football scoreboard in left field and told Mantle: "Geez, you might be able to hit one out of here today."

Noren knew all about the stadium's prevailing winds, when the ballpark held the heat and when the breeze blew through the open grandstands. He knew where the ball carried and where it had never gone before. He had played two years for the Senators before being traded to New York in May 1952 as insurance against Mantle's infirm right knee.

Noren knew that Babe Ruth had hit a ball into the graceful crown of an oak tree on the other side of the center field wall; that Larry Doby had hit a ball over the thirty-one-foot wall in right field, prompting an irate call to the Senators' front office: "Someone from your stadium just threw a ball onto our house and woke up my children, and now I can't get them back to sleep."

He also knew that no one in major league history had ever hit a ball over the thirty-two rows of poured-concrete bleachers erected in left field just in time for the 1924 World Series, the only one the Nats ever won. Noren thought that Mantle might be the man to do it and this might be the day. "I played there two years. I knew the ballpark pretty good. The wind was blowing out a little—not a gale. And I always thought he had more power right-handed."

April 1953 was as kind to Mantle as the previous spring was cruel. Then he was expecting his father's death; now he was awaiting the birth of his first child. His draft status—and his place as DiMaggio's heir—had been resolved. He arrived at spring training as a World Series star who had batted .345 against the Dodgers, driving in the winning runs in game 7, and had outwitted Jackie Robinson on the base paths. "That young man's arms and legs and eyes and wind are young, but his head is old," Branch Rickey had said. "Mantle has a chance to make us forget every ballplayer we ever saw."

On April 9, in an exhibition game in Pittsburgh, he hit a ball onto the roof of Forbes Field, a 450-foot effort that duplicated Babe Ruth's last major league home run. The Babe would have been impressed: the night before, Mantle, Martin, and Ford had managed to miss the train from Cincinnati while cavorting across the river in Covington, Kentucky. They had paid a taxi driver $500 to drive them to Pittsburgh, arriving just in time for batting practice.

Three days later, at an exhibition game in Brooklyn, Mantle was chatting with the home-plate umpire at Ebbets Field when the public address announcer greeted him with news from the stork in Joplin, Missouri: "Mickey doesn't know it yet, but he has just become the father of an eight-pound, twelve-ounce baby boy."

They named him Mickey, Jr., although his full name was Mickey Elven, after Mutt, not Mickey Charles. Later, Merlyn would regret saddling her oldest son with his father's name. But in the spring of 1953 Mantle was not yet fully who he would become, and neither he nor Merlyn could envision how much of a burden that "Junior" would become. Five days later the proud papa, who did not meet his son until June, was shagging flies in the Griffith Stadium outfield with Irv Noren and Johnny Sain. "It was a gray, overcast day with a wind blowing directly out to center field," said Bob Wolff, the Senators' broadcaster, who got paid to notice such things. "And the flags were unfurled in that direction."

Chuck Stobbs, a lefty newly acquired from the Chicago White Sox, was the Senators' starting pitcher, a last-minute decision by manager Bucky Harris, who had noted the Yankees' opening day loss to another left-hander but had forgotten Mantle's game-winning grand slam off Stobbs the year before. "Stobbs is pitching," Noren told Mantle. "And you can hit him pretty good."

The weather had played havoc with the first week of the season, forcing cancellation of the Senators' home opener set for Monday afternoon, April 13. President Dwight D. Eisenhower, away playing golf in Georgia, did not intend to make the traditional first heave. *Washington Times-Herald* reporter Jacqueline Bouvier was dispatched to the stadium to query the Nats about the presidential snub. They expressed more interest in her. "She's going with some senator," a companion informed them. The president returned to D.C. in time for Thursday's rescheduled festivities. By Friday afternoon, Ike, Jackie, and the capacity opening-day crowd had disappeared. Paid attendance was 4,206, an embarrassment camouflaged by a throng of 3,000 boys in the upper deck along the third base line who had gotten in free. It was Patrol Boy Day, an annual event in the nation's capital.

With the Yankees leading 2–1 in the top of the fifth inning, Stobbs committed pitching's cardinal sin: he walked Yogi Berra with two outs and the bases empty, bringing Mantle to the plate. Later, in his coaching life, Stobbs always admonished young pitchers, *no two-out walks.*

Stobbs was a three-letter man at Granby High School in Norfolk, Virginia, where baseball was his third-best sport. He was just eighteen, the youngest player in the major leagues, when he made his debut with the Red Sox in 1947. Now he was in his seventh season, playing for his third team and making his first start for the Senators. He was only twenty-three, but he had gotten old early. He didn't throw very hard, and that spring he hadn't thrown very much. Shoulder stiffness had limited his innings during spring training.

Mantle wasn't feeling up to par either—he had pulled a muscle in his left leg the day before, one of those early-season injuries that would plague him throughout his career. Charley horse was the official diagnosis. He stepped to the plate, as he often did, with a borrowed bat, a 34-ounce, 34½-inch Louisville Slugger belonging to teammate Loren Babe, who hit a total of two home runs in his major league career.

As Mantle settled into the batter's box, he was greeted by the friendly visage of Mr. Boh 460 feet away, in deep left field. Mr. Boh was the one-eyed, mustachioed mascot of the home brew, National Bohemian, which could not be sold in the stadium by municipal ordinance.

Left field in Griffith Stadium was as vast as center field in Yankee Stadium—405 feet down the line. Though the fence would eventually be brought in to accommodate a beer garden and the visitors' bullpen, it would always be a pitcher's ballpark, allowing 41.7 homers per season compared to the major league average of 81.5. That forgiving acreage was the reason Bucky Harris thought a control pitcher like Stobbs would flourish in Washington.

Griffith Stadium's construction was minimalist: exposed steel girders, concrete, and brick. The ungainly structure was the architectural equivalent of a mismatched suit. Its charms, such as they were, were sensory and supplied by the surrounding neighborhood—the smell of bread rising at the Wonder Bread factory on Seventh Street, where the stadium vendors purchased hot dog rolls (bringing them back warm between games of a rare sold-out doubleheader), and the joyful sound of African-American spirituals from Elder Solomon Lightfoot Michaux's church greeting dispirited fans after another desultory loss. There was a small rectangle of land gouged out of center field because the owners of the abutting homes refused to sell when the ballpark was built in 1911. That tall, stubborn oak stood sentinel behind the center field wall.

Mantle took Stobbs's first pitch for a ball. In the Yankee dugout, benchwarmer Jim Brideweser turned to coach Jim Turner and said, "You know, I bet this kid could hit that big scoreboard."

"Naw," said Turner. "Nobody could do that."

The second pitch was either a fastball or a slider, Stobbs told reporters later, he couldn't remember which. Either way it was right over the plate. As Stobbs went into his windup, a gust of wind blew through the open facade behind home plate. "Straight out to left field," recalled Bill Abernathy, a patrol boy who had opted to sit with his father in the vacated presidential box.

Sam Diaz, an observer at the Weather Bureau, would later report: "Between 3 and 4 p.m. there were gusts up to 41 miles per hour in the direction of the bleachers at Griffith Stadium . . . the lightest at 20 mph."

Bill Renna, a spare Yankee outfielder, watched attentively. "They threw him a changeup, I think. He moved forward but kept his bat back and took a short step and held back because it was a changeup. Then it released: body, arms, bat came around, really synchronized. Everything went in a smooth swing. Everybody in the dugout got up and moved forward onto the steps. We just kept walking forward in unison watching the flight of the ball."

The ball left his bat traveling at an estimated speed of 110 miles per hour. Clark Griffith, the namesake and grandson of the Senators' owner, was sitting in the family box behind the third base dugout, having cut class at Sidwell Friends School for an afternoon of baseball. "It went up and got caught in the jet stream," he said. "It took on a life of its own."

The thwack of contact resounded through the empty stands. The sound would stay in the memory of Roy Clark, the musical son of a Washington square dance bandleader, sitting with his father along the first base line. "It just echoed in that ballpark," Clark said. "Even before it was halfway to its destination, you knew that it was gone. Looked like it was in the air for five minutes."

The ball kissed Mr. Boh's cheek, clipping his handlebar mustache above the word "beer" as it headed out of the ballpark toward Fifth Street, NW. The visiting bullpen down the left field line offered an unimpeded view. "You're waiting for it to come down, to go into the crowd," backup catcher Ralph Houk said. "The next thing it's over the crowd and out of the stadium. There's a moment of silence. Everybody is looking that way—even all the infielders on the opposing team and the left fielder. He's looking for it, and he can't believe it went out."

Mantle rounded the bases with his customary modesty, head down as he touched each bag. Contrary to later reports, he did not giggle while rounding first base, not within earshot of first baseman Mickey Vernon, anyway. The ball was hit so high that Mantle was at second base by the time it came down, testified second baseman Wayne Terwilliger. On the mound, Chuck Stobbs hung his head and dropped his glove, Abernathy recalled. He may have been the only person in the ballpark paying attention to the pitcher. "His glove fell off his hand," Abernathy said. "He just looked down at the mound."

It was Mantle's first home run of the season, the first of twenty-nine he would hit at Griffith Stadium. Returning to the dugout, he smiled in a way that acknowledged his debt to the wind. "Yes, he knew it," Gil McDougald recalled. "He didn't get all of it. I mean, Billy Martin, for God sake, hit one on the label and it went for a home run."

Years later, after Roy Clark had become a renowned country singer and Mantle's good friend, he told him he had been at Griffith Stadium that day. "He looked like a kid on Christmas morning—just a big grin," Clark said. "And then he immediately said, 'That wasn't the hardest ball I ever hit.'"

Upstairs in the press box, Arthur E. Patterson, the Yankees' director of public relations, regarded the ball's disappearing act from the vantage point of opportunism. Marketing The Mick was his job. Patterson was an old sportswriter who had spent twenty years at the *New York Herald Tribune* before making the seamless transition from hack to flack, first for the Yankees and then for the Dodgers. He knew a good story when he saw one—even if he couldn't actually see what happened. The left field bleachers cut off the view from the press box.

"That one's got to be measured!" Patterson declared. Or so legend says. He dashed from the stadium on a gust of inspiration. "To his dismay the baseball already landed when he arrived, so he picked out the spot where it might have come down," Red Smith wrote years later. "To this day, the measurement is regarded as exact."

Actually, measuring the thing was the suggestion of New York *Daily News* beat writer Joe Trimble, and he meant it as a joke. But Patterson immediately saw entrepreneurial potential. Some innings later, he returned with a baseball and a story he would tell in a variety of iterations until his death in 1992. He had arrived on Fifth Street, a residential block lined with row houses and billowing oaks, to find "a surprised and delighted Negro lad" (*The Sporting News*) named Donald Dunaway running down the street with a bruised baseball, its cowhide scraped like a child's knee. They quickly entered into a mutually beneficial arrangement—the boy, who lived around the corner at 343 Elm Street, NW, would show him where the ball had landed in exchange for whatever money Patterson happened to have in his pocket. The sum was variously reported as 75 cents, a dollar, $5, or even $10.

The deal struck, Patterson testified, Dunaway led him to the backyard of 434 Oakdale Place, a modest two-story attached brick row house on the south side of a one-block street that dead-ended at Fifth Street and the back of the left field wall.

Hustling back to the press box—afternoon deadlines were looming—Patterson breathlessly reported that the ball had traveled 565 feet, making it the longest home run ever measured. None of the intrepid residents of the press box ventured out of the stadium to interview the mystery boy or to make an independent attempt to verify Patterson's claim. He never said he had employed a tape measure; nor did the word appear in any of the morning papers. The not-so-sweaty literati knew better, as Bob Addie wrote in the *Washington Times-Herald* three days later, in a column taking his readers "behind the scenes to show you how these records are determined."

" 'Here's the dope,' panted Red. 'The fence is 55 feet high to the beer sign. I walked 66 feet from the 391 mark to the back where Mantle's ball cleared the bleacher limit. That would be 457 feet. Now I paced off 36 strides, which means three feet a stride or 108 feet to where the ball eventually landed in the backyard on Oakdale St. It's a small backyard so the ball didn't have a chance to bounce much. So add them all up it's 565 feet.' "

The announcement was piped into Bob Wolff's broadcast booth and he dutifully and enthusiastically reported it. "He was the Yankee PR man, so you accepted what he said." The next morning, Mickey Mantle and Donald Dunaway were front-page news in every sports section in America. "The magnificent moppet of the Yankees today hit the longest home run in the history of baseball," Trimble declared, even after subtracting three feet from the total.

"Other things happened," Louis Effrat reported in

the *New York Times,* "but no one appeared to be interested."

The *Washington Post, Washington Star,* and New York *Daily News* published panoramic photos of the stadium with an arrow tracing the ball's breathtaking trajectory. Ten feet longer than the Washington Monument is high! "The neighbors thought it was a flying saucer," *The Sporting News* reported.

When the Yankees were rained out the next day in Philadelphia, New York's prolific scribes had plenty of time and column inches to expand upon the feat. The beatification continued unabated all spring. In June, *Time* magazine stationed him on the cover. A corner banner proclaiming "Coronation, Four Pages in Color" actually referred to Britain's new queen, but Yankee fans interpreted it otherwise.

Over the next two months he hit baseballs out of three more ballparks: April 28, right-handed, Busch Stadium, St. Louis; June 11, left-handed, Briggs Stadium, Detroit; July 6, right-handed, Connie Mack Stadium, Philadelphia.

The Spalding Company felt compelled to deny excessive liveliness in its baseballs. *The Guinness Book of Records* was rewritten. Mantle agreed to write his first autobiography. He also signed deals with Wheaties, Camels, Gem razors, Beech-Nut gum, Louisville Slugger; and endorsed a whole wardrobe of Mickey Mantle–sanctioned clothing, Esquire socks, Van Heusen shirts, Haggar slacks.

There was one downside to the publicity. Several Washington sports columnists including Addie received irate telephone calls from aggrieved parents whose sons had been drafted by Uncle Sam: "If that boy can hit that long a drive, why isn't he in the Army?"

The Hall of Fame called to request the bat and ball. Patterson promised to send them after a suitable viewing at Yankee Stadium. A display case was constructed and placed in the Stadium lobby. Sometime between Friday night, May 29, and Sunday, May 31, the day before the sacred relics were due to make their pilgrimage to Cooperstown, the ball was stolen. "Apparently the bat was wired too securely to the back of the case and couldn't be wrenched loose," the Associated Press reported. "The Yankees quickly absolved Chuck Stobbs."

Mel Allen took to the airwaves to appeal to the conscience of the thief. The ball was returned. Stobbs was demoted to the bullpen, having lost five of his first seven starts. He would be remembered for one pitch in a fifteen-year major league career. His initial good humor—"He really hit the heck out of it, didn't he?"—eroded when he returned to the park one day to find a large, white painted ball on the spot where Mantle had knocked the smile off Mr. Boh's face.

It was all anyone ever wanted to talk to Stobbs about—except perhaps a wild pitch that landed seventeen rows up into the grandstand, also a major league record. Every April 17, his friend Bob Kleinknect, who worked the concession stand behind home plate, and missed the home run, sent an anniversary card. Sometimes he signed Mantle's name, sometimes he added a note, *Thank you for what you did for me.*

Long before his death from throat cancer in July 2008, Stobbs stopped talking about the home run that defined his career as much as it did Mantle's. It was no different from the "other big ones," Mantle said later, "except that Red Patterson attached a number to it." In so doing, he elevated Mantle to a new level of hype: "Young Man on Olympus," *Time* called him.

Ordinary language could not contain him. A new term was coined: the Tape Measure Home Run! The first reference to the putative tape measure may well have been in Shirley Povich's annual Christmas column in the *Post* cataloguing Santa's largesse. Among 1953's better gifts, Povich listed "that tape measure."

Three years later, between games of a July 6 doubleheader, the Northern Virginia Surveyors Association presented Mantle with a 600-foot, gold-plated tape measure, which now resides behind glass in Mickey Mantle's Steak House in Oklahoma City: "Presented to Mickey Mantle for hitting 585 ft. home run at Washington, May 1956," an engraved plaque attests, mangling the truth even further.

By then, Patterson had decamped for the Dodgers. He had done his job.

Rarely has so much ridden on one ball. The modern era's obsession with clout, the language of home-run power (going deep, dial 8 for long distance), the nightly recapitulation of each day's blasts in smoothly edited highlight packages accompanied by percussive thwacks and cracks and booms can all be traced to April 17, 1953.

Left in the ball's wake were three unsolved mysteries: where did it go, how did it get there, and what became of Donald Dunaway?

2.

In 1953, Griffith Stadium was a white man's palace—albeit a homely one—standing on the edge of a black neighborhood called LeDroit Park. The ballpark occupied the former site of a Civil War hospital dedicated to the care of freedmen and sat just to the south of Howard University, the predominantly black college chartered by the federal government at the war's end. In 1873, a Howard professor and trustee named Amzi Barber purchased 40 acres of college-owned land and built a "whites-only" gated community conceived as a pastoral village located on one of the city's main commuter trolley lines.

The Gothic cottages, stately row houses, Italian villas, and Victorian mansions shared a common green. The neighborhood was protected by restrictive covenants and by a fence, brick and iron on the southern border, wood on the northern boundary adjacent to Howard Town. The barrier was an indignity and inconvenience for residents, who had to walk a mile out of their way to get to public transportation. Twenty years of "fence wars" eventually resulted in the destruction of the wall, the exodus of the white population, and the rise of LeDroit Park as the nexus of African-American culture in the nation's capital. It was the logical place to look for Donald Dunaway, the only eyewitness to the denouement of the Tape Measure Home Run. If he was alive. If he ever existed.

There were no Donald Dunaways listed in the Washington, D.C., phone book for 2006–2010; in the 1954 D.C. directory there were no Dunaways at the 343 Elm Street address he gave to the newspapers at the time. None of the seventy-two Don, Donald, or

Donnie Dunaways or Dunnaways of his approximate age listed on www.whitepages.com had ever heard of him. There was no Donald Dunaway (one *n* or two) in D.C. public school attendance records for Lucretia Mott Elementary School, the "colored" school attended by most LeDroit Park children (Gage-Eckington, an elementary school at the end of his block, was still a year and a world away from Supreme Court–imposed desegregation). Nor was he enrolled at nearby Garnett Patterson Junior High School.

The Hall of Fame and the Yankees' front office had no updates in their files. Letters and newspaper clippings left in every mailbox on Elm Street and Oakdale Place elicited no reply. Many of the phone numbers for Elm Street addresses were disconnected.

I hired a private eye. She found no trace of Donald Dunaway in Social Security death records or U.S. military service records. She suggested a deed search for 343 Elm Street. There was no Dunaway on the deed. Neither the current owner nor the owner before her had ever heard of the family.

The men gathered under the food tent at the annual LeDroit Park reunion had trouble placing the name. *Dunaway? Dunaway. Yes, there was a Dunaway. No, there wasn't. I thought Albert Taylor caught that ball.* Bobby Lane, the unofficial neighborhood historian, put an end to the discussion: "There ain't no tape measure, and there ain't no Dunaway."

Brad Garrett, renowned former FBI special agent, was engaged. Garrett spent five years tracking—and finding—Mir Aimal Kansi, the perpetrator of the 1993 CIA murders, and obtained confessions from the D.C. Sniper and the first World Trade Center bomber. But Donald Dunaway eluded him. After consulting secure databases and racking his brain, he said: "This guy is harder to find than Kansi."

He recommended shoe leather.

LeDroit Park was added to the National Register of Historic Places in 1974, but in the summer of 2007 its renaissance had not yet reached Dunaway's old block. Some houses had "for rent" signs spray-painted on plywood doors. A hand-lettered sign in the second-floor window of 343 Elm Street advertised its availability. Some neighbors were never at home; others spoke no English. A man at the end of the street who gave his name as Clarence said, yes, he knew Donald. They had gone to school together, parked cars together, broken windows out of the stadium together. "He passed away, but I don't remember what year."

Then a gracious woman named Sandra Epps appeared on Oakdale Place, offering to make introductions to longtime residents. After two years of shoe leather, mailbox stuffing, and unanswered phone calls, doors opened. Miss Rosa Burroughs invited me into her parlor, directly across the street from the Dunaways' former home.

Yes, Miss Rosa said, she knew the family. He was slight and had a light complexion. Yes, Donald was alive. She had seen him at the bus stop at Fourteenth and P Street just the year before. But she did not know how to get in touch with him. Perhaps her friend Miss Sarah would. She would ask next time she saw Miss Sarah at bingo.

Six months later, Miss Sarah, who had been feeling poorly, returned to bingo. Oh, yes, she remembered Donald and his sister Maxine, the wife of Elder Walter McCollough, the pastor at Bishop C. M. "Daddy" Grace's United House of Prayer. The church provided a home phone number.

One night, Maxine McCullough returned my call. "Yes, he caught the ball," she said. "But why don't you ask my brother?"

3.

"You couldn't have been looking very hard," Donald Dunaway said when he answered the telephone in his apartment less than two miles from where he had found the ball.

Approaching his seventieth birthday, he was not

in the best of health. A small man, perhaps five feet, five inches tall, he had been laid low by diabetes and "the arthritis." A pair of metal spectacles magnified his rheumy eyes, and his fingernails were long and curled. A gray-speckled beard and a watch cap pulled low on his brow failed to obscure his good-humored smile. But he did seem a bit stung to hear that the boys from the old neighborhood did not recollect his name. "But they remembered Duckie, didn't they?"

Duckie was his street name, bequeathed by an uncle who thought he waddled like a duck as a toddler. Although he had been apartment- and wheelchair-bound for most of a year, he gamely agreed to take a trip back in time to his old stomping grounds. Walking with a cane, he shuffled to a waiting car. I drove. He provided the directions. "There," he said when we reached the intersection of Fifth and Oakdale. "Right *there*."

His account of April 17, 1953, was different from the codified version of events. Sometimes he even contradicted himself. But he was adamant and consistent about the big things. He did not see the ball land on the fly in the backyard of 434 Oakdale Place. And he never showed anyone where he had found it.

He was fourteen years old—not ten, as Red Patterson had reported—a sixth-grader at Bundy Elementary, a school for hard cases such as himself. "I was mischievous," he said with a mischievous smile. He was a member of the LeDroit Park gang and proudly wore the red and black colors. "Like the Bloods and Crips now," he said. Only without knives and guns. "They knew back then how to fight. We always used these two hands right here."

School was not a priority. He had already repeated two grades when he was transferred from Mott to a school twenty minutes away. "Because I was bad. Bad enough to go to Bundy School. It was for slow learners."

On Friday afternoon, April 17, he said, "I hooked. I snuck out of school at recess and kept going."

He walked down New Jersey Avenue to the ballpark, which was less than two blocks from the rented two-story row house he shared with his mother, grandmother, sister, and half brother. The stadium lights, installed in 1942, cast a beneficent glow over the neighborhood and the Dunaways' front porch. Duckie was a Senators fan and would swing on the porch listening to the game on the radio. Like a lot of neighborhood kids, he had an entrepreneurial interest in Griffith Stadium. Sometimes he sold scorecards, sometimes he ran errands; mostly he hung out in the parking lot behind home plate and waited for foul balls. If he got lucky and the security guards didn't chase him away, he might catch as many as three on a good day and sell them to departing fans for $1. "It was what you call my hustle," he said.

Usually he sneaked into the ballpark—it wasn't hard. There was a lumberyard across the way on Seventh Street and plenty of boards lying around to help scale the back wall opposite the Freedmen's Hospital Morgue. But on April 17, 1953, he was a paying customer. "Had me some money," he said, savings from selling the *Afro-American* and the *Pittsburgh Courier* inside the stadium.

He bought himself a 75-cent ticket for the left field bleachers and took a seat on the concrete benches a row or two above the left field fence. "Down low, close enough to touch the ballplayers."

He marked his seat for me on a photograph that appeared in the *Post* the next morning. He had a fine view. He saw the ball head in his direction, saw it hit off the beer sign perched above the bleachers, watched as it headed out of the yard on a trajectory that might have carried it into his own backyard two blocks away save for the intervening row houses and the laws of physics. "I could see when it hit," he said. "I turned my head around and saw the flight."

He stayed in his seat long enough to watch Mantle cross the plate. Then: "Out of some perverse instinct,

I said, 'Let me go see if I can find it.' I lucked up and found it."

He started down Oakdale Place, a narrow block of low-slung attached brick row houses developed right after Griffith Stadium was built. He walked down one side of the street and up the other, searching every garden and under every parked car, more than once. Dan Daniel reported in *The Sporting News* that "a Negro woman hanging out one of the windows" had directed Dunaway and Patterson to the site. Not true, Dunaway said. He was alone.

In 1953, a row of six similar houses on Fifth Street faced the left field wall of the stadium. A swath of grass—"a cut-through," Dunaway called it—ran behind the houses and parallel to the sidewall of 434 Oakdale Place. "Something told me to look in the back. I went through the little cut."

In the fall of 2008, the lot once occupied by the Fifth Street row houses was empty, enclosed by chain-link fence with a sign proclaiming Howard University's intention to rebuild on the site: COMING SOON, NEW HOMES AT HISTORIC LEDROIT PARK. The chain link made it impossible for us to reach the backyard of 434 Oakdale Place where Patterson said Dunaway had led him to the ball. The yard was enclosed by a picket fence; the back section ran parallel to an alley, where another wooden fence once kept the black residents of Howard Town from entering LeDroit Park.

I asked Dunaway to show me how far back behind the house he had found the ball. "Under the window," he said, pointing to a second-story side window, which was at least twenty-five feet closer to Oakdale Place and to the stadium than Patterson's declared location. Which meant the ball had never reached the backyard of 434 Oakdale Place at all.

"No," he said. Never said it had.

Turning, he pointed to the now-empty corner lot where each of the demolished row houses had had a small fenced-off yard. "I looked in everybody's yard 'til I went to this particular yard and found it," he said. The ball was sitting against the back of the house— probably the second or third from the corner—in a pile of dead leaves. "Standing out," he said.

He took the ball to an usher he had spoken with on his way out of the ballpark. "He said, 'You found that ball? Damn, you is kidding.' He said, 'C'mon, I'm goin' to take you 'round to the clubhouse.'"

In *All My Octobers*, Mantle wrote, "I think the kid just showed up at the clubhouse wanting to sell the ball or get an autograph."

That's pretty much what happened, Dunaway said. The usher escorted him to the visitors locker room along the third base line, where they were greeted by a clubhouse attendant, who summoned someone official-looking. Dunaway assumed he was a reporter because he had a pad and pencil and wrote down everything he said. The usher provided the bona fides. "He said, 'This is the fella who caught Mickey Mantle's ball,'" Dunaway said. "I gave him the whole outlook."

But he never mentioned 434 Oakdale Place. Nor did he take anyone to the spot where he had found the ball. "I told him I found it on Fifth Street behind a guy's house. I told him where to go look for it himself."

What did the man look like? "White," he said. "He shook my hand. He said it might have been one of the longest ever."

Dunaway handed over the prize and was promised a ball autographed by the slugger that never arrived in the mail. "He gave me a hundred dollars," he said. "The guy told me how famous I would be for catching the ball. I was more excited about the money than about being famous. That was a lot of money back then."

Newspaper reports of a much smaller bounty made him indignant. "Not no dollar. He gave me a ball then, too. It was autographed by four or five players. I gave that ball to one of my grandnephews. He was playing baseball with that as he was growing up."

A photographer took his picture with the ball before he relinquished it. Friends told him later that their fathers had seen it in out-of-town papers. He looked for it in the *Washington Post* the next morning and was disappointed not to find himself there. (No photograph could be located in the archives of the *Post*, the *Washington Star*, the *Afro-American*, the Historical Society of Washington, D.C., or the Library of Congress.)

That afternoon, his uncle Willie, a railroad porter who lived with the family when he wasn't working, took him shopping on Seventh Street. He bought himself some khaki pants, socks, shirts, tennis sneakers, and a couple of beers for his uncle. When he got home he had "some 'splainin'" to do. He spent the rest of the money on candy and pinball and taking neighborhood girls to Tom Mix shows at the Dunbar Theater.

He finished the school year, but sixth grade was his last. That summer, he was caught stealing and was sent to Blue Plains in southeast Washington, D.C., once called the Industrial Home for Colored Children. He ran away that fall. It wasn't any harder than sneaking into Griffith Stadium.

He didn't go back to school, and he didn't go back to the old neighborhood. He didn't want any part of the truant officers looking for him in LeDroit Park. He lived on the streets and with an uncle in Anacostia who let his mother know he was all right. He spent afternoons at a police boys' club on Florida Avenue where no one asked questions, worked at a local shoeshine stand, and set pins at the Lucky Strike bowling alley.

Trouble led to more trouble and incarceration in a facility for juvenile offenders in Ohio, where he learned how to do laundry, to say the rosary, and that he didn't want to spend any more time in jail. He was twenty years old and had five months of parole to serve when he was released after five years.

He worked in hotel laundries around the nation's capital until the mid-1980s. He survived on frugality, good luck—he hit a couple of $5,000 Four Ways in Atlantic City—and "the grace of God," his brother-in-law Elder Walter McCollough said.

He donated money to every Catholic charity that asked for help and filled his apartment with the rosary beads he received. He never married, which he regretted, and had a daughter, whose name he could not spell. Most of his friends from the old neighborhood were dead and gone when we met. But the memory of the day he hit the jackpot at Griffith Stadium remained vivid. "One big day," he said.

The best day of his life.

He died on March 3, 2010, due to complications from diabetes, gout, and pneumonia. He was seventy-one years old.

4.

In the spring of 2008, Alan Nathan, professor emeritus of physics at the University of Illinois–Champaign-Urbana and chair of the Society for American Baseball Research's Science and Baseball Committee, succumbed to my entreaties and clambered to the roof of Howard University Hospital to test the prerogatives of memory and place against the hard discipline of science. Armed with my new laser range finder, historical photographs, Google Earth images of the neighborhood with dimensions of the old stadium superimposed on a grid, Sanborn Insurance maps from 1953, building permits for the demolished Fifth Street row houses, newspaper accounts, and a very good head for math, Nathan set out to establish the most plausible fate of the ball. Unlike Red Patterson, he also had a tape measure.

The myth of the Tape Measure Home Run was consecrated and perpetuated by Mel Allen's 1969 recreation recorded by for an album celebrating *One Hundred Years of Baseball History*, which was replicated on a 1973 Yankees record—*Fifty Years of Sounds*—and given to every fan at the last game before the Stadium closed for renovations. It was later rebroadcast on *This Date in Baseball History* and *This Week in Baseball*. Neither his call of the game nor Bob

Wolff's D.C. broadcast had been recorded. Allen's bellowed re-creation became accepted fact: "We have just learned that Yankee publicity director Red Patterson has gotten hold of a tape measure and he's going to go out there to see how far that ball actually did go."

But when Bill Jenkinson, a baseball historian and author with a penchant for the long ball, confronted the spinmeister on the matter of the tape measure in the early 1980s, Patterson cheerfully admitted he had never had one. He also told Jenkinson he had never claimed the boy told him he saw the ball land on a fly. He had paced off the distance with his size eleven shoes. Marty Appel, who later became the Yankees' director of publicity, told Patterson his shoes belonged in the Hall of Fame.

Previous scientific attempts to ascertain the actual flight of the ball have not been universally well received in the world of Mantleology. In 1990, Yale professor Robert K. Adair published the bible of baseball science, *The Physics of Baseball*, devoting an entire chapter to the Tape Measure Home Run. Twenty years later he stood by his original conclusion that the ball had traveled 506 feet, with a margin of error no more than 5 feet. "The number 565 is pure fiction," Adair told me. "It was where they picked the ball up after it rolled across the street."

Jenkinson conducted his own analysis, relying on both anecdotal accounts and scientific data. His conclusion: the ball traveled no farther than 515 feet and probably 10 feet less than that. When his findings were published in 2008 by Yahoo! Sports, the baseball blogosphere reverberated with cyberhowls. Mantle devotee Randall Swearingen denounced Jeff Passan's column as "a malicious prosecution" of The Mick. MICKEY'S HISTORIC HOMER ON TRIAL!

From the roof of the hospital, one thing was immediately apparent to Alan Nathan. The ball Mantle hit off Chuck Stobbs could not have traveled 565 feet in the air or rolled or bounced that far from home plate. Nor could Patterson have walked a straight line in his size elevens from the backyard of 434 Oakdale Place to the back of the stadium, as he told reporters. According to the 1953 Sanborn map, the houses on Fifth Street and the fences behind them would have obstructed both his path and that of the ball.

Figuring out what *could* have happened was more challenging, given the protean nature of the urban landscape. Blueprints of the stadium could not be located. Clark Griffith, the surviving member of the stadium's baseball dynasty, had no records. The exact dimensions of the scoreboard were in some dispute. No building permit for its construction or for the concrete bleachers was on file with the District of Columbia. The exact location of home plate was unknown. "Mickey's tree," the landmark oak beyond the center field fence, had been sacrificed to progress—a hospital parking lot.

However, the corner of Fifth and V streets and the building behind the third base line were unchanged. Using my range finder and those fixed points, Nathan located home plate on the 1953 map "to within a circle of radius of about three feet."

Our hope had been to use the range finder to measure the distance from home plate to the reported landing zone. But the hospital's roof made that impossible. Resorting to old-fashioned arithmetic, Nathan calculated a straight-line distance of approximately 557 feet from home plate to the backyard of 434 Oakdale Street. But he had already rejected Patterson's claim as "crazy" and now discounted Adair's analysis because it underestimated the effect of the spin of the ball and failed to account for the Fifth Street row houses.

Unimpeded, Nathan concluded, the ball could have traveled as far as 535 to 542 feet—"A lot further than Adair says." But the houses were in its way.

Our next step was to try to find a scenario that matched Dunaway's description and that also had a basis in science. For more than a year, Nathan consulted other physicists, including Adair and SABR members devoted to the study of demolished ballparks. He reviewed all the anecdotal evidence and weather data: Bill Abernathy's recollection that the wind picked up just as the pitch was thrown; Lou Sleater's dugout observation that the ball went "straight up" like "it was going to be a pop-up to the shortstop and just took off"; and third baseman Eddie Yost's testimony that the ball hit halfway up the Mr. Boh sign, above the words OH, BOY, WHAT A BEER, and then veered right at a 20- to 30-degree angle. Nathan considered—and rejected— several hypotheses. Could the ball have ricocheted off Mr. Boh, then bounced on the pavement of Fifth Street and *over* the row houses into the backyard of 434 Oakdale Street? No; the proximity of the buildings and the angle of descent precluded that. Could it have pecked Mr. Boh on the cheek on the way out of the ballpark, landed on Oakdale Place, and taken a hard right into the cut-through? Nope, not unless it was more magical than the Kennedy bullet. Nathan concluded that "the only reasonable scenario is one whereby the ball hits the roof of a house on Fifth Street on the fly, then bounces off into the backyard."

Proving it was another matter. The quest for exactitude was compromised by insufficient information about the wind speed 60 feet above field level and the

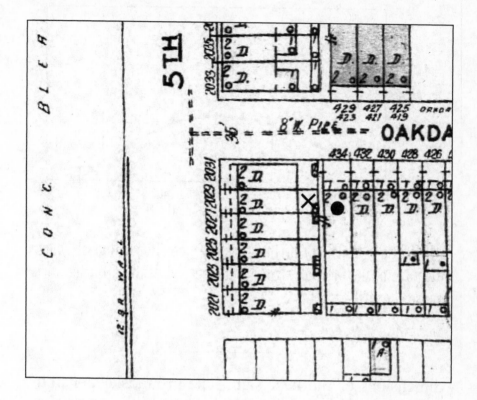

Certified 1959 Sanborn Map showing the corner of Fifth Street and Oakdale Place, behind the left field wall at Griffith Stadium, Washington, D.C. A fence that once kept black residents out of the all-white community of LeDroit Park ran east-west in the alley behind 434 Oakdale Place.

- Red Patterson: backyard of 434 Oakdale Place
- × Alan Nathan: backyard of 2029 Fifth Street NW, now demolished

Credit: Environmental Data Resources, Inc.

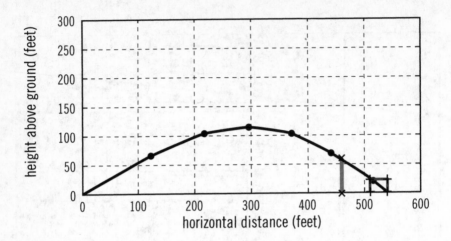

Hypothetical trajectory of the Tape Measure Home Run showing the edge of the Mr. Boh sign and the second of the attached row houses on Fifth Street NW

Credit: Alan M. Nathan

dimensions of the beer sign. The groundskeeper and the club secretary offered conflicting reports: one said it measured 55 feet high, the other said 48 feet. Both agreed that the ball had hit at least 6 feet above the back wall of the stadium.

So Nathan did what scientists hate most: he assumed. He knew the distance from home plate to the beer sign—460 feet. He calculated that the ball had left Mantle's bat traveling at about 113 miles per hour at a launch angle of 30 to 31 degrees. He assumed that the wind had been blowing 20 miles per hour, less than the highest reported gusts; that the ball had been deflected but not severely; that it had nicked Mr. Boh at a point 55 to 60 feet above street level but retained enough horizontal speed to traverse the 52-foot distance from the edge of the sign to the roofs of the houses on Fifth Street.

According to Nathan's calculations, the front of the second house was 512 feet from home plate. The back of the house was about 540 feet from the plate. Its tin roof slanted six inches to the rear. "If my estimate of where the sign was is correct, then very little deflection would be necessary to get the ball to about where Dunaway said he found it," Nathan said. "It could have hit the roof of the second or third house and rolled down into the backyard."

In short, according to the laws of physics, the fate of the ball as described by Donald Dunaway is scientifically plausible. The best hypothesis is that he was telling the truth.

7

November 2, 1953

FISH BAIT

1.

On Saturday, October 31, the day he was scheduled to check into Burge Hospital in Springfield, Missouri, for surgery on his damaged right knee, Mantle called in sick. Bad stomach, Merlyn told Mantle's minder Tom Greenwade, who had made the arrangements. It wasn't just a case of preoperative nerves, though there were some of those. There was a matter of family honor to settle—a footrace with his brother Ray to determine who was the fastest Mantle. "I thought I had better find out before coming up here," he told Kenny L. Brasel, a reporter from the *Springfield Leader & Press*.

Ignoring his doctor's caution against strenuous exercise, Mantle challenged Ray to a 100-yard dash. He won, then got sick to his stomach.

His queasiness was understandable, given his well-founded apprehension about going under the knife. He knew what his doctors could not—he was never the same after the 1951 World Series. Straightaway, he could still fly down the line, his fanny getting lower with each step. But he knew he had lost lateral movement, the ability to change directions, to cut and run. And he didn't tell Frank Sundstrom, the young surgeon who took the preoperative history, how much pain he was in until after the operation.

Mantle arrived at Burge Hospital on Sunday, November 1; surgery was scheduled for the next afternoon. He checked into a two-room suite—one for the patient, the other for Greenwade, the designated spokesman and security guard. Mantle didn't want to go under the knife—what twenty-two-year-old who made his living with his body would? He especially didn't want to do it in New York, Springfield papers reported, where his convalescence in Lenox Hill Hospital in the fall of 1951 had been disturbed by the "aggressive attitude of Easterners" who ignored the "No Visitors" sign on the door.

Yankee general manager George Weiss had referred

him to a Springfield physician named Bertram Meyer, who had introduced him to Dan Yancey, a local orthopedist with a national reputation. Mantle had reinjured his knee on August 8, Ladies' Day at Yankee Stadium, chasing down a hard hit ball to left center field in front of 68,000 people. "By lightning work, the Oklahoma kid held the blow to a single," the *Times* reported, "but as he stopped short to make the pivot for the throw to second base he gave his right knee a severe jolt."

"Sprained ligament," said team physician Sidney Gaynor. "Yankee Stadium jinx" was Mantle's diagnosis. Lying on a clubhouse couch, with his cap perched on his still-sweaty brow and an ice bag balanced on his knee, he bemoaned his fate. "It's the same sort of an injury I received in the 1951 World Series. I've never been hurt in a baseball game until I joined the Yankees."

Fitted with a bulky brace that laced above and below the knee, he talked his way back into the lineup on August 18. When he failed to reach a ball he should have caught, Stengel sent him back to the bench, where he remained, save for four pinch-hitting appearances, until August 29.

And so a season that had begun with a bang in April, with a league-leading .353 batting average in June and four 4 RBI games by mid-July, ended with respectable but less-than-stellar totals: .295 BA, 92 RBI, and 21 home runs, only 5 of them after August 8.

On September 4, the Yankees received an anonymous death threat postmarked Boston, warning Mantle if he played at Fenway Park over the Labor Day weekend his career would "come to an end with a .32."

Though the FBI concluded that the letter was most likely the work of a crank or perhaps a very young Red Sox fan, police hauled Mantle off the train in Hartford, Connecticut, en route to Boston and escorted him to Fenway with a security detail. He hit a home run and later told reporters that he had never circled the bases faster.

His two World Series home runs (a two-run game-winning home run in game 2 and a grand slam in game 5) helped the Yankees vanquish the Dodgers in yet another interborough championship. He also helped Dodger pitcher Carl Erskine set a new World Series record for strikeouts by contributing 4 of a record 14 in game 3. "He would take a good fast ball pretty much over the plate and swing at the curve ball out of the strike zone," Erskine said. "Stengel was screaming at him from the dugout."

Having created—and profited from—the myth of the Tape Measure Home Run, Yankee officials now wondered why Mantle swung at each pitch as though he wanted to detonate the ball. They were more worried about his physical condition.

The previous fall, the Army had again revisited his draft status. Again he was ruled 4-F but not because of the osteomyelitis that had previously disqualified him. A new Selective Service guideline—the Mantle rule, it was called—mandated that anyone who had not received treatment for the disease in the previous two years was eligible to serve. On November 3, 1952, the Army surgeon general in Washington ruled that Mantle was excused because of a "chronic right knee defect resulting from an injury suffered in the 1951 World Series." The report cited the routine "rejection of persons with dislocated semi-lunar cartilages or loose bodies of the knee which have not been satisfactorily corrected by surgery."

That grisly center field injury—not the sprained ligament in August—was the condition that brought Mantle to Burge Hospital two years later. With the technology available in 1953, Mantle's doctors simply couldn't see how badly he had damaged his knee. Today's orthopedic essentials, magnetic resonance imaging (MRI) and arthroscopic surgery, were still decades away. The best sports orthopedists in the world had no way to visualize the extent of the injury or to predict the sequential degeneration that would follow.

They had no reason to doubt that torn cartilage fully explained the swelling, locking, and buckling of his knee. They had justifiable confidence in their sunny prognosis—that the damage was limited and the ligaments were intact. True, his knee would never be as strong as it had been prior to October 1951, but they believed he could regain 95 percent of its preoperative strength. The surgery, Yancey promised, would prohibit further "slipping of the knee" and eliminate Mantle's tendency to favor it. He would report to spring training on time and run without inhibition or the need for a brace.

His patient was less sanguine: "I don't know what I'm getting into, but this is my own idea," he said as he was wheeled into the operating room.

2.

That morning Frank Sundstrom, Yancey's young associate, took a patient history and did a preop physical. He held the retractor during surgery and stitched up the 2½-inch wound. He did not see any pre-operative indication of a previous ligament tear or prior knee surgery. He found no significant abnormalities resulting from osteomyelitis. He noted in Mantle's chart what so many other medical professionals observed: he was an astonishing physical specimen. "One of the best athletic bodies I had ever seen, and that was even before steroids," Sundstrom said fifty years later. "He had such beautiful, strong, well-defined muscles."

That singular male beauty was noticed in the Yankee locker room as well. "Mickey's muscles, in spite of their size, zip with the looseness and speed of a lightweight boxer's," trainer Gus Mauch said. "When I massage his arms and shoulders, they transmit some sort of extra something which I never experienced before in over thirty years of handling athletes."

But, Mauch believed, this apparent perfection masked an imbalance between muscle, bone, and connective tissue. As his successor, Joe Soares, explained: "Mantle had a severe, congenital condition. His muscles were so large, but his joints—wrist, knees, ankles— were frail. This discrepancy between the awesome muscles and the weak joints caused the vast majority of his muscle tears and injuries, and exercise wasn't going to help this constitutional defect."

Sundstrom concurred: "A ligament will withstand so much tension on it before it deforms or strains or tears," he said. "I suspected that he would suffer ligament damage or tendon damage because of the horsepower in his muscles."

Bunny Mick, a lifer in the Yankee organization, saw

evidence for Sundstrom's theory as Mantle took batting practice one day. "He screamed in pain, grabbing his chest as if he had had a heart attack. He'd swung so hard he ripped the muscle in his chest."

In short, Mantle's strength was his weakness. He tore himself apart. This flawed medical logic would become an essential element of Mantle mythology. But it would not survive the test of time and medical scrutiny. Mantle wasn't brought down by the way he was built or his cavalier attitude toward off-season conditioning and rehabilitation. It was this simple: given the existing state of sports medicine, nothing could have prevented the degeneration of his right knee, short of another line of work.

When Yancey opened Mantle's knee, he found, as expected, a bucket-handle tear of the medial meniscus, a piece of cartilage shaped like a crescent moon that fills and cushions the space between the femur and the tibia. The only surprise was the extent of the damage; the cartilage was split at one end and separated at the other, suggesting the wear and tear already occurring inside the joint. He removed approximately 30 percent of the meniscus, a piece of tissue "about the size of your little finger," Sundstrom said, using the latest in orthopedic knives. Greenwade was summoned to the operating room to examine what *The Sporting News* called "vagrant cartilage." Looked like "a piece of pork rind fish bait," he declared.

Yancey patiently explained to the press that cartilage functions "like a shock absorber." Though it does not grow back, he confidently—and erroneously—assured reporters that fibrous tissue would eventually fill the void. The surgery was unremarkable, the postoperative course smooth, the future unimpeded. "It was a fairly common thing," Sundstrom said. "If you had to have something wrong with your knee, you'd want this to be it."

3.

Reconstructing Mantle's medical history is almost as difficult as it was to diagnose the extent of his injuries in 1953. His family had none of his medical records, and the Yankees hadn't retained them. Gaynor stipulated in his will that all of his remaining files were to be destroyed after his death and his daughter, Deborah, complied with his wish. There are no records remaining at Burge Hospital, and Lenox Hill Hospital in New York would not acknowledge whether any records remain in the archive, citing privacy laws. When Mantle's son Danny contacted the Mayo Clinic, where his father had had subsequent checkups and surgeries, he was told there was nothing in the medical records pertaining to his father's care.

Compounding the problem is the fact that Mantle's accounts of his medical history were often inaccurate, inconsistent, and incomplete—beginning with the erroneous assertion that he had his first knee surgery at Lenox Hill Hospital in October 1951. The date was never challenged. Why would it be? Who forgets his own medical history? And so the phantom surgery of 1951 was added to the list of surgical events solemnly updated with each of Mantle's successive physical disasters.

The fact that no surgery was performed until November 1953 does not lessen the impact or severity of his original injury. On the contrary. It is extraordinary that Mantle played two full seasons following the World Series injury on an unstable, unrepaired knee.

Various accounts (newspaper, magazine, and book) described Mantle's initial injury as torn cartilage, torn ligaments, or torn tendons. A later iteration described a fractured kneecap, the bone protruding through the skin, and blood sullying his uniform. This may well have been an extrapolation based on Mantle's gory recitation. *I could feel my leg snap. I really thought my leg had fallen off at the knee.* None of the three teammates who first rushed to his side—Houk, Silvera, Kusava—

remember blood in the outfield; nor is any evident in photographs taken when he was examined in the locker room.

Just how fateful was that moment in center field? To answer that question, I assembled a "case history" from newspaper clippings, contemporaneous photographs, and interviews with teammates, trainers, and physicians. I presented the evidence to Dr. Stephen Haas, who was team physician for the Washington Wizards and Capitals and is now the medical director for the National Football League Players Association. I asked him to make an educated forensic diagnosis, a plausible case history for what happened to Mantle's knee.

"It appears that the most likely critical event was an acute combination of torn medial collateral and anterior cruciate ligaments and a medial meniscal tear," Haas said.

In modern vernacular, he blew out his knee.

This "unhappy triad" of injuries was named and identified by Haas's mentor at the University of Oklahoma, Dr. Don O'Donoghue, in 1950. Knowing what it was and doing something about it were two entirely different things. "It is not surprising that the ACL tear was not seen since the surgical techniques at that time made it difficult to fully visualize that structure," Haas said. "I am sure the meniscal tears, and loose bodies, got all of the attention at surgery because they are not as subtle and relatively easily treated. It probably satisfied the surgeon as to the cause of his symptoms."

Don Seger, who became the Yankees' assistant trainer in 1961, had worked on Mantle in spring training in the mid-Fifties. Even then he was struck by the degeneration of Mantle's knee. He too is convinced that Mantle suffered the unhappy triad. Before the advent of MRI, doctors used what's called the "anterior drawer test" to diagnose a ruptured ACL. "It pulls your leg forward while you're sitting on a table and your leg's at ninety degrees and you're relaxed," Seger said. "You pull it back and forth forward to see how much play you have in it, how far your leg would go away from your knee. That's the 'drawer sign,' like pulling a drawer out. He'd sit there and wiggle his knee and make it go forward. I almost winced to watch him do it."

Mantle turned this sickening flexibility into a kind of parlor trick. "Watch this," he said when his minor league teammate Keith Speck visited the clubhouse in Minnesota one day. "He pushed that one bone in his lower leg, it would come out two inches," Speck said. "He pushed out behind his knee with his thumb, and the bone would go right straight out."

"Watch this," he'd say to his boys as they headed out

the door before school. "The whole top part just went like this," said Mantle's son David, demonstrating how his father had twisted his knee as if opening a pickle jar. "It was like it could come apart," Danny Mantle said.

Mantle never talked to Seger about his knee. The trainer concentrated on those gleaming, problematic muscles. "I massaged his thighs and legs frequently and for long periods. A lot of things you couldn't do with his knees. He inflamed so badly when you put him on a weight program or something of that nature. He would become so inflamed he just couldn't walk."

Mantle disdained the bulky, double-hinged brace the Yankees provided for increased stability. Had he used the brace, Seger says, it would have compounded the deterioration in his knee. "To attempt to put a knee brace on him, it would grind those articulating surfaces," Seger explained. "It compressed them so together. That would set up an inflammatory reaction."

O'Donoghue, long considered "the father of orthopedic sports medicine," reported that 25 percent of acute athletic knee injuries resulted in the unhappy triad. The numbers have escalated with the increased speed and size of athletes, amateur and professional, under competitive duress. It happens when wide receivers "plant and cut" on turf. It happens when basketball players land awkwardly on a straightened leg after making a shot, or blocking one. It happens when base runners slide into second base with a hyperextended knee. It happens without warning and often without contact. According to a 2009 report by Dallas Mavericks' team physician Tarek Souryal published by WebMD in emedicine.com, 200,000 Americans sustain an ACL injury every year.

Of the four ligaments charged with the responsibility of holding a human leg together, the anterior cruciate is the most crucial and the most central, winding its way north to south through the middle of the joint. Its job is to make sure the knee doesn't come through the front of the leg, which is exactly how Mantle described his injury. When an ACL ruptures, it does so with a bang. In a gym, it can cause an echo, like a stack of books landing on a hardwood floor. When Mantle went down, Phil Rizzuto said he could hear the pop at shortstop.

And when an ACL goes, it usually takes the medial meniscus with it, leaving loose fragments of cartilage in the joint, irritating and corroding the surface of the knee and getting stuck in all the wrong places. When they get stuck, the knee locks up with stabbing ice-pick pain.

Unlike the medial collateral ligament, the ACL does

not repair itself and cannot be stitched back together. It must be reconstructed, surgery that was rarely attempted in 1953 because the procedure was so invasive and the outcome so uncertain. "He never had a reconstruction, only removal of torn cartilage and scar tissue," Haas said. "He would have been out at least a year if he had a reconstruction, probably longer based on the techniques done in that era."

How did Mantle play with a torn ACL? It can be done, Haas says. "Mickey Mantle can be classified as a 'neuromuscular genius,' one of a select few who are so well wired that they are able to compensate for severe injuries like this and still perform at the highest levels, overcoming a particular impairment at a given moment. It is a phenomenon comprised of motivation, high pain threshold, strength, reflexes, and luck."

In 1968, the last spring of Mantle's career, Soares observed: "Mickey has a greater capacity to withstand pain than any man I've ever seen. Some doctors have seen X-rays of his legs and won't believe they are the legs of an athlete still active."

4.

The world may have gasped at the way he outran balls in Yankee Stadium's capacious outfield, how he flew from first to third on a single to right field, how he beat out routine infield ground balls. But at field level, those who had seen him run before the injury saw subtle changes in his gait—a slight glitch when he tried to accelerate out of the left-handed batter's box, compromised lateral movement and a loss of speed that was hard for most observers to fathom, considering how fast he could still run. "He lost a lot of speed," Ralph Houk said. "I would say going from home to first, he lost a full step."

When Houk became Yankee manager, he followed Stengel's example and refused to allow Mantle to steal, hoping to protect Mantle's "delicate, almost feminine" knees. Trainers taped him together with long sheaves of Conco athletic tape that wound from ankle to thigh. The thick foam bandages were wrapped so tightly they cut off circulation and mummified his flesh, turning it white and puffy. When the bandages came off, Charlie Silvera said, the skin was creased with "bulges where the tape had maybe come loose a little bit."

Sometimes it was worse than that. "There would be blood oozing out," said Virgil Trucks, who joined the Yankees in 1958. "Not like you've stabbed somebody, but it was ample."

In the clubhouse Mantle never complained. But Merlyn confided his private agonies to some of the other wives. "After a ball game, he would just stretch out on the sofa and moan in pain for hours," Lucille McDougald said. "She said, 'You have no idea how much pain he was in.'"

Stoicism could mask only so much. A grimace became a feature of each left-handed at-bat. "You could see the pain in his face every time he swung, even at age twenty-three or twenty-five," Bunny Mick said. "He couldn't let his body rotate because of the knee. He couldn't follow through on the swing for the rest of his career."

Silvera said, "After a doubleheader he couldn't get out of the car. He'd have to take one leg at a time and swing it over."

With Merlyn back home awaiting the birth of Mickey Jr. in the spring of 1953, Mantle shared a room in the Edison Hotel with Irv Noren, who observed his pain close-up in their less than sumptuous accommodations. Even at age twenty-two, Mantle took mornings slow. "I had bad knees," Noren said. "His legs were worse than mine. In the morning, he'd say, 'Irv, you awake?'

"I said, 'I can't get out of bed.'

"He said, 'Me, neither.'

"We'd both be hobbling around the room 'til we could get our legs straightened out."

Sometimes even then, Noren said, Mantle returned to the dugout with tears in his eyes. It would never get any better. Over the course of his career, Mantle missed 255 games due to injuries—more than a season and a half of baseball.

Each successive knee injury predisposed him to the next, in what Haas called "cascading episodes of instability." The compromised ligament made him prone to more cartilage tears. Thorns of cartilage eroded the protective coating around the joint, causing severe premature arthritis. The hamstring muscles, acting as a secondary line of restraint, had to work harder to compensate for the lack of stability, which led to recurrent muscle pulls. The vehemence of his left-handed swing, with his right knee locked and extended and his foot rolling over on the ankle on the follow-through, put enormous stress on the already unstable joint.

The result of this degeneration is visible even in a grainy reproduction of an X-ray that appeared in a 1988 sports medicine trade magazine announcing Mantle's endorsement of the non-steroidal anti-inflammatory drug Voltaren (diclofenac sodium). Dr. George Ehrlich, then head of medical affairs for Ciba-Geigy, the manufacturer of Voltaren, interviewed Mantle when he and Whitey Ford enrolled in the clinical trial in 1987. Looking at that 1988 X-ray twenty years later, Ehrlich came to the same conclusion that he reached then: "When you take an X-ray of a healthy knee, you can see the space between the femur and tibia—that's where the cartilage is. In this X-ray, there is no space. The femur is directly on top of the tibia. It's bone on bone."

That night over dinner, Ehrlich chatted with his seemingly enthusiastic patient about the benefits of the drug. "My suspicions were raised at the time," Ehrlich said. "Later, the people in the medical department told me he never followed the instructions. I thought he did. He told me he was better. He didn't follow the directions. I don't think he followed anyone's directions. He was a great athlete, a very poor patient."

But a very obliging pitchman. He even got into a New Jersey swimming pool to demonstrate the wondrous properties of the anti-inflammatory medicine. "We'd like to pay you to watch Mickey," the public relations man told a young lifeguard, Rahmin Rabenou.

"I'm on duty," the conscientious teenager replied.

"No, we *really* want you to watch Mickey."

An offer of $30 backed up the urgency of the request. "He can't swim."

After watching Mantle affect a kind of dog paddle, Rabenou, who became a physician, thought, "Maybe he *should* have the lifeguard with him."

Voltaren couldn't make Mantle float, but it might

have alleviated much of his pain. Viewed in the context of his other health issues, he had good reason not to take it, having already been diagnosed with liver disease, which is incompatible with such a medicine. In fact, Ehrlich said, "A damaged liver would have precluded his being in any drug study."

Dave Ringer was Mantle's personal physician in Georgia in the last years of his life. One day, a drug rep for Ciba-Geigy visited his office, brandishing miniature Mickey Mantle bats and golf balls along with the usual medical samples. "He said, 'Mickey Mantle uses Voltaren and it's excellent for the osteoarthritis of his knees and he loves it.'

"I said, 'No, he doesn't.'

"And the drug rep, young guy, looks at me and he said, 'Yeah, he does. He uses it regularly. He advertises for us. He said it works very well.' "

Ringer sent his nurse to get Mantle's chart. "See, I told you he doesn't take Voltaren," Ringer told the man from Ciba-Geigy. " 'He's got liver problems.' That drug rep just about fell on the floor."

At Mantle's next appointment, Ringer told him about the visit. "He said, 'Oh, my God, don't tell anybody. I'm getting $200,000 a year for that. I don't take it. I can't take that drug, but I don't want them to know that.'"

5.

Wife, mother, in-laws, and infant son were waiting for Mantle when he came out of knee surgery in November 1953. He remained in the hospital for eleven days, the first of which he spent in groggy sedation due to severe pain. He wasn't allowed out of bed for five days (at which point he was permitted to swing his leg while seated on the edge of the bed) and did not leave his room until he was discharged.

Today, professional athletes who have arthroscopic knee surgery—using the thin tool that requires only tiny incisions—routinely return to the playing field in a matter of weeks. Mantle's recuperation was expected to take a minimum of three months.

Greenwade did everything possible to ensure a swift recovery. T-bone steaks were delivered to Mantle's room every evening. Special nurses were brought in; Mantle admired them greatly. Other than sanctioned visitors—minor league teammate Al Billingsley was admitted after first being summarily turned away—he healed in splendid isolation while get-well cards piled up in the hospital mail room.

He talked his doctors into releasing him eighteen hours early so he could attend his twin brothers' football game between undefeated Commerce High and their rivals from Grove, Oklahoma. He was ordered to return to Springfield for office visits two or three times a week for the next three weeks. He left the hospital on crutches after spending thirty minutes signing autographs in the Polio Cottage with strict instructions to be "judicious with his activities."

This was a prescription Mantle was unable or unwilling to follow. Two days before Mantle left Burge Hospital, Billy Martin returned to California from an All-Star exhibition tour of Japan with $4,000 in his pocket and a haul of loot—cameras, binoculars, silk robes, and samurai swords. He was flush but lonely. His wife had left him, taking their infant daughter with her. Mantle invited him to winter in Commerce. "I knew he had nothing left in Berkeley."

They had torn up American League cities all that season. Mantle was still mourning his father; Martin was mourning his marriage. Merlyn was back home nursing a fussy baby. Martin's divorce dashed her hope for conventional off-season domesticity.

Martin loaded up his new Cadillac convertible and headed east. By the time the visit ended, he had totaled the car while drag-racing against Mantle in his Lincoln Continental. They were en route to a Commerce High School basketball game. Martin had Ray and Roy, the team's star players, in the backseat. Mantle and his Lincoln were unharmed.

Their exploits that winter and in Martin's future off-season visits were well known and not particularly well received around town. Old-timers in Commerce can still point out the spot where Mantle turned his car over on the road later renamed in his honor. He pointed out a hole in the "Home of Mickey Mantle" sign at the edge of town to Irv Noren one off-season. "There was a rock thrown through it," Noren said. "Mickey said, 'This is what they think of me in my own damn hometown.'"

Mantle's patron Harold Youngman, the prosperous owner of a Missouri road-building company, handled his investments and indulged his taste for hunting, fishing, and basketball. He procured cushy off-season jobs for Mantle and Martin. Their responsibilities consisted mostly of flying around the countryside in his twin-engine Beechcraft, chatting up town commissioners who awarded the contracts for new construction.

Youngman also sponsored Mickey Mantle's Southwest Chat All-Stars, a basketball team that took on all comers including the Harlem Globetrotters. Mantle was the coach and occasional point guard. The Yankees were none too pleased about his minutes on the court and sent a telegraph telling him to cease and desist. "He just kind of shrugged it off," recalled Paul Churchill, a member of the team.

After Mantle played in the championship game two nights later, Churchill said, he got another telegram: REPORT TO NEW YORK CITY.

This was Martin's first but not his last sojourn in Commerce. He got fat on Merlyn's forbearance and Lovell's biscuits and gravy. "Eating them beans and shit," Mantle told me. "Merlyn didn't like him," Max Mantle said. "Didn't have a reason to."

Martin gave permission to Mantle's worst instincts, not that they needed a lot of encouragement. Merlyn was left alone with an infant who was as allergic to milk as her husband was to domesticity. "All I saw of the two of them that winter was their backs going out the door," she wrote in *A Hero All His Life*. "If they did all the hunting and fishing they claimed they were doing, the fish and quail population of Oklahoma and Missouri took a fearful beating."

Elbow bending was their preferred sport. One off-season, the wise proprietress of Billie's bar bogarted the keys to their car when the miscreants stopped to say goodbye on their way to spring training. She declined their IOU and the promise to send the money as soon as they got to spring training. "She said, 'When you do, I'll send you the keys,'" recalled preacher Jerry VonMoss. "They had to call Casey Stengel and he had to promise to wire the money. She still wouldn't give 'em the keys until she got the money, and Western Union was in Miami, four miles away."

Not all their scrapes were as easily resolved. Angie Greenwade, Tom's daughter, recalled being awakened by the telephone in the early hours one winter morning. "This sheriff called to say that Mickey and Billy had been in a fight in a bar," she said. "This sheriff had them in jail. He wanted to know what to do. Dad told him to keep them there. He got in his car, and away he went to get them out of jail.'"

They'd drive around the countryside in Mantle's convertible with the top down, the country music turned up high, and a bottle of Jack Daniel's on the dashboard. "They'd shoot wildlife from the car, which we called 'hot shooting,'" recalled Picher native Frank Wood. "The game ranger arrested them. He hauled them into the Ottawa County Jail. The sheriff let Mick go. But he wasn't gonna let Billy Martin go free. Put him in the pokey for three days!"

On one such outing that winter, Mantle stepped into an abandoned prospecting hole. "That's how he twisted his daggum knee," said Wood, who got the story from his father. "Lots of times they'd plug them up and grass would grow up on it. If you had a hunting dog, you had to be careful they didn't fall in."

When the telephone rang at Dan Yancey's Springfield home in early February 1954, his children immediately noted his agitation. His prize patient had developed a Baker's cyst behind his right knee, a swollen, fluid-filled bulge that restricted his motion and set back his recovery. "The only time we ever saw Father get upset with a patient was Mickey Mantle," said his daughter Alice Yancey.

Youngman flew Mantle and Cousin Max back to Springfield on his private plane. When he examined Mantle, Frank Sundstrom was as appalled as Yancey. "We learned he had been quail hunting. We took him back to the operating room, turned him over, and removed his Baker's cyst."

The surgery was performed on February 4, 1954, three weeks before the start of spring training. The procedure took thirty-five minutes. By the end of it, Yancey had recovered his good humor. He assured reporters that Mantle would run as fast as ever and would be able to report to camp on time.

Mantle had yet to sign his 1954 contract. On February 27, he agreed to a one-year deal for $21,000, ending a brief headline-making holdout, which was common in the days before unionizing came to the national pastime. A holdout was a player's only leverage. But this did not improve his relationship with the Yankee higher-ups, who announced that Greenwade would be "put in charge of Mantle until he reported for training." No more bird dogs or outings with Billy Martin.

Greenwade took Mantle home to his house in Willard, Missouri, where his daughter, Angie, proudly relinquished her upstairs bedroom to The Mick. Sportswriters in need of copy for season previews poked and prodded. "Is He Really *Tree*-mendous?" one bold-faced headline inquired. "Casey Stengel has said he is but Mickey is *still* a problem."

The accompanying story made much of the Famous Bubble Gum Incident of 1953, when Mantle was photographed blowing boyish bubbles in center field, not the preferred image of baseball's most sober-minded franchise. (It was a lucrative misdemeanor. He got a $1,500 endorsement deal.)

When he arrived in St. Petersburg on March 2, he was unable to straighten his leg. He spent more time in Gaynor's office than in the batting cage, thwarting Stengel's plan to tutor him in the art of bunting. The Ol' Perfessor wanted to teach him "a butcher boy swing," hitting down on the ball to minimize fly balls and strikeouts. While Mantle limped, Stengel fumed. When he failed to show up for scheduled bunting sessions with Frank Crosetti, Stengel was livid. Later, in an interview with Bil Gilbert of *Sports Illustrated*, Mantle conceded, "I never learned anything."

Even Martin—*the fresh kid*, in Stengelese—listened and learned. Stengel saw himself in Martin, who made the most of his limited ability with competitive ire and fire. He saw in Mantle the player he had wanted to be. To squander such a surfeit of ability was unthinkable, irresponsible, maddening. *What's the good of telling him what to do? No matter what you tell him, he does what he wants.* This angry, frustrated refrain became the subtext of managerial disappointment.

In 1954, Stengel was still bearing the scars of derision that had greeted his appointment as Yankee manager in 1949. Five consecutive pennants, one more than the sainted Joe McCarthy and Stengel's mentor John McGraw, had earned him the right to be taken seriously, he thought. Instead, he was greeted with this from Dan Daniel in *SPORT* magazine: "Is Stengel GREAT—Or Is He Lucky?" A sixth straight pennant would put that question to rest and elevate him to the Mount Rushmore of New York baseball managers.

For that, he needed a whole Mickey Mantle. Every torn muscle, every missed opportunity and game was a rebuke to Stengel's legacy. He couldn't see past his own ambition. "He couldn't see past the fact that he was hurt," said Robert W. Creamer, author of the definitive Stengel biography.

Stengel was an emissary from the old school; Mantle was the face of postwar America. Yet there was fondness between them and more shared experience than it appeared. They came from the same part of the world, and when Stengel had to send him to the minors in 1951, it was to the team he had signed with out of high school, the Kansas City Blues. Like Mantle, Stengel was a mischievous, rambunctious boy, shy in company, hot-tempered on occasion, and an indifferent academic. But unlike Mantle, Stengel was a serious and dedicated student of the game. He had played for the storied McGraw for three seasons, even living with him one year. He had watched McGraw's careful, protective grooming of Mel Ott, nurturing an unrealized talent to greatness. The Ol' Perfessor figured he knew how to raise himself a star.

But Mantle was unwilling. "I don't think he knew what the definition of constructive criticism might be, because I don't think he'd ever listen to it," said his brother, Larry, the high school football coach. "Mickey developed this defensive mechanism to shut people out; he didn't listen to criticism. I think his short temper was probably a defensive mechanism. He didn't achieve everything, and he didn't wanna hear people say things.Now, he mighta listened to it from Daddy, but I can't imagine anybody else, maybe Casey when he was young."

Mantle never openly defied Stengel's instruction; he simply ignored it. Like a recalcitrant teenager, he could be sullen and volatile. He took out his frustrations on inanimate objects and turned his anger on himself. "Here," Stengel said, handing him a bat after one dugout tantrum. "Why don't you bang yourself on the head with this."

Born in 1890, Stengel was Grandpa Charlie's contemporary, but he wasn't paternal. The nurturing, cuddly Stengel was a sportswriters' invention. He coddled writers, not players. He spoke to Jerry Coleman only twice in nine years. The first time he said, "Nice work." The second time, after hearing Coleman call himself "horseshit," Stengel growled, "No, you're not."

"Casey was not a very pleasant guy to play for," Coleman said. "He was very tough, very hard-nosed. He didn't care if your name was Smith or DiMaggio. His goal was singular—winning the pennant. Who did it for him were the people who could play best that day. You look: the name's on the board, you play. Name wasn't on the board, you didn't play. No explanation, no nothing. He never called anyone into the office and said, 'I'm not going to do this because of this.' He did it. That's harsh treatment for players who expect more."

Mantle often said that Stengel was like a "second father to me." If so, Creamer says, he was an angry father and Mantle was a stubborn son. Ryne Duren, the myopic flamethrower who became a Yankee when

Martin was banished to Kansas City in 1957, saw clearly
what was lacking in Mantle's relationship with Stengel:
"Casey should have been the father image rather than
what he was, and he resented him for it. *Why can't you
treat me decently instead of being such an old bastard?*
I think Casey took advantage of his vulnerability. Casey
didn't see the little boy in him that needed a father."

Bunny Mick, a longtime Stengel lieutenant, sighed,
thinking what might have been "if he had just taken
that kid in his arms."

With Mutt gone and Stengel stubbornly remote, Greenwade was deputized in loco parentis by the front
office. But their relationship was infused with resentment Mantle harbored from being "outslickered" when
he signed with the Yankees. "From early on Dad tried
so hard to fill in for Mutt, tell Mickey the safe way to
live and how to take care of his body," Angie Greenwade said. "And Mickey didn't take advice. He really,
truly thought he was invincible, I suppose. Dad would
say, 'Well, he's just going to end up in terrible shape or
come to no good.'"

When Billy Martin was recalled by the Army
midway through spring training in 1954, the Yankee
wise men seized the opportunity to introduce a stabi-
lizing influence—Bill Dickey informed Coleman that

he had a new roommate. "I didn't drink, so they put me together with him to steady him," said Coleman, who found his new roomie to be shy, quiet, boyishly unassuming—delightful. On road trips, Mantle carried Coleman's luggage up from the lobby. During spring training, they took turns driving to the ballpark. "One day I look out the window. It's twenty of nine. He's there in his car. I said, 'Mickey, why didn't you honk the horn?' He said, 'I didn't want to bother you.'"

Mantle had been sitting outside for twenty minutes.

The Yankees tried surrounding him with family, putting three more Mantles in pinstripes. They signed his cousin Max to a minor league contract in March 1954 and invited the twins, Ray and Roy, to take batting practice at Yankee Stadium, fueling the fleeting prospect of an all-Mantle outfield. Mantle always said they were better natural athletes. But they lacked his drive and his luck in receiving the fierce attentions of their father.

It was worth the gamble and the small financial investment. The twins were assigned to the Yankees' Class D team in the Sooner State League. Greenwade drove them to McAlester, Oklahoma, in a black Cadillac just like the one in which he had signed their big brother.

Max was released a month before the twins arrived.

They stuck around long enough to play in a Florida instructional league the following spring. Ray was drafted into the Army and never played again after he was discharged. Roy suffered a career-ending injury running out a base hit.

The twins' visit to New York proved to be the highlight of Mantle's year. At the end of April 1954, he was batting .175. In June, he was still unable to play both games of a doubleheader. When Stengel blew out the candles on his birthday cake, celebrating his sixty-fourth birthday on July 30, he bravely declared that he should be fired if the Yankees failed to win the pennant.

A month later, with his team losing ground to the Indians and the season growing short, reporters noticed a newly published book on his desk. Its title: *The Year the Yankees Lost the Pennant,* a picaresque fantasy in which a Mantle look-alike unable to resist temptation makes a pact with the Devil to bring down the Yanks.

It was a gift, Stengel said. Hadn't read it. Wasn't going to.

"It's fiction," he growled.

8

September 26, 1954

NO OTHER TIME

1.

On the last day of his fourth major league season, Mantle played the first meaningless game of his career. One hundred and three wins, four more than they had needed to win the 1953 pennant, and the Yankees still finished eight games behind the Cleveland Indians— "them plumbers," in Stengel's derisive lexicon.

He fixed the blame on Mantle's slowness to heal from his off-season knee surgeries and his knuckle-headed refusal to act like an adult. "He's gotta change a lot," Stengel declared. "He's gotta change his attitude and stop sulking and doing things he's told not to do.

He'll have to grow up and become the great player he should be when he reports next spring."

Perhaps that's why he sent Mantle out to play his first and last complete game at shortstop, the position of his callow youth. Around the horn beside him, Stengel stationed an unlikely second baseman, Moose Skowron, and an improbable third baseman, Yogi Berra. "My tape measure lineup," Stengel called it. He did not point out that Mantle had hit only nine home runs since the All-Star break.

That sleepy Sunday was a slow news day in a year of momentous events: Joe DiMaggio married and divorced Marilyn Monroe; the Miss America pageant went prime time; Joseph McCarthy was censured by the U.S. Senate; the Supreme Court handed down its landmark decision in *Brown v. Board of Education*; and for the first time in six years, the Yankees failed to win the American League pennant.

The also-rans in the Bronx were knocked off the back page by a better story. Willie Mays had returned from two years in the Army to lead the Giants to the National League pennant and to complete the kingly triumvirate of New York center fielders—Willie, Mickey, and The Duke.

The coincidental ascendancy of Mays, Mantle, and Snider announced a golden era in baseball and ratified New York's sense of itself as the center of things. "There was no other time," said Yankees infielder Andy Carey. "No other ten years—Mays, Mantle, and Snider."

There was no better time to be a baseball fan or a boy growing up in New York. Pete Cava, who still calls himself "a Mickey guy," argued ardently on behalf of The Mick every day in study hall at New Dorp High School on Staten Island. His pals, Glenn Cafaro and Greg Bischoff, were equally partisan about Willie and The Duke. Bischoff spilled blood on Snider's behalf, in a ritual card burial with his Mickey-loving best friend. "It was an alignment of planets," said Cava. "If Homer is alive, Ajax is playing in the Bronx, Ulysses is in center field for the Giants, and Achilles is defending the field in Brooklyn."

When the season opened, The Duke was baseball's sure thing (the National League leader in runs and slugging percentage in 1953, when he tied with Eddie Matthews, at .627). Mantle's right knee made him almost as much of a question mark as the Giants' returning GI. If his unrealized potential was a vexation, Mays was a revelation—speed, grace, *and* power. When last seen in a major league uniform, nine days before reporting to the Whitehall Induction Center in May 1952, Mays was batting .236. Gladys Gooding, the organist at Ebbets

Field, serenaded him after the game with her rendition of "I'll See You in My Dreams."

In his absence, Snider and Mantle dominated the footlights, starring opposite each other in the 1952 and 1953 World Series (Duke's Dodgers played the role of perennial foil). Oracles had confidently predicted another Yankee-Dodger World Series in 1954. So much for oracles.

Mantle rebounded from a desultory spring and was named the starting center fielder for the American League in the All-Star Game; Snider was his opposite. (Mays and his thirty-one home runs be damned.) But in the second half of the season, he and the Giants asserted themselves, overtaking the Dodgers to clinch the pennant on September 21.

By then, Stengel had grown irritable with Mantle's imperfections. On September 10 in Chicago, Mantle went 0 for 4 and struck out twice, his ninety-third and ninety-fourth strikeouts of the season. His batting average fell below .300. Worse, he failed to run out a ground ball.

Casey Raps Loafing Mantle—*New York Post*

Casey Stengel Incensed at Mickey Mantle's
Lack of Hustle

He bawled out the star on the bench last night when Mick failed to run out a grounder. White Sox players also said that Mantle had loafed in the field."

Mantle had been working out at shortstop since July. Stengel bristled when reporters asked whether he was putting Mantle in jeopardy by exposing him to the uncertainties of a 6–4–3 double play. "He can get hurt in the outfield too, can't he?" Stengel snapped. "Every time he's been hurt, it was out there. Maybe I'll never make the move, but he'll be ready. I got him workin' out here now so he'll be used to the different throw. I don't want him killing any customers in the grandstand."

By the time the Yankees took the field on the afternoon of September 26, everything of consequence in baseball had been decided—the pennants won, the World Series starting pitchers named, the betting line posted—the Indians were 8½-to-5 favorites over the Giants.

Only the National League batting title remained unresolved. Quite improbably, Snider and Mays had arrived at the last day of the season in a virtual tie for the batting championship, separated from each other and Giants' right fielder Don Mueller by hundredths of a point. "Close enough to be covered by the same handkerchief," Arch Murray wrote in the *New York Post*.

"We were aware of each other," Snider allowed.

How could they not be? "They had us side by side in the paper every day."

They kidded each other around the batting cage but it quit being fun when Dodger owner Walter O'Malley began posting their stats on the scoreboard, a galling rebuke to a player schooled by Branch Rickey not to focus on individual statistics. "If you're in first place, the numbers will be there," the Mahatma preached. It was the prevailing ethos in an era before players' agents prepared thick binders of statistics to market multimillion-dollar clients.

Baseball was in the air that afternoon as it had been all season, passed from stoop to stoop and borough to borough on competing AM frequencies. In the visiting dugout in Philadelphia, Giants' manager Leo Durocher parroted Vin Scully's play-by-play from Brooklyn. The Duke had earned The Lip's disdain by sitting out against the Giants' tough lefty, John Antonelli, the day after they clinched the pennant in Brooklyn.

If New York's elders were attuned to the voices of the game, city boys like Denny Minogue and his crew from upper Manhattan preferred to spend the afternoon pretending to be the players they aspired to become. Minogue had sung at two Masses by game time in the Bronx, Brooklyn, and Philadelphia. Liberated from the liturgy, he raced to the playground at P.S. 152 to meet his pals Mike Green and Bobby Cook for an afternoon of stickball and bravado. Mike was a short Jewish kid. His affinity for The Mick wasn't hard to understand. "Cookie" was a center fielder and his own man—you had to be if you were a Duke guy in upper Manhattan. Denny was an Irish Catholic choirboy who wanted to be black because of Willie Mays. On the playground at P.S. 152, they wouldn't let him pitch because he threw too hard, but they couldn't stop him from singing the praises of the Say Hey Kid.

When the sun went down that evening on the last day of the regular season, they headed to Freddie's corner store, as they had every day, to learn how their guys had done. All over the city, boys like them were loitering in congested candy store aisles, sucking down an egg cream or a black-and-white, arguing until the newspaper trucks rumbled down the street loaded with heavy bundles of certainty to tell them who had earned bragging rights for the winter.

> Spooner Fans 12 and Wins, 1–0
> —*New York Herald Tribune*

Snider went hitless, eclipsed by the Dodgers' pitching phenom, Karl Spooner, who struck out twenty-seven in his only two starts of the season. The Duke was stymied by Jake Thies, an unprepossessing rookie for the Pittsburgh Pirates, who described himself as "a sidearm, underhand pitcher." Duke offered no comment.

Mays' 3 Hits Nip Don for Title, .345–.342
—New York *Daily News*

Mays was subdued after winning the batting title in his first full season in the majors. "Sure, I was thinking about it," he said. "Leo wouldn't let me forget it."

Mueller had nothing to say to the press. But the right fielder had a question for Mays. "Is it true you're the best center fielder in baseball?"

"The best right fielder too," Mays replied.

Yanks Lose Finale to A's 8–6 Despite Power Shift
—New York *Daily News*

Only 11,000 showed up at Yankee Stadium to see Stengel's power lineup. "Mantle drew three passes to make 100 for the year, fanned to make 107, and singled sharply to right in his last try for the .300 finish," Joe Trimble wrote in the *News*. He also played his old position, shortstop, without killing any paying customers.

Subway series: Winner's grin spreading over the face of the Yankees' twenty-year-old star, Mickey Mantle. In the visiting clubhouse at Ebbets Field, Mantle was elevated and venerated after hitting the game 7 home run that decided the 1952 World Series. He had every reason to smile. It was a smile "quite unlike any other," Tim McCarver said, "almost a measure of a man."

In the fifth inning of game 2 of the 1951 World Series, New York's other rookie center fielder, Willie Mays, hit a tweener to right center field. The collision of fates was almost operatic: Mays, Mantle, and DiMaggio triangulating the future of the game. Mantle caught a spike in a drain trying not to run into Joe "Fuckin'" DiMaggio. It was the turning point of a life that had just begun. A day later his father was diagnosed with non-Hodgkin's lymphoma. Mantle's knee and heart were never the same. He never looked that young again.

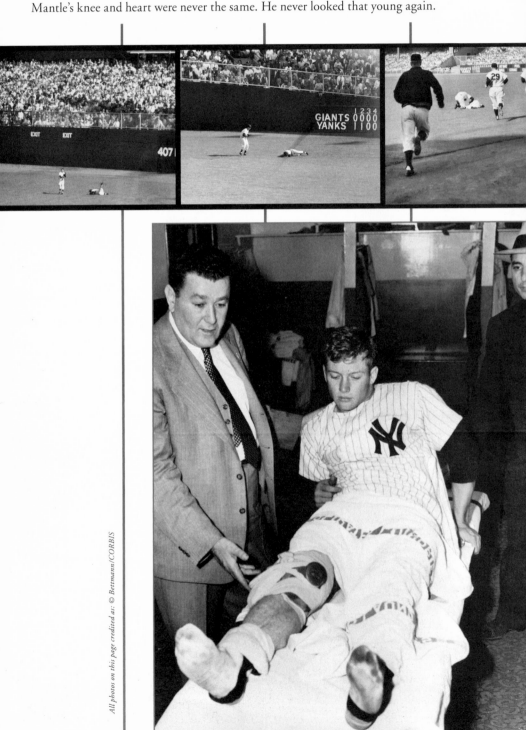

To New York wise guys, Commerce, Oklahoma (*top*), sounded like a made-up name. *Above*: The mining shack where the Mantles lived for a decade, the corrugated metal shed where he and Mutt practiced every day at 4 P.M., and the toxic legacy of the lead and zinc mines. *Left:* The sign announcing the renovation of Mantle's childhood home was stolen off the front porch. *Below*: Dying, Mutt told his son all he wanted for Christmas in 1951 was a wedding. In the last family photograph, taken two weeks before the marriage of Mickey and Merlyn, Mantle plays the hand he's dealt.

★★★★
FINAL

DAILY ☐ NEWS

NEW YORK'S PICTURE NEWSPAPER ®

LARGEST
CIRCULATIO
IN AMERIC

New York 17, N.Y., Saturday, April 18, 1953

4¢ IN CITY LIMITS 5¢ OUTSI CITY LI

28 Pages

MANTLE SLAMS RECORD HI
GIANTS SPLIT; YANKS WIN

—Stories Pages

Donald Dunaway said he found the Tape Measure Home Run ball in a pile of leaves opposite the window of 434 Oakdale Place, not in the backyard.

Holding a Grand Slam. Mickey Mantle holds ball which, despite its scuffed appearance, will make for his mantlepiece. Mickey, batting righthanded in fifth inning in Was
ball 562 feet for record homer. Ball was ripped when it hit house after clearing wall at Griffith Stadium. Yanks won, 7-3. —

Surrounded by adoring fans, or team-mates, or reporters, or by the four young sons who rarely saw him, Mantle was seldom alone. In the Yankee club-house, where he was his happiest and best self, "the loneliness of greatness" was still palpable.

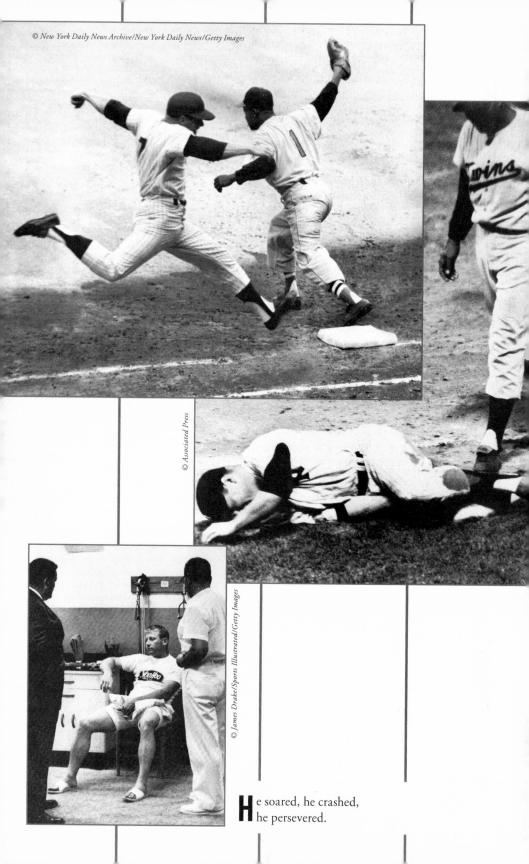

He soared, he crashed,
he persevered.

He swung and he missed and he hurt. Later, he admitted how much he had hurt himself.

Top: The drawing of Mantle at the Pearly Gates appeared in the *Dallas Morning News* the day after his death. His wife later outbid one of his drinking buddies for the original art, cartoonist Bill DeOre said, because she blamed them for contributing to his downfall. *Bottom*: Mantle and pals, Billy Martin and Whitey Ford, at Old Timers' Day at Shea Stadium in 1975. He was the Last Boy in the last decade ruled by boys. "He never grew up, and it ruined him," teammate Jerry Coleman said.

Stengel was in a more generous mood after the game, lauding Mantle's patience at the plate—maybe the young fella was getting to be a more careful hitter after all. "He fell off in the last month an' that bad knee of his had something to do with it," Stengel opined.

Mantle's comment was buried deep inside a page-two story in the *Herald Tribune*. He said his trick knee would probably prevent him from playing shortstop regularly.

2.

That evening, New York's three center fielders gathered at Lou Walters' Latin Quarter for the 8 P.M. broadcast of *The Colgate Comedy Hour*. Just the year before, Mantle, Martin, and Ford had celebrated the 1953 pennant at the nightclub by signing Dan Topping's name to the tab. But this was Willie's time, and Willie was running late. Ed Sullivan's variety show on CBS, *The Toast of the Town*, aired opposite *The Colgate Comedy Hour* Mays *was* the toast of the town. So while Joe E. Brown warmed up the crowd at the Latin Quarter, Mays opened for Sullivan five blocks up Broadway at the CBS-TV Studio 50, then dashed downtown in time to be introduced with Mantle and Snider as members of *Look* magazine's All America Team. It was the first time New York's three leading men shared center stage in their four-year run on the Great White Way.

After the show, they went their separate ways. Mays tagged all the bases of newfound fame. He was paraded in an open car up the Canyon of Heroes in New York's traditional ticker-tape embrace. He appeared on *The Today Show* with Dave Garroway and his cohost, J. Fred Muggs, the all-star chimp; and he was Steve Allen's guest on the national premiere of *The Tonight Show*.

Snider headed west with the Eddie Lopat All-Stars for a barnstorming tour of the Hawaiian Islands. Mantle's trick knee prevented him from joining them as planned. His roommate, Irv Noren, agreed to take his place on Lopat's roster. Noren was leaving for California that night and offered Mantle a ride to Commerce in his brand-new, four-door Chevrolet. Mantle packed his bags. He couldn't get out of town fast enough.

Noren's pal Julie Isaacson, a union guy with all the right connections, told him to leave his Chevy on the sidewalk outside the Latin Quarter. After the show, he found two of New York's finest—mounted police— standing guard over the car packed with all their belongings. "Nobody touched Mickey and Irv in New York City when Julie was around," Noren said.

They stopped in Union, New Jersey, for a bite to eat

before hitting the road. Their friend Frank Petrillo had a car dealership there—and a wife who prepared Mantle's favorite Italian dish—spaghetti with "that cheese that smells like feet." The lady of the house hit him over the head with her dish towel after that remark. "What the hell do you know about good Italian food, you midwestern hick?"

Frank, Jr., age nine, and his baby brother, Jay, age two, were allowed to stay up late to see their favorite ballplayers. Mantle and Noren were in a playful mood. "Mickey and Irv were tossing my kid brother around," Frank, Jr., said. "Jay's head hit the ceiling. Anytime after that my brother did something wacky, my mother blamed it on Mickey and Irv."

They took off after midnight in an autumn drizzle. There were no four-lane interstates; 55 miles per hour was an aspiration, not a constraint. Commerce was 1,271 long miles away. They didn't stop except to eat and to stretch their aching legs. Noren didn't let Mantle get behind the wheel until they were two miles outside of town. He had heard too many stories about Mantle's driving.

They didn't talk much baseball as they headed west. They didn't listen to the radio broadcast of the first game of the World Series. "Listened to that damn cowboy music the whole way," Noren said. "There was one song that was pretty tragic. This guy had hit his two kids, and they were saying, 'What did you hit us for, Daddy?'

"Didn't kill 'em or nothing. Mickey had tears in his eyes, and I did, too."

Finally, Noren said, "Turn that damn thing off."

They didn't hear the radio announcers Al Helfer and Jimmy Dudley describe the top of the eighth inning from the Polo Grounds when the Indians' Vic Wertz redirected Don Liddle's first pitch toward the gully in the deepest recess of center field and Mays turned his back on expectation, setting off in pursuit of a ball no one thought he could catch—except him.

They didn't stop to watch the first game of the last World Series broadcast without color commentary in stark black and white. They didn't see The Catch that would become the defining moment of Mays's career. They were on a dusty midwestern ball field on Route 66. Mantle had spotted a bunch of schoolboys shagging balls. "Hey, let's have some fun," he told Noren. "We've been driving all day and night."

Noren pulled the car over and parked by the grandstand. He grabbed a bat and a few balls from the trunk and approached the biggest kid on the field and gestured to Mantle. "I told him I wanted to see if I could strike this guy out," Noren said. "I told him, 'When we get through I'll give you three brand-new balls.' Then I told all the kids to go out four hundred feet."

The last ball he threw left a jet trail across the afternoon sky. It flew over the school and over a copse of trees beyond that. "Boy, you can't strike that guy out," the kid told Noren when he returned with the ball ten minutes later.

"You know why? That's Mickey Mantle."

"And I suppose *you're* Yogi Berra," the kid said.

They were still laughing when they hit the road again. "Kid probably went home and told his dad, 'Mickey Mantle drove up in a car and hit the ball over the schoolhouse,'" Noren said.

"And his dad probably told him, 'I told you never to lie,'" Mantle replied.

3.

Who's better? Willie, Mickey, or The Duke?

Duke says the argument started in 1954, their first full season together in the majors. It began in earnest the moment Giants' third baseman Hank Thompson claimed the final out of the Series from the sky.

You could get a fat lip in any saloon in town by starting an argument as to which was best.

Red Smith wrote the line on the occasion of Mays's

retirement in 1973. But he was talking about 1954. That year, New York was home to 7,891,957 true believers and kibitzers with at least that many opinions, loudly expressed.

Mantle makes balls disappear!

Mays plucks line drives out of thin air!

Snider is as smooth as Houdini!

Passions were fierce, often hereditary, and fightin' words were backed up. Opinion was obligatory. New Yorkers argued on Brooklyn stoops, on Bronx fire escapes, in poorly lit stairwells of Queens housing projects, and in the pages of seven major (English-language) daily newspapers (not to mention the *Brooklyn Eagle*, *La Prensa*, *El Diario*, and the Jewish *Daily Forward*).

Baseball men argued between, during, and after helpings of rubber chicken on the winter banquet circuit, with Leo "The Lip" leading the way. "Leo Durocher needed about ten seconds after the final 1954 statistics were posted to realize the publicity potential in the feud," *SPORT* magazine reported. " 'Snider is wonderful,' he said. 'But my boy Willie just happens to be the greatest there ever was.' "

"Old Ty Cobb was one of those who rose to Leo's bait," *SPORT* continued. "Tyrus irritably barked, 'How many years has Mays hit .300 or more for the Giants? In one season, I believe. How many years has he batted in 100 or more runs? Hell's bells, at this point the discussion is ridiculous!'"

Neither "The Lip" nor "The Georgia Peach" made mention of Mantle.

Over the next three seasons, as their fortunes and their batting averages waxed and waned and the statistical evidence accumulated, the argument intensified. At the dawn of the last season for New York's three franchises, the debate remained unresolved.

> The nature of the Mantle-Mays-Snider argument is
> such that it changes every year. So who's the best?
> Come around in 15 years when all the
> statistics are dry and we'll know.
> —April 17, 1957, *Newsday*

And still there is no agreement.

Indians pitcher Bob Feller: "Mickey was a better all-around ballplayer. Mays was a better actor."

Dodgers outfielder Tommy Davis: "I don't care if Willie was pink, blue, yellow, beige, he would still be the best."

Dodgers pitcher Carl Erskine: "The only thing Mays had over Mantle was durability. Duke was royalty. They never had much royalty in Brooklyn."

Giants first baseman Willie McCovey: "We knew

Willie Mays was head and shoulders a better player. We knew *he* played in New York."

Yankees first baseman Moose Skowron: "They talk about Willie Mays. Willie Mays was in two World Series. I go by winnings."

Leave it to Yogi Berra to make the most sense: "I'd like to have *that* outfield."

The point of the argument was having it, not solving it. It was sustaining and somehow defining. Who you chose as your guy told you something about yourself—who you wanted to be and *how* you wanted to be.

It was easy to want to be Willie. He was so blithe, levitating above green fields with weightless joie de vivre. To chase his example—to run with liquid abandon—was to cast off white propriety and prejudice in favor of freedom of expression. Who wouldn't want to try on the accoutrements of incandescence?

Consistent, reliable, orthodox, The Duke never disappointed. He aspired to Joe D.'s understated grace and dropped the ball the only time he tried Willie's basket catch. He made his major league debut two days after Jackie Robinson broke the color barrier. He hit more home runs in the Fifties than Mantle, Mays, or anyone else—forty home runs in each of five straight years. But it was his fate to be overlooked. He was the perfect hero for an underdog borough and every underdog kid.

If DiMaggio had grace, Mantle had insouciance. It informed his posture, the way he leaned on a bat, hip jutting one way, pelvis the other, in a casual frieze of male swagger. And he got better-looking with age, growing into his role and his musculature, offering hope to prepubescent boys everywhere. To assume his limp was to master how to function with disability and disappointment. To mimic his swing was to channel his power.

Mantle brought danger and expectation to the plate. With every swing there was a chance of seeing something "so fucking ridiculous," said Billy Crystal. Or so fucking awful.

Though they appealed to different constituencies, Willie, Mickey, and The Duke had more in common than center field; they shared the same psychic real estate. All three were sons of semi-pro baseball players with large ambitions for their boys. All three knew what it was to be overwhelmed by expectation and major league pitching. All three spent time in the bushes their rookie year.

But they were peers more than friends. They coexisted in box scores and headlines, crossing paths chiefly on the base paths. Umpires enforced the no-fraternizing rule with $50 fines when $50 meant something.

Snider and Mays played each other twenty-two times a year, every year. There was little love lost between the Dodgers and Giants and less between their fans, as Mays learned after hitting two home runs in Brooklyn one Sunday afternoon. All four tires on his car were slashed when he returned to the Ebbets Field parking lot. He took the subway back to Manhattan.

Mantle and Mays played over each other's shoulders in ballparks on opposite shores of the Harlem River, but they faced each other only twice, in the 1951 and 1962 World Series. Mantle and Snider played each other in every World Series—or so it seemed. But fate had a way of bringing them together in moments of drama and definition. Mantle's first New York home run sailed over Snider's head in an April 1951 exhibition game. Seven months later, Mays hit the fateful tweener at the Stadium that Mantle called "the ball that caused all my trouble." When he told the story on the *Warner Wolf Show* in 1981—the first joint TV appearance of Willie, Mickey, and The Duke, which they did for free—Mays replied, "I thought you were playing center field."

Who would believe that four years later, in game 6 of the 1956 World Series, Snider would catch a spike in the same drain, and in the same way. He even used the same language to describe the injury: "Something snapped."

Snider saw Mays's best catch, not The Catch that got all the attention "because it was in the World Series."

The one Mays made on opening day in Brooklyn in 1952, when he levitated himself in pursuit of a line drive and launched himself headfirst into the outfield wall. Knocked himself out cold but held on.

Snider witnessed Mantle's best catch, in game 5 of the 1956 Series, when he outran Gil Hodges' drive to the deepest reaches of Death Valley, enabling Don Larsen to throw the only perfect game in World Series history. "Probably the only good catch I ever made," Mantle would say later. "I wasn't that good of a fielder."

His home run in the bottom of the inning broke a scoreless tie. It was his best World Series game, and nobody noticed.

After the diaspora, when the Dodgers and Giants decamped to the West Coast in 1957, Mantle had New York to himself. Arriving at the Los Angeles Coliseum for the Dodgers' first home game in California, Snider found Mays waiting at the batting cage, pointing to the right field wall 440 feet feet away. "Hey, Duke, they're burying you, man."

Mays reigned supreme over DiMaggio's hometown, but his accomplishments were fogged in by the elements at Candlestick Park and by East Coast newspaper deadlines. He feared he'd been forgotten. But when he returned to New York for the first time in 1961 for the Mayor's Trophy Game at Yankee Stadium, Bob

Sheppard's introduction was drowned out by the roar. Mantle turned to his teammate Al Downing and said, "Boy, they really love him here, don't they?"

By 1962, The Duke of Flatbush had become as superfluous in Los Angeles as his nickname. One day, he overheard Dodger manager Walter Alston complain, "I can't wait 'til I get rid of these old guys." The Dodgers sold him to the laughable Mets on April Fools' Day 1963 for $40,000. At the end of the season, Dodger general manager Buzzie Bavasi wrote him a personal check for $10,000 and admitted the Yankees had wanted him, to back up The Mick.

That fall, Mantle and Mays met in a World Series that was supposed to answer the question "Who's better?" for once and for all. Their stunning haplessness prompted a shrill bleacher critic to ask instead: "Who's worse?" When Mantle returned to center field after yet another futile at-bat, the arbiter offered this verdict: "Hey, Mantle, you win."

Mays shone like a solitaire for five years after Mantle retired. In 1972, the Giants, who had promised never to trade him, sent Mays to the Mets for a belated New York encore. But it was too late. In October 1973, America winced as Willie staggered under a fly ball on a sun-drenched World Series afternoon. Like Mantle and Snider before him, the Say Hey Kid had become just another old ballplayer who had overstayed his body's welcome. "I did," Snider said. "We all did."

The argument didn't end when the record books closed. Daily box score tallies gave way to "what-ifs." What if Willie hadn't played two years for Uncle Sam? What if Mickey hit into that Candlestick gale? What if Mays had played center field for the New York Yankees? What if Mantle had played on two legs? What if Duke hadn't had that short right field fence in Brooklyn or the damnable outfield wall at the L.A. Coliseum?

"Willie, Mickey and The Duke" became the occasion for endless confab and the subject of Denny Minogue's 1981 anthem to the Fifties— "Talkin' Baseball." "I'm talkin' Willie, Mickey, and The Duke (Say hey, say hey, say hey)." By then, Minogue had surrendered his major league aspirations and had adopted the nom de tune, Terry Cashman. The lyrics to his now ubiquitous melody evoke a singular era before fate dictated how history would sing their praises. On September 26, 1954, no one knew who would be judged best, but everyone cared about the answer.

By the time the song hit the airwaves, the terms of the debate had changed. The mathematical calculations of a simpler world, the five-tool judgments made by stopwatch-bearing scouts, gave way to a generation of stat-wielding revisionist historians with competing formulas designed to assess and compare the relative value of individual performance, past and present. Armed with a profusion of new and ever-improving metrics, Bill James and the heirs to his sabermetric revolution relocated the argument over Willie, Mickey, and The Duke from the street corner to sports radio, and then to cyberspace, where partisans armed with new ammunition continue the fight. "It's kind of like the Civil War," said Pete Cava. "Just because they signed the treaty at Appomattox doesn't mean the war is over. The debate still goes on."

When the New York *Morning News* published the first primitive box score in 1845, no one could have imagined it would lead to this: Runs Created (RC), Runs Created Above Position, (RCAP), Runs Created Above Average (RCAA), On Base + Slugging (OPS), On Base + Slugging Adjusted (OPS+), On Base Percentage × Slugging (SLOB), Win Shares and Wins Above Replacement Player (WARP 1–3), Value Over Replacement Player (VORP), Offensive Winning Percentage (OWP), Marginal Value Above Replacement Player (MORP), and Defensive Wins Above Replacement (DWARP).

Branch Rickey, baseball's most original thinker, planted the seeds of this new math by hiring the first team statistician in 1947—the same year he and Jackie Robinson defied baseball's color barrier. Seven years later, he and his stat man, Allan Roth, pioneered the formula for on-base percentage. This arithmetic innovation was met with predictable disdain: "Baseball isn't statistics," groused Jimmy Cannon. "Baseball is DiMaggio rounding second."

Jim Bouton seconded that opinion years later: "Statistics are about as interesting as first base coaches."

Not even the visionary Mahatma could have predicted the statistical blitzkrieg that has changed the way players are measured and remunerated, how front offices assemble teams, and how managers manage games. By 2001, James, paterfamilias of the stat-geek generation, had conceded that clarity had been all but lost in the numerical dust storm of mutating calculations and shiny new algorithms. But he remained unequivocal in the assessment first published in the 1985 edition of *The Bill James Historical Baseball Abstract*: "Mickey Mantle was, at his peak in 1956–57 and again in 1961–62 clearly a greater player than Willie Mays— and it is not a close or difficult decision."

John Thorn, baseball historian and numbers cruncher extraordinaire, concurs—up to a point: "Mantle is superior to Mays—with a bat in his hands. If one is to judge them as all-around players Mays is superior because he was so much better in center field. For pure offense, Mantle is it."

(For a statistical comparison of Mantle and Mays, see appendix 3, page 715.)

Mantle often joked about how many years you'd have to deduct from his eighteen big league seasons if you subtracted the number of times he struck out. The answer, according to Dave Smith of Retrosheet, is three. Mantle also made light of the number of times he batted without hitting the ball. Do the math, he advised: 1,734 walks plus 1,710 strikeouts means he batted 3,444 times "without hitting a ball. Figure 500 at-bats a season, and that means I played seven years in the majors without hitting the ball."

But, according to the new math, he overvalued the strikeouts and undervalued the walks. In the sabermetric universe, an out is an out is an out. Strikeouts carry more emotional weight than statistical import. To the gimlet eye of a modern stat geek, walks are the key to Mantle's superior on-base percentage and the reason he fares so well in a preponderance of the new offensive metrics.

Nor is Mantle's self-flagellation over his lifetime .298 batting average warranted. "He was measuring himself with the yardstick he grew up with in the 1930s," Thorn said.

Little wonder that teammates fulminated at all the

"what ifs." "I hate it when people say how much he wasted," said Clete Boyer. "Jesus Christ, how much better could he have been?"

4.

On a flawless spring training day in 2006, arms folded over a slight pinstriped paunch, Reggie Jackson turned away from tracking the flight of one hundred batting-practice hacks to consider the question of Mickey Mantle and white-skin privilege. Forty-five minutes into Jackson's disquisition, Derek Jeter jogged over to find out what was holding Mr. October's attention. "We're just talking about how Mantle would have been remembered if he was black," Jackson said.

Jeter, a post-racial hero who has perfected the art of public speaking without saying anything at all, executed the patented midair pirouette usually reserved for hard-hit balls in the hole and headed in the opposite direction. It is the one "what if" nobody wants to talk about. If Mickey had been black and Willie had been white, what kind of conversation would there have been? Would there even have been a conversation? How much does race influence the way they are remembered?

"I'm glad you asked me that," Jackson said. He

was a distinct minority in that regard. He considered briefly the tormented life of Jack Johnson, the first African American heavyweight champion, who inspired Howard Sackler's Pulitzer Prize–winning drama *The Great White Hope*. "He lived a life on the dark side with alcohol and women. I don't know that we condemn men for that. In this country we accept it as part of being a man. I don't think Mickey would have been disparaged if he had been black."

"Would he be forgiven?" I asked.

"We're going to forgive Doc Gooden; we're going to forgive Darryl Strawberry. I think character flaws bring compassion for all colors. Now certainly, I don't know that if Willie Mays were white there would be any question that this guy's the greatest ballplayer of all time."

So, if the races were reversed, would there have been an argument? "No," Jackson said. "I don't think so. Mickey was super-talented and he played for the Yankees in New York and he was white."

That's why former pitcher Jim Kaat believes, "Mantle was so much more of a national figure," than Mays. "Most of the kids that played baseball or followed baseball were country boys from Oklahoma, North Carolina—good old country boys. In that era, most of the kids playing baseball were white. And that's why Mickey was the perfect hero."

Ironically, because of race, Mantle may be disproportionately adored and historically undervalued. "This sounds strange to say, given all the adulation that surrounded him, but in a certain sense I think Mantle is now underrated and may be even more so as time passes and fewer current fans actually saw him play," said Bob Costas, the erudite sportscaster, who idolized Mantle as a child, but was clear-eyed about his hero by the time he eulogized him as an adult. "When he retired he was third on the all-time home run list, but 536 doesn't seem as monumental today. In addition, he was essentially shot at age thirty-three but played four more years. Those years took his lifetime average from .309 to .298 and his slugging percentage from .582 to .557. It's easy and understandable for some people to say, 'Sure, Mantle was great but because he played for the Yankees and was a charismatic white player at a time when most of his greatest contemporaries were black or Hispanic, he was overglorified.'

"And a cursory look at raw numbers supports that notion. Mays and Aaron eventually went way past him. Frank Robinson had fifty more home runs and his lifetime average was only four points less. Lesser Hall of Famers like Willie McCovey, Eddie Matthews, and Ernie Banks are actually pretty close to him on the home run list. Roberto Clemente was a different kind of player, but he was passionate and stylish and his life had great meaning beyond baseball.

"So what's the big deal about Mickey? Here's the big deal—Mays is the greatest all-around baseball player I've ever seen. Aaron is the true home run king and his career achievements are staggering. And yet at his best, in his prime, from the mid-Fifties to mid-Sixties, Mantle was pretty much their equal and at times was better. That's even with all the injuries. Who knows what he would have been had he been healthy, since he was the greatest combination of natural power and natural speed the game has ever seen. But even as it was, look closely at the numbers—especially OPS, the combination of on-base and slugging percentage, which is the best measure of a hitter's value.

"In his prime, Mantle's OPS was higher than Mays's and significantly higher than Aaron's in any stretch of their careers, as were his home runs per time at bat. Bill James has always ranked Mantle's peak value as the highest of his contemporaries and virtually every one of the new baseball metrics is very favorable to him.

"Sure, there is some nostalgia and mythology connected to Mickey, but the real baseball truth is this— the more knowledgeable you are about the game and its history, the higher you rate him.

"Race certainly played into how some people *identified* with Mantle and Mays, just as it later did with Bird and Magic. But it doesn't change the fact that for a decade Mantle was virtually Mays's equal, just as Bird was no worse than a very close second to Magic, no matter the racial perceptions.

"Those who say there was no real comparison between Mays and Mantle are considering it from the end point based on career totals, but that's not where the whole Mantle-Mays debate took shape. It took shape in the Fifties and early Sixties when they were the game's most dynamic players and when you could make a fair case for either one. Overall careers? Mays hands down. In their primes? Pretty much a toss-up. And that would have been true if Willie was white and Mickey was black or if you painted both of them blue."

Monte Irvin was Mays's first roommate and became Mantle's friend when he was invited to his fantasy camp in the Eighties. Like Jackson, he believes there would not have been any discussion about "who's better?" had the races been reversed and Mantle would not be "quite as loved and adored. The reason Mickey was liked so much was number one, color; number two, talent; number three, he was personable."

Like McCovey, Irvin believes that the press protected Mantle, shielding his indiscretions because of his race. "I don't think they would have covered up some of the things they did if it was the other way around," McCovey said. "That's the only way I feel he benefited from being a white guy."

Irvin watched with admiration as Mantle improved

himself first as an outfielder and later as a public person. He listened when fantasy campers asked Mantle *the* question. " 'Well,' he'd say, 'Mays was a better fielder. I had more power and hit the ball farther.' He came out and told the truth.

"He made people like him. I told Willie, 'You should be a little more personable. They'd like you the way they like Mickey.'

"But he never did."

PART TWO

A Round with The Mick

ATLANTIC CITY, APRIL 1983

I had been to the Claridge Hotel once before, on Valentine's Day, 1965. It was my father's idea for a romantic, if belated, celebration of my parents' twenty-fifth wedding anniversary. He hadn't counted on the chlorine fumes seeping through the air vents. "What a dump," my mother sniffed.

The Claridge opened for business in December 1930, which was one kind of gamble. Set back from the boardwalk, it fronted on Brighton Park, where a fountain illuminated by thirty tinted lamps threw off a wild array of pinks, yellows, and greens. Twenty stolid stories high, with a gleaming gold cupola perched at the top, the Claridge was known as the "Skyscraper by the Sea," the best place in town to ride out a hurricane. Its thrusting architecture, supposedly inspired by the Empire State Building, has earned it a spot on a list of the world's ten most phallic buildings, according to a Web site that purports to measure such things.

The hotel's name, appropriated from the elegant London establishment, was spelled out in marble script on the lobby floor. A grand spiral staircase dominated the gilt and marbled entryway. Generations of newlyweds posed there like figures atop a wedding cake. My parents were not among them.

Lyndon Johnson sweated out the 1964 Democratic Convention at the Claridge, which lacked air-conditioning. It went downhill from there. The last of the grand hotels built along the boardwalk, it was also the last to open for gambling after the casino trade arrived in 1976. When I showed up in April 1983 for my 11 A.M. interview with the new Director of Sports Promotions, I was greeted by a doorman decked out in full Beefeater regalia; barmaids introduced themselves as wenches. The spiral staircase had been ripped out to make room for additional slots. But "a little bit of Britain" just wasn't pulling in the nickel-slot-playing daytrippers schlepping down the Garden State Parkway from Manhattan—"bus people," as they're known in

THE LAST BOY • 243

the trade. Management needed a way to compete with the bigger, glitzier gambling emporiums springing up along the boardwalk.

In 1980, Willie Mays had signed on with Bally's Park Place Hotel and Casino and was promptly banished from baseball by Commissioner Bowie Kuhn. So Mantle knew what to expect when he accepted the job at the Claridge. Lee MacPhail, his old Yankee boss, said, "I wish you wouldn't do it." Kuhn advised following DiMaggio's caffeinated example: be a Mr. Coffee. "I would, but nobody ever called me," Mantle replied.

On February 8, 1983, Mantle signed a $100,000-a-year contract that called for him to spend a minimum of sixty days a year at the Claridge, "the smallest and friendliest" casino on the strand. All he had to do was show up and be nice.

Mickey was late. I waited for him in the London Pavilion, a glass-enclosed dining room, formerly a veranda, where Sinatra and the rat pack once soaked up the rays and took in the ocean air. I sat by the window and gazed at the slate gray ocean, thinking about what Howard Cosell had told me about my hero and trying not to: "Mickey Mantle should be in jail. He's a drunken whoremonger." I thought about the questions I wanted to ask him and the ones I was dreading to ask. The dog-eared pages of the notebooks from that week-

end bear witness to my anxiety. Ask re: Dad. Ask re: son Billy. Ask: Who's better? Mickey Mantle or Willie Mays?

By the time Mickey arrived, the dining room was empty and the powdered eggs in the stainless-steel buffet trough had returned to their original state.

"Hi, I'm Mick," he said, sticking out his hand.

"Hi, I'm nervous."

"Why?" he drawled. "Scared I was gonna pull on your titty?"

Had I been thinking quickly, had I been thinking at all, I would have replied, "Yes."

"Coffee," he said. "Can't I get a cup of fuckin' coffee?"

There wasn't time for a jolt of joe. A blue stretch limousine was waiting in front of the hotel, along with a reporter from the New York Daily News, the hotel's public relations director, and Bill Greenberg, one of six high rollers selected by the casino's marketing department to play a round of golf with Mickey Mantle.

The chauffeur put pedal to the metal and five people with nothing in common except the occasion slid all over the plush velour interior trying not to collide with one another. For fifteen minutes no one said a word except "Excuse me."

High roller Bill wore his prosperity around his waist

and a thick gold chain around his neck; white chest hair billowed from his golf shirt and caught in his jewelry. His tone was familiar and obsequious. "So, Mick," he said, "you gonna play in that leukemia thing in Wilmington?"

"The leukemia thing" was a golf tournament. Whitey Ford was scheduled to be there.

"Very good lobster in Wilmington," Mickey replied.

"And good crabs," I added.

"I led the league six straight years in the crabs," Mickey said. "Did you know that? Major league record."

"Still hold it?" Bill asked.

"Still hold it. And my wife was second four times."

The engine thrummed. Vodka sloshed inside a faux cut-glass decanter. I mentioned my recent interview with Boss Steinbrenner in which he had floated the notion of turning Yankee Stadium's Monument Park into a water park for disadvantaged youths from the South Bronx. Mickey laughed. "It was 480 in center field when I played," he said. "It's 420 now, and he's talking about bringing them in further. I said to him, 'They ought to let them throw the ball up and hit it.' That pissed him off."

I told him about the time my grandmother's ample cleavage failed to conceal my Sammy Esposito glove from the serious-looking men in yarmulkes guarding the entrance to Yom Kippur services at the Concourse Plaza. "We got thrown out," I said. "They thought I was being disrespectful. I thought they were being disrespectful."

Mickey laughed. "Your grandmother must have loved you very much."

Our common history in the old stomping grounds established, he asked me, "What was that tall building? You could see it over the center field fence?"

"The Bronx County Courthouse."

A "golden fortress," New York mayor Fiorello La Guardia had called it. The courthouse presided over the intersection of 161st Street and the Grand Concourse, across the street from the hotel where I chose the voice of the Yankees over the word of God. "Well, anyway, somebody used to get up there with a big mirror and shine it in the visiting team's eyes. It was a long ways off. Hell, it was like lightning. First of all, you couldn't hardly hit it over the center field fence. I only hit two over in eighteen years, and that was the only two I ever saw hit over."

Bill interjected: "You got a big write-up in the paper the other day. I meant to bring it. Billy Martin really chewed up Bowie Kuhn. He said, 'Mickey's like a brother to me.' You got the same temperament, right?"

"I wouldn't say that," Mickey replied. "He got traded because he was a bad influence on me. They found out it wasn't Billy. It was Whitey. Billy left, and it was the same old shit."

He trotted out one of his stock Billy the Kid lines: "I always tell him if it hadn't been for him, I could still be playing. Him and Whitey."

Bill asked what Whitey was up to. Mickey shrugged: horses, golf, a few investments. "He does good on the outside," he said, as if Ford had been paroled.

Mantle wasn't doing so well when former Yankee co-owner Del Webb introduced him to Bill Dougall, an executive in the Del Webb Corporation, on the golf course at the Sahara in Las Vegas. Webb hinted that maybe Dougall could find something for Mantle to do. Some years after Webb's death, when Dougall was named president of the Claridge, he thought of The Mick. "I made a deal with him, but the deal I made with him was nothing like the one that Willie Mays made," he assured me.

Dougall didn't want anyone confusing The Mick with a hotel ambassador like Mays or a Vegas greeter like Joe Louis or with the furniture. No, Mantle would be at the Claridge only "when he's active and participating in an event," Dougall declared. "We wouldn't have Mickey sitting like a statue, for Christ sake."

The Del Webb Corporation was a major player on the Vegas strip but new to Atlantic City. The Claridge was a trial run before committing big dollars to the East Coast gaming industry. Hiring The Mick as a front man was a natural—what better way to mitigate reservations about legalized gambling than to bring in the All-American Boy? "We're not trying to make money off of him," Dougall told me. "The main thing is national exposure."

He did not tell me that Mantle's contract contained a clause prohibiting him from drinking in public while on duty.

Mantle's excommunication by Kuhn was immediate. Publicly, he brushed it off. "Fourteen years, all I'm doing is going to spring training as a batting instructor, which I wasn't, hanging around, same as here, signing autographs." His unconscious wasn't as easily mollified about the foreclosure of any meaningful role in the game that had been his life's work. He began to see a chiseled tombstone in his dreams. HERE LIES MICKEY CHARLES MANTLE: BANNED FROM BASEBALL.

Mickey shrugged. "He had to do it. He did it to Willie. He made his mistake when he did it to Willie. In the back of my mind it bugs me a little. It sounds worse than it is. A guy or two said, 'Jesus Christ, you were my boyhood idol, now you're banned. You must have done something bad.' I feel really kind of bad no one took up for me. It's, like, 'Well, fine, he's gone.'"

Mays took up for him, sort of. "He's never gonna harm baseball or anybody else," Mays said. "The only one he ever harmed was himself."

It's not as though he was in the casino plugging silver coins in a nickel slot, which is where his mother, Lovell, parked herself when she visited the twins at the Vegas casino where Webb had gotten them jobs as pit bosses. Lovell did not believe in coming home with money in her pocket. The kids chipped in to buy her a glove to protect her lucky hand.

"Time magazine came around, and they wanted to get a picture of me dealing craps and blackjack," Mickey said. "Well, there ain't no way I would do that. One thing I said I wouldn't do is be a shill or somethin'. I would never do anything degrading."

He decided to lighten the mood with a joke. "God calls Saint Peter over, and he says, 'Saint Peter, I was down on Earth and I made this man and this woman and I forgot to put their sexual organs on them. You take this pecker and this pussy down there and put 'em on them.'

"Saint Peter says, 'Okay.' And he's getting ready to leave, and God says, 'Be sure to put the pussy on the short, dumb one.'"

The PR woman noticed the weather. "Getting nicer."

"Brightening up," Bill said, as the limo approached the gate of the Linwood Country Club.

"I hold the club record here," Mickey said. "Twelve lost golf balls. Nine holes."

The Linwood Country Club opened for business in 1921 in response to what the Atlantic City Press called "overcrowding of the existing courses." The overcrowding was most apparent to Jews blackballed by country clubs catering to Atlantic City swells. The club, built on the site of a planned racetrack, was surrounded by marsh and dune grass and adjacent to the posh Atlantic City Club.

The course sits back from the ocean in view of East Egg Bay and the Atlantic City skyline. A cruel, coastal wind was blowing hard, rattling the cattails and mussing Mickey's thinning hair. But he was there to announce his first official event for the hotel, the Mickey Mantle Invitational Golf Tournament. The fancy title Dougall had bestowed on him—Director of Sports Promotions—meant creating sporting events for him to promote.

Of the fifty or so shivering souls assembled outside the caddy master's shack—cameramen, soundmen, network correspondents, public relations flacks, hotel executives, caddies, and reporters—only eight were actually scheduled to play golf, six high rollers plus Mick and Bill Dougall. The point of the outing was not so much to play golf but to announce that Mickey would be playing golf, which required seeing him play golf.

The press release announcing the tournament that had come across my desk promised a day with The Mick, unlimited cocktails, a filet mignon dinner, a free T-shirt with a big number 7 on the sleeve, and a one-on-one interview. Golfers committing to play in the June event would receive a Mickey Mantle Invitational Golf Tournament sweater with a baseball embossed over the left breast. Mickey took one look at the design and grumbled, "Hell, it's like wearing your ball cap downtown."

A caravan of golf carts was waiting at the first tee: four for the golfers, four for the assembled media—the New York Daily News, Los Angeles Times, The Washington Post, CBS Sunday Morning, and local camera crews—and one for assorted beverages. The network guys wired Mickey for sound and told him not to venture into any water. "I play 'bout half my golf in the water," he replied.

As Mickey climbed into the lead cart, his right leg already stiff from the cold, I asked whether he ever walked the course. "I don't walk," he said. "I can't walk."

On the first tee, Mickey took a club from his bag and stretched, the way he used to unlimber with a bat behind his back. Reporters craned their necks, soundmen jostled for position. He addressed the ball to a symphony of camera shutters. His first shot went wide right, ricocheted off a tree, and came to rest in swaying beach grass. Mickey set off in pursuit, driving in circles. The media caravan drove in circles behind him. "We want to get him looking for the ball," a correspondent informed his cameraman with an urgent golf whisper.

"If you use this shit, it's going to be the dullest TV show ever put on the air," Mickey said.

His second shot found a trap. Kicking the damp sand back into place, he said, "This'd be a good job for Billy. He could get a lot of practice here."

He meant Martin, of course, whose managerial afterlife required kicking a lot of sand at home plate and in sand traps. Mickey played a lot of golf, too, chiefly at the swank Preston Trail Golf Club in Dallas. Now he was getting paid to do what he did every day. Every morning he checked the box score the way he always did and waited for a game the way he always did. "Get up at six A.M., get the paper, check the score, see if

Billy got fired. Call the office to see what's going on. Go to the office 'til noon to see what's going on. Go to Preston Trail, have lunch, play golf, go home."

There was no shortage of playing partners. But he cleared his schedule when Martin came to town. "Billy called in Texas to play golf. This was Friday, just before he went to the ballpark and got kicked out and suspended. He said, 'Let's play golf tomorrow.' So I go to Preston Trail at nine A.M. Everybody tees off. I'm sitting up there reading the paper, watching the TV. Twelve o'clock, he never has shown up. I never have heard from him since. I finally went out and hit some balls."

"Sounds like Billy," somebody said.

Mickey's third shot hit the green and his mood lightened. "Hey, did ya hear Steinbrenner gave Yogi a new million-dollar contract?" He waited, lining up his putt and the punch line. "Dollar a year for a million years."

The tongue-tied boy from Commerce, Oklahoma, had perfected the patter. It is an acquired skill, a kind of celebrity ventriloquism. Mickey had learned to crack wise in measured morsels of sound for the amusement of unctuous strangers. But it was hard to maintain good-ol'-boy bonhomie in the face of an Arctic spring gale.

"You know Bob Cousy?" one of the high rollers

asked on the fourth tee. "Well, I've played golf with him. And Hawk, Hawk Harrelson. I've played with him, too."

"Oh," Mickey said, shivering. After nine numbing holes, he hung a right and detoured off the course. The caravan came to a halt at the clubhouse door and Mickey made a quick getaway. When he returned some ten minutes later, he seemed disappointed that we were all still there. "I was hopin' somebody would call it off while I was inside."

There was no reason he had to play the entire round, except that the press release promised he would. The cameramen had all the footage they needed after the first hole. But to quit would have been to admit that the whole thing had been orchestrated for the benefit of thirty seconds of airtime. The solipsistic logic of Media Day dictated that he play on. The thermometer on the clubhouse wall read 37 degrees.

My teeth were chattering. Mickey noticed. "Can't we get this girl a fuckin' sweater?" he said. "She's gonna fuckin' freeze."

A sweater materialized. It reached my knees and warmed my heart. He asked if I wanted to ride shotgun on the back nine.

Turning the cart into the wind, he floored it, zooming over a wooden bridge at the eleventh hole, heading straight for a water hazard, with a parade of hard-charging media types following in his wake. Just shy of the water, he made a hard left, looking back over his shoulder with glee as the press detail scrambled to stay out of the drink.

Recklessness was always part of his charm, his cheerful, who-gives-a-fuck élan. But with each increasingly precarious turn threatening to upend the cart, with every vicious twist of his lower body as he swung through the ball, his limp became more pronounced and the consequences of his wildness more patent. He never did learn to cut down on his swing.

"Here, feel this."

He took my hand and placed it on the most famous knee in baseball history. It felt like jelly. "There's a ball rolling around in it, a calcium deposit. When it gets caught . . ." He shrugged. "I've got no cartilage. So when I swing . . ."

He moved his hands through an imaginary plane. "They get stiff when it's cold and damp, stiff and sore. It's like a real dull toothache. That's the way it is all the time when I play golf."

"When was the last time they didn't hurt?" I asked.

"When I was eighteen."

The doctors were recommending knee replacement. "I've been dreadin' to do it. I didn't want to have any more operations. I thought I'd be dead by now."

By the seventeenth hole, he was playing polo with

his golf ball, swooping down on the green with all the horsepower the battery-powered cart could muster, and using his putter as a mallet. As he and Dougall approached the eighteenth tee, a well-nourished seagull unloaded on the hotel executive. "That's par for this whole day," Mickey said.

He limped off the course with a low score of 79.

His chest hair defrosting in the limousine, high roller Bill offered an encomium to Mutt Mantle. "Coming from a fan's opinion, the opinion of people about you and your relationship with your dad, which was very, very strong . . ." He paused, trying to regain a grammatical foothold, and took another sip of vodka. "You can name one thousand other ballplayers, and the association does not prevail which it does here, father and son."

"That's all he lived for, was to see me make the major leagues."

"How old were you when he started you?"

"From the time I was four or five years old. He named me after Mickey Cochrane before I was ever born."

"Is it true he took twenty-two dollars out of his paycheck to buy you a Marty Marion glove?" I asked.

"Yeah, and he only made thirty-five dollars, too."

"You must miss him," I said.

"Oh, yeah. He died my second year with the Yankees. He never did get to see me get that good, but I think he knew I was going to."

Mickey dedicated The Quality of Courage, a collection of athletic encomiums fashioned on John F. Kennedy's Pulitzer Prize–winning Profiles in Courage, to Mutt: "the bravest man I ever knew."

I bought it when it came out in 1963. "I read it over and over."

"It was kind of about my dad, people he admired and people I learned to admire later on."

The people his dad admired—posthumously selected by Mickey's ghostwriter, Robert Creamer—included Jimmy Piersall, Ted Williams, Red Schoendienst, Jackie Robinson, Nellie Fox, and Roger Maris, who played his first game in the majors five years after Mutt's death. In deference to Mutt and the willing suspension of disbelief, Billy Martin didn't make the list. The putative author expressed enthusiasm for the cash advance and exerted editorial control over only chapter 1, the one about his father.

I asked what kind of ballplayer Mutt had been.

"My dad was pretty good, but there was no way to judge because he never did play pro ball. But everybody tells me he was the best semi-pro player in Oklahoma."

"*Good enough to play in the majors? As good as you?*"

"*Shit, no. I could outrun him. Had a better arm. Nobody was better 'n me.*"

He briefly considered the rhythm of his given name, the syllables as smooth as river rocks, the best baseball name ever. "*My dad named me Mickey Mantle. It sounds like a made-up name.*"

A stage name, and Mickey had gotten good at acting the part. "*I'll tell you what,*" he said. "*It's amazing my name's been as good as it has been. When people think of Mickey Mantle, they think of Jack Armstrong, the All-American Boy. That shit.*"

9

May 30, 1956

A BODY REMEMBERS

1.

On opening day in the nation's capital, Mantle relocated two Camilo Pascual fastballs beyond the center field fence in Griffith Stadium—past the flagpole, over the thirty-one-foot wall and into the boughs of the beloved backyard tree that caused the ballpark to be built around it. The first of the two gargantuan left-handed efforts sailed over the tree and landed on the roof of a house beyond the 408-foot sign and bounced across Fifth Street. The second disturbed birds nesting in the bower. Most papers said it was an oak left over from virgin forest; others called it a maple. Pascual called it "Mickey's tree."

Pedro Ramos, his voluble countryman and team-mate—generously characterized by Mantle as "one fucking bright Cuban"—waved a white towel at Pascual as Mantle rounded the bases again. "He hit one *into* the tree and the next one went *over* the tree," Ramos said, the force of the two 500-footers having conflated his recollections. "They are still looking for those balls. That tree *remembers* Mickey."

"*Tree*-mendous," Casey Stengel declared. "They tell me that the only other feller which hit that tree was Ruth. He shook some kids outta the tree when the ball landed. But the tree's gotten bigger in twenty-five years, and so I guess have the kids The Babe shook outta it."

In the sixth spring of his major league career, Mantle had arrived at the tipping point. The 1955 season had been a good one. He led the American League in triples (11), home runs (37), walks (113), runs (129), slugging (.611), and on-base percentage (.433). That May he had the only three-home-run game of his career, hitting two left-handed, one right-handed. But on September 16, he pulled his right hamstring trying to beat out a bunt and made only two more regular-season appearances, both as a pinch-hitter. He was limited to ten at-bats in the World Series remembered in Brooklyn for sweet redemption and in the Bronx for Mantle's absence from the lineup. When the Yankees gathered in St. Petersburg the next spring, sports columnist Dan Daniel posed the question everyone in baseball had been mulling all winter: "Which way will his career turn?"

The exhibition season offered tantalizing clues and cautionary omens. Mantle's hamstring was still weak, and he quickly reinjured it. But he struck out only once (not until the twelfth game) and hit six home runs, two of which found their way into Tampa Bay beyond Al Lang Field. A third prompted Stan Musial to say, "No home run has ever cleared my head by so much as long as I can remember."

The new Mantle announced himself in the first game of the regular season. "Mickey attained maturity on opening day," Jerry Coleman said. "It was—boom! boom!—and he had two home runs without even trying."

Mantle said he had quit trying to hit homers. "I'm beginning to learn that easy does it," he told *Times* columnist Arthur Daley.

Within a month, this new, laid-back slugger had churned up a tide of dread in opposing pitchers.

May 5 vs. Kansas City: 3 for 4, 3 RBI, 2 home runs, one of which threatened to leave Yankee Stadium.

May 18 at Chicago: 4 for 4, 2 RBI, 2 home runs (one

left-handed, one right-handed), a double, and a walk. His ninth-inning home run tied the game, which the Yankees won in the tenth inning.

May 24 at Detroit: 5 for 5, 1 home run, 1 RBI.

Five days later, little Billy Crystal of Long Beach, Long Island, attended his first game at Yankee Stadium. At age eight, he was young enough to believe that Miller Huggins, Babe Ruth, and Lou Gehrig were buried beneath their monuments in center field.

Rain dampened the uniforms of the Marine color guard as the flag was raised in center field before the first game of the Memorial Day doubleheader. The soggy forecast also diminished the expected holiday crowd. There were not quite 30,000 fans in the ballpark when the players assembled along the baselines for a moment of silence. A mist shrouded the copper filigree of the Stadium's frieze and hovered over the bullpens where the starting pitchers, Pedro Ramos and Johnny Kucks, were trying to get warm. A bugler played taps.

Jack Crystal owned the Commodore Record Shop in Times Square, the jazz emporium in New York. Louis Armstrong had given him his tickets. A priest was seated in the row before them along the third base line. So when Mantle came to the plate in the fifth inning and Ramos came in with a waist-high fastball on a 2–2 pitch, Billy didn't see the sweet left-handed swing or the collision between ash and cowhide. Nor could he see the trajectory that carried the ball where no other had gone before. "The priest stood up and blocked my view of the ball hitting the facade," he said. "Though I do remember standing up on my chair, 'cause everybody else just went 'Aaaaaaaah.' It was just huge. It just went up and up and up, and it just settled down at the last second. And the priest actually said, 'Holy fucking shit!'"

In the Yankee bullpen, Tommy Byrne gazed at the heavens. "You just keep lookin' and you keep wonderin', 'Well, how far is the damn thing goin' to go?'"

As Mantle rounded third base, Pascual stood on the dugout steps, waving a white towel at his compatriot on the mound. "Look what he did to you! He rocket up in right field!"

Between games, team officials consulted the archives and the blueprints and determined that the rocket had traveled 370 feet, hitting the facade 118 feet above field level, 18 inches from oblivion. Ramos thought the ball had left New York. Indeed, it ended up in Eddie Robinson's Baltimore restaurant, a gift from The Mick to the Yankee first baseman.

Much to Pascual's regret, Mantle declined Stengel's offer to rest his aching hamstring and take the rest of the afternoon off. In the fifth inning of the second game, he hit his third home run of the year off Pascual, the 141st of his career. It was a modest effort that landed only halfway up the right field bleachers. At the end of the day he was leading the majors in six offensive categories: runs (45), hits (65), total bases (135), home runs (20), RBI (50), and batting average (.425). He had struck out only twenty-one times. Even the usually imperturbable Harold Rosenthal of the *Herald Tribune* was moved to excess: The "Merry Mortician" was burying the rest of the league.

By the time the Detroit Tigers arrived two days later, the facade home run had been memorialized in front-page photographs adorned with soaring arrows. Outfielder Harvey Kuenn eyed the distance and demanded corroboration from a young sportscaster named Howard Cosell who had witnessed the clout. "Did he really hit it up there? *Really?* His strength isn't human."

Whitey Herzog, the future Hall of Fame manager who played left field for the Senators that day, had been traded to Washington on Easter Sunday. Summoned to the manager's office after church services, Herzog learned his fate. "You're pretty good but you're not as good as the guy I got," Stengel said.

"Shit, I know that," Herzog replied.

How good was Casey's guy?

"Nobody could play baseball better than Mickey Mantle played it in 1956," Herzog said.

For once he wasn't sabotaged by physiology. He was batting .371 with 29 home runs—ahead of Babe Ruth's 1927 pace—against the Red Sox on the Fourth of July. The Yankees held their collective breath after he charged a ball hit his way in the eleventh inning of the first game of a doubleheader. He thought he could prevent the winning run from scoring. Then he felt a familiar twinge in his right knee. "Sprained ligament on the outer aspect," said team physician Sidney Gaynor. Mantle missed the next four games.

The pain went "all around the leg" but it did not derail him for long. He won the Triple Crown, leading the American League in home runs (52), RBIs (130), and batting average (.353). He was the *Sporting News* Major League Player of the Year and the Associated Press Male Athlete of the Year. He received the Hickok Belt, awarded to the top professional athlete of the year, as well as the first-ever Babe Ruth Sultan of Swat crown as the major leagues' top slugger.

Tangible evidence of Mantle's strength was ample and astonishing. A spring training baseball bag shredded by the force of repeated collisions. Tony Kubek said you could see the sawdust fly. Tin soda cans crushed between his thumb and index finger. No one in his family knew how he he had gotten so strong. His twin brothers, Roy and Ray, whom he called Rose and Rachel, were taller, but neither had his forearms. "Seemed to me like they was this wide," Larry Mantle said, cupping his hands to make a circle eight inches in diameter. His shoulders seemed "like fifty-three inches wide" to his son David.

It was the muscle running down the *back* of his 17½-inch neck that intimidated and distracted Cleveland pitcher Mike Garcia. When that thing got to twitching, Garcia said, that's all he could see. "Built like a concrete wall," said the totemic slugger Frank "Hondo" Howard.

Hondo and his contemporary Boog Powell are two of the only people on the planet who know what it feels like to hit a ball as hard as Mickey Mantle. It feels like nothing else and it feels like nothing at all, Howard says. It is the answer to baseball's own Zen riddle: How do you feel the absence of tension?

"Everything is in unison," said Powell. "Your whole body, your whole swing, everything is right together. Everything he had in his body was coming out of that bat."

Clark Griffith, the grandson and namesake of the patron of Washington, D.C., baseball, saw Mantle up close throughout his childhood. "He had magnificent rotation, the way his back spun around, the leverage in his shoulders. I loved his stance. I loved the way he finished when the bat was wrapping around his body. To take a vertical bat and to get it moving, you have to have very strong hands to move it into the plane of the pitch just before contact."

Mantle's might inflicted damage on bats, balls, and egos. After watching Mantle hit a blistering home run off "Sudden" Sam McDowell in 1968, Yankee pitcher Stan Bahnsen asked the batboy if he could inspect the wood for bruises. "There were three or four seam marks in the barrel a quarter of an inch deep," he said. "Those seam marks were *buried*."

Billy Pierce, then pitching for the White Sox, recalled a July night at Yankee Stadium in 1959 when Mantle KOed a rookie outfielder with a line drive. "The right fielder went to catch the ball, and it hit him right in the chest," Pierce said.

"Just to the right of the breastbone," said Jim McAnany, who survived the bullet that came half an inch from occasioning his obituary. "I just went down like I was shot. It knocked me off my feet."

McAnany was prominently featured in the next morning's *New York Times*, photographed sitting on the outfield grass surrounded by concerned and incredulous teammates. According to the caption, the ball had glanced off his glove. In fact, it glanced off him. He never saw it coming. It was the first inning of his first game at Yankee Stadium, and he lost the ball in the big-city lights. He was charged with a two-base error, adding insult to potentially lethal injury. "When I came into the dugout, I said, 'I think I got a hole in my chest.'"

The next day, Mantle inquired after his health. "Sore," McAnany said, which was to say the least. The X-rays were negative for breaks—no internal injuries, but The Mick had left his mark. The seams of the ball were imprinted on his chest. "The American League too," he said.

Mickey Lolich, who grew up in Oregon idolizing Mantle from afar, had a near-death experience pitching to him in Tiger Stadium. "I threw him a sinking fast ball down and away, and he swung at it. All I heard was a buzzing sound, which means the ball was hit tremendously hard. I've heard the sound a few times in my life. It's actually the seams grabbing the air as it goes by. I turned and saw the ball going into center field on a low bounce. All of a sudden, I felt a pain. He had hit me on the inside of my right leg, high up on the thigh, just below the very important family jewels. I had faced a moment of death and never even seen it!! My legs went to total jelly. My stomach went up into my throat. My catcher comes out—it was Bill Free-han—he says, 'Did that ball hit you?'

"I says, 'Yes, it did.'

"He says, 'Holy shit! Are you okay?'

"I says, 'No, I'm not!'

"The umpire comes out and says, 'What's goin' on out here?' And Freehan says, 'That ball hit Mick!'

"The umpire says, 'My God. You okay?'

"And I says, 'No!'

"He says, 'Well, you take just as long as you want.' "

No wonder Bob Turley welcomed his trade to the Yankees: "He shocked the shit out of pitchers." If they were lucky.

Jim Kaat of the Minnesota Twins sought divine intervention when he fell behind on The Mick. "Two-and-oh on Mantle, Earl Battey would wave his arms and make the sign of the cross."

Catchers were in a uniquely vulnerable position. "He could make a bat hum over your head, *hoooee*," said Ed Bailey.

Infielders laughed when managers tried to wave them in on the grass. "Especially with nobody on," Powell said. "You do your best imitation of a Mexican bullfighter, you just olé everything."

One time Mantle squared to bunt, Clete Boyer remembered, "and he hits a fucking line drive to right field." Another time, a ground ball knocked the glove off shortstop Joe DeMaestri's hand in Philadelphia. "It spun my hand so hard that it turned my hand around," DeMaestri said. "The glove went right through my legs and into the outfield. The ball rolled out, and my glove did too. It was like a normal ground ball but hit like 180 miles per hour."

A thud often signals contact with what Ted Williams called "the joy spot" of the bat. "The guys that hit the balls the farthest, there's a click that goes along with it," former catcher Tim McCarver said.

The sound of the ball coming off Mantle's bat was distinctive. "With your back turned, you knew it was him," Powell said. "It was a ring. It was more like a musical note."

Three weeks after Mantle failed to hit the ball out of Yankee Stadium, he hit one out of Briggs Stadium in Detroit. The score was tied in the eighth inning when manager Bucky Harris paid a visit to his pitcher Paul Foytack. "All he'd ever say was 'Steady in the boat, now,'" said Virgil Trucks, who was on the Tigers bench. "Bucky had just sat down, and he hears this crack of the bat. He looks up, and that ball went out of Briggs Stadium, landed in Trumbull Avenue at the back of the stadium and bounced, and they found it on the roof of a cabstand. That ball had to travel 600 feet."

Into a stiff wind. "All Bucky said was 'Mmm mm, that would bring tears to the eyes of a rocking horse.'"

Between them, Pascual and Ramos gave up 4.3 percent of Mantle's 536 home runs—Pascual 11, Ramos 12, tying him with Early Wynn for Most Victimized by The Mick. So Ramos didn't take any chances when he faced Mantle a month after the close call with the Stadium facade—he walked him four straight times. When Mantle hit another disappearing-act home run off Ramos at Griffith Stadium in 1957, he sent the clubhouse boy to the Senators' locker room with an autographed ball and a message: "Tell him to get a cab, and if he can find that ball I'll sign that one for him, too."

In 1956, Mantle became a measuring stick for teammates and opponents, some of whom, in defiance of major league nonchalance, came out to the ballpark to watch him take batting practice. St. Louis Cardinals' manager Tony LaRussa says Mantle remains a standard of comparison: "When a guy runs really well, does he run as well as Mantle? If he has power, does he have Mantle's power?

Dodger pitcher Ed Roebuck saw the speed and the power in 1956. In spring training. "He bunted on me," Roebuck said. "I fielded the ball. It sounded like a bunch of wild horses running by."

That fall, in the sixth inning of game 4 of the World

Series, Roebuck threw him a sinker that "hung out over the plate." Not for long. Duke Snider admired the parabolic view in center field. "I don't mind you not charging it," Roebuck told him later. "But you don't have to stop to see how far it went."

Roebuck understood: length is a guy thing. Size matters. "Seriously," he said, laughing. "That's what made the male regard Mantle that way. Forget God. Mickey Mantle can hit the ball farther than anybody."

2.

Mantle was not a baseball scholar. Kneeling in the on-deck circle, he might inquire of a retreating teammate, "What's he throwin'?" Doubtless he would have understood Ted Williams's splendid reply to protégé Mike Epstein when he asked The Splinter about hitting. *Well, you just do it.*

One hot July afternoon late in Mantle's career, Epstein encountered his childhood hero in the Yankee dugout during early batting practice. "The last two balls I hit were in the upper deck in right center or in right field," Epstein recalled. "And I hear this voice in the dugout, and he says, 'Boy, we could use that power over here. We got none!'

"And I looked over—there's Mantle. He's standing there in his undershorts and his shower shoes. So I went over and I said, 'When you stride, do you feel your body doing anything?'

"And he said, 'Well, what do you mean?'

"I said, 'Do you feel your body moving in a certain direction or doing something that you can talk to me about?'

"And he said, 'Mike, honestly, I don't know nothin' about hitting.'

"You know, he drawled it out. He said, 'I just watch the other hitters.'

"In Yankee Stadium in those days, the bat rack was by the home-plate side of the dugout, and that's where we were. So he just reached in and took a bat—left-handed, in the dugout—and he's just taking some strides. I said, 'You feel anything?'

"He said, 'Nah.' And then he said, 'You know, actually, I sort of feel my body going backwards as I'm striding forward.'

"So I said to him, 'Well, do you feel the same thing right-handed?'

"And he did it about five or six times, and he looked at me and he says, 'No.' "

Mantle had no idea what he did right or wrong or differently batting right-handed and left-handed. More than likely he would have had little truck with present-day baseball pedagogy. Today's students of the game have PhDs in physics and industrial engineering; applied engineering, applied math, applied psychology, applied biomedical engineering; kinesthesiology and a new area of inquiry called biological cybernetics. They don't talk baseball; they discuss the "relationship amongst the sweet spot, COP, and vibration nodes in baseball bats," the topic of a treatise published in *Proceedings of the 5th Conference of Engineering of Sport.*

Mantle had no answer to the question "What makes Mantle Mantle?" There was good reason for that. It has to do with the biology of memory, the subject that earned Eric R. Kandel the 2000 Nobel Prize in Physiology or Medicine. "I think your question is not dramatically different than asking 'What makes Mozart Mozart?'" Kandel said.

The answer requires an understanding of how a body remembers.

Kandel is the director of the Kavli Institute for Brain Science at Columbia University. Athletics are not prominently featured on his résumé. In gyms and locker rooms "muscle memory" is a catchphrase for the ability to recall and replicate a perfected motion, such as a baseball swing, in the freedom of infinite space. In Kandel's "New Science of Mind," muscle memory is an idea, specifically "an idea of exactly what groups of muscles to move in response to a particular stimulus"—

a fastball, for example—and the ability to recruit "the family of muscles that have to be moved to accomplish a particular task."

Kandel, a physician trained in psychiatry and neurobiology, explained: "There are two kinds of memories. They're called implicit and explicit. Explicit memory is a memory of people, places, and objects. If you think of the last time you sat in a baseball stadium and remember who you were with, you're doing explicit memory storage.

"Implicit memory storage is hitting a tennis ball, hitting a baseball, doing anything that involves sensory motor skills."

With sufficient reiteration, an explicit memory can become implicit, literally moving from one storage center in the brain to another. Once a task is mastered, it can become automatic, almost reflexive. "When you first learn how to drive a car, you're just paying attention all the time," Kandel said. "You're terrified. You're saying, 'Now I shift, now I don't.' After you learn, you don't tell yourself when to shift. You do it automatically."

Muscle memory is a form of implicit memory that is recalled through performance, Kandel said, "without conscious effort or even the awareness that we are drawing on memory."

Which is why Mantle could not explain what he did

or how he did it. It's also why the greatest athletes usually make the lousiest coaches.

When Rob Gray tests baseball players in his Perception and Action Lab at Arizona State University, forcing them to articulate what they do and how they do it, their performance deteriorates. Compelled to surrender what he calls "expertise-induced amnesia"—in short, to make an implicit memory explicit again—"they start thinking about what they're doing and mess everything up," he said.

Until recently scientists believed that the brain was fixed in its anatomy: you would lose brain cells with age but you couldn't alter the architecture. Turns out they were wrong, Kandel said. The brain can bulk up, too. Repeated experience can form new synaptic connections, especially if you start building up those implicit memories before puberty. The right genes nurtured the right way—meaning early enough and often enough—creates the potential for a particular kind of genius. "You and I would call it implicit intelligence," Kandel said. "That's what you're looking at. That's beautiful."

Mantle was an Einstein of implicit intelligence. If he played today, technology would be able to explain to him what he could not explain to or about himself. Doctors would measure the firing pattern of his muscles. Coaches would gauge the speed of the barrel of his

Louisville Slugger with Doppler radar and attach infrared markers to the tip of his bat. They would be able to answer Harvey Kuenn's cry to the heavens: "How can a man hit a ball that hard?"

Greg Rybarczyk, a mechanical engineer who studies and measures every major league home run hit every year on his Web site Hit Tracker (www.hittrackeronline.com), has grappled with that question for three decades. "Mantle represents a unique synthesis of strength, artistry, and an almost magical priming of his body," he said. "Just as painters prime a surface and farmers prime a pump, Mantle primed his body to function with ballistic efficiency. A blasting cap is also known as a primer, a small explosive device that sets off a larger charge, an entirely apt reference to Mantle in that at the moment he begins to bring the bat forward, it's as if a blasting cap is going off that detonates the rest of his swing.

"He's magical because so many men over so many years have tried to hit a baseball with the grace and power of Mantle and so few have been able to do it. Artistic, because the end result of Mantle's swing is so much more than you might expect from a man who stood only five feet, eleven inches tall and because no scientific examination of his swing has ever really pinned down exactly how he produced such amazing power. And it was much more than just strength, it was great strength plus technique of the absolute highest order."

In an effort to pin down how Mantle generated such power, I asked Preston Peavy, a techno-savvy hitting coach, to analyze Mantle's form, using the visual motion-analysis system he created for his students at Peavy Baseball in Atlanta. He converted film and video clips of Mantle into a set of kinematics, moving digital stick figures that show the path of each part of the body as it moves through space. For an analysis and comparison of Mantle's left-handed and right-handed swings, see appendix 2, on page 701. (To view the kinematics, go to www.peavynet.com or www.jlace.com.)

3.

A 90-mile-per-hour fastball doesn't leave much time for thought. Traveling at a rate of 132 feet per second, it makes the sixty-foot, six-inch journey from pitcher to batter in four-tenths of a second. The ball is a quarter of the way to home plate by the time a hitter becomes fully aware of it. Because there is a 100-millisecond delay between the time the image of the ball hits the batter's retina and when he becomes conscious of it, it is physiologically impossible to track the ball from the pitcher's hand to the catcher's glove. David Whitney, the director of the Vision and Action Lab at the University of California, Davis, explains, "A 100-millisecond delay doesn't seem very signifigant. But if a baseball is traveling at 90 mph, that translates to around fifteen feet. If we perceive the ball fifteen feet behind where it's actually located, the batter has to start his swing very early on in the baseball's trajectory."

Neurologically speaking, every batter is a guess-hitter. That's where implicit memory comes in. The ability to infer the type of pitch and where it's headed with accuracy and speed is inextricably linked with stored experience—the hitter has seen that pitch before, even if he can't see it all the way. Add the reflexes to respond to that memory and a visual motor system that allows the batter to react on the fly to a change in the trajectory of a flying object, the right DNA, and Mutt and Grandpa Charlie out by the shed throwing tennis balls, and you have Mickey Mantle.

Every at-bat is a dance of double pendulums. The pitcher leads, using his body as a kinetic chain to deliver energy from his legs through his trunk into his shoulder, arm, and, finally, the ball. The batter follows, reacting in kind. The converging and opposing forces may or may not be equal, but the goal is the same—to turn potential energy into kinetic energy as efficiently as human physiology allows.

The pitcher has the inherent advantage of foreknowledge—he knows what he's going to throw—and he has the downward slope of the mound to generate momentum. With only flat ground and muscle power at his disposal, the hitter creates force by twisting his upper and lower body in opposite directions like a rubber band. When that human rubber band is stretched taut and is ready to snap, it uncoils, propelling the bat through the strike zone.

This deceptively simple act is an intricate biomechanical task requiring the coordinated mobilization of virtually every muscle in the body in less than a second. "Everything but the chewing muscles," said Dr. Benjamin Shaffer, a specialist in orthopedic sports medicine and head physician for the Washington Capitals. "Unless you grit your teeth."

Nobody gritted more than Mantle. Lefty or righty, he swung with felonious intent.

Yogi Berra once called him "naturally amphibious." In fact, he wasn't. His right-handed swing was unstudied. His left-handed swing was learned behavior, constructed with thousands of backyard swings. "I don't think he had any weakness that you could exploit," the Indians' ace right-hander "Rapid Robert" Feller attested. "If he did have, they'da done it."

(Mantle batted .500 against him.)

Ultimately, Mantle's injured right knee gave righties a vulnerability to exploit. Everybody knew the book on him. "From the right side, he had no holes in his swing," Clete Boyer said. "Left-handed, he had a little blind spot up high because of his knees."

"I threw him titty pitches," said Jim "Mudcat" Grant. "Up around the chest, around the nipples, a little bit off the plate, inside. Of course, we knew he had bad legs, and there were times we went for the legs in terms of pitching inside. We didn't throw to hit him. We wanted to move him a little bit, cause some discomfort."

Frank Lary of the Tigers was widely reputed to throw at Mantle's knees, a reputation the "Yankee Killer" denies. "Nah, that was Jim Bunning," he said.

Normally a model of on-field decorum, Mantle had to be restrained one day after Bunning hit him in the leg with a pitch.

Claude Osteen, the estimable lefty and longtime pitching coach, remembers the turning point of his career, the defining at-bat when Mantle convinced him he'd become a pitcher. It occurred in 1964, when he was with the Senators. "The report was, you had to jam him," said Osteen, then in the eighth of his eighteen years in the majors. "If you've got a guy like Mantle who's standing miles away from the plate, where there is so much daylight between the inside corner and his hands, it's frightening. It's right down the middle of the plate for a guy like that."

Visions of a tape measure home run or worse—a line drive hit back up the middle—tormented pitchers in their sleep the night before they faced the Yankees. But Osteen knew that pitching inside was what he had to do to get Mantle out. "I went in there. It was a fastball, right on the black," he said. "Right away he went straight into the ball and closed that daylight up."

Osteen held his breath and watched Mantle's hands. "The bat looked like it was forty-two or forty-four inches long," he said. "It could have been fifty inches long, and you knew he could swing it."

Mantle swung—and hit a nice little grounder to the shortstop. He had brought his hands out too far to get the barrel of the bat on the ball. That's when Osteen knew: "If I can get this guy out, I'm capable of getting anybody out."

He also knew how easily it could have gone the other way. The margin of error was a sliver of daylight. Which explains why when he was asked how he had pitched to Mantle, the late Frank Sullivan said, "With tears in my eyes."

10

May 16, 1957

RETURNS OF THE DAY

1.

"Oh, *that* night," Carmen Berra said, recalling Billy Martin's twenty-ninth birthday, the last one before you get old. Mickey and Whitey had planned a night on the town. Gil and Lucille McDougald were invited but had made other plans. Elston and Arlene Howard couldn't get a babysitter. Andy Carey and Jerry Coleman declined. "Who's coming?" Coleman asked. Hearing the guest list, he said, "I think I'll pass."

Bob Cerv and Irv Noren, former teammates in town with the Kansas City A's, joined Carmen and Yogi Berra, Joan and Whitey Ford, Hank and Charlene Bauer, Merlyn and Mickey Mantle, Johnny Kucks, and the birthday boy for dinner at Danny's Hideaway, where he was toasted often and liberally. When everyone else headed to the Waldorf-Astoria for an after-dinner drink or two and Johnnie Ray's 10:30 P.M. show, Cerv and Noren went home to bed.

The pastry chef at the Waldorf baked a birthday cake, which the Yankees took with them when they decided to go see Sammy Davis, Jr., at the Copacabana. "I was the one that insisted we bring Billy's cake," Carmen said.

It was the era of mink stoles, pink gardenias, and floor shows. Legs were gams; bands were big. The Copa was a mainstay of New York café society, *the* nightspot for "stay-outs and their pin-ups—three shows a night, seven nights a week, at 8, 12, and again at 2." The Copa billed itself as "the hottest club north of Havana." Located at 10 East 60th Street, just off Central Park, the sober limestone exterior with the decorous burgundy awning gave no hint of the prevailing Latin attitudes and latitudes in the basement of the sedate Fifth Avenue apartment building.

The Yankees and their baked goods arrived in time for the 2 A.M. show. Jules Podell, who ruled the club with an iron fist and a massive gold pinky ring, took care of them. He was a Yankee fan. "They put a special table for us up front," Carmen said. "We were the kings and queens of New York."

Being of such regal stature meant that you could disappear below the city streets into a fantasy world (capacity 670) populated with bold-faced names and the gossip columnists who lavished them with ink. Leonard Lyons, Walter Winchell, Ed Sullivan, and Dorothy Kilgallen mingled with talent scouts, casting agents, sports stars and the wise-guy colleagues of the club's very silent owner, mob boss Frank Costello.

The headliners—Frank Sinatra and Jimmy Durante, Dean Martin and Jerry Lewis, Tony Bennett and Sammy Davis, Jr.—usually got more play in the morning papers than the swells seated in the plush semicircular banquettes. But not the morning after Billy Martin's birthday party.

Also celebrating that night at two large tables nearby were members of an upper Manhattan bowling club, the Republicans, who had begun their evening with dinner at Mama Leone's. There were nineteen in their party, among them Edwin Jones, forty-two, of 600 West 188th Street, who went over to pay his respects, draping a familiar arm over Martin's shoulder.

What happened after that remained a matter of dispute. Did the Yankees resent the intrusive bonhomie? Did the bowlers take umbrage at the VIP treatment the ballplayers received? Accounts vary. One thing they all agreed on: everyone had had a lot to drink. Words were exchanged between the tables, and between the bowlers and the stage. Then they traded fisticuffs. Leonard Lyons, making rounds for his "Lyons Den" column in the *Post*, wrote: "The great battlefields include Bastogne, Verdun, Gettysburg and the kitchen of the Copacabana. The nightclub fracas, front page yesterday, was preceded by a racial slur, directed at Sammy Davis Junior's farewell show."

The *Journal-American* story quoted an anonymous Yankee wife: " 'What started the whole thing was that someone at the bowlers' tables—and they were a noisy group—called Sammy Davis a name that sounded like Sabu.' That angered the singer and he shouted, 'Will the person who called me that come forth?' "

In 1944, Harry Belafonte was banned from the club just before shipping out with the U.S. Navy. The following year, Lena Horne used her clout in contract negotiations to desegregate the main dining room downstairs. But racial tensions persisted in America. "They were calling Sammy Davis, Jr., Little Black Sambo," Merlyn Mantle told me. "Four guys came to his aid," and asked the bowlers "to tone it down."

Among the four, Hank Bauer: "We've got ringside seats, great big round table, and we're drinking B & B and coffee. And this big fat Jewish guy came walking by me. He said, 'Don't test your luck too far tonight, Yankee.'"

Bauer assumed Jones was Jewish after reading that he owned a delicatessen. "I give him my best vocabulary—two words. And now he's down at the end of the table, him and his son-in-law, I think it was. The son-in-law went back to the men's room."

Martin, an improbable peacemaker, got up to have a word with him. Mantle followed. "So Ford says, 'You better go see what the hell's happening,'" Bauer said. "My wife, Charlene, says, 'It ain't none of your business.'

"I say, 'Yes, it is.' I went back there, and I opened up the door. I saw nothing but tuxedos. And Yogi and Johnny Kucks ran into me and said, 'Get the hell out of here.'"

Edwin Jones, the convivial bowler, was unconscious on the floor.

One of New York's finest arrived on the scene, summoned an ambulance, and reported the incident to the station house on East 51st Street. Detective Chris Coyle, who was assigned to investigate, decided, "This case is too hot to handle," the *Journal-American* reported, and referred the matter to the Manhattan district attorney.

Lyons elaborated in his May 19 column: "Davis was

at the drums when the commotion started and so none of us could hear, or in the darkened club see the action in the rear involving the Yankee ballplayers. On my way out, I saw the victim being carried into an ambulance, and so I returned to the Copa Lounge and there met the ballplayers. Yogi Berra noticed I was asking questions. He feigned utter innocence, walked toward me and greeted: 'Hello, what's new?' "

Lyons kept the rest of the conversation off the record but shared the details with his son, film critic Jeffrey Lyons. "Yogi and my father were about the same height," Jeff Lyons said. "He couldn't see easily over Yogi's shoulder. Yogi shifted left and right to keep my father out. My father said, 'You give me an exclusive and I'll tell you the way out the secret passage when it was a speakeasy.' "

Thus, the New York Yankees made their getaway.

At 3:16 A.M., Jones was admitted to the emergency room at Roosevelt Hospital, where he was treated by a young medical resident named Cedric Priebe, whose report noted: "Nose broken (but not displaced); ribs, scalp and jaw bruised; x-rays inconclusive."

Jones didn't know who or what had hit him—he said he remembered nothing until he woke up at the hospital, which was odd since he walked in on his own two feet. The next morning, he was well enough to receive reporters at his upper Manhattan apartment, where he professed love and affection for Hank Bauer. "I'm not going to make a case of it," he said.

To which his lawyer, Anthony Zingales, replied, "You be quiet."

By the time the Berras got home to New Jersey, the Yankees had dispatched their private investigators. "Yogi called Johnny Kucks at 5 A.M. and said, 'Don't open your door,'" Carmen said. "Because he was young, they didn't want him to get in trouble."

At the station house, Leonard Jones, the victim's brother, filed a complaint against Bauer, the ex-Marine, known to his teammates as "The Bruiser," charging him with felonious assault.

The morning papers had already gone to bed by the time the Bauers got back to their apartment in the Concourse Plaza. "About four o'clock in the morning, the phone rang," Bauer said. "It's a writer. 'Hank, what are you going to do about this?'

"I says, 'Now what the hell are you talking about?'

"'Well, this guy claims you hit him.'

"I said, 'I'm sorry. I didn't hit anybody.'

"Now we're up, and the bellboy calls me and he said, 'Hank, don't come down here.'

"I said, 'Why not?'

"He said the lobby's full of writers, TV cameras,

and everything. He says, 'When you get ready to go to the ballpark, you give me a call, and I'll take you down on the freight elevator.'

"Now, at the ballpark, the first guy I run into was Casey. 'What happened?' I said, 'Case, I'll tell you the truth. I wanted to hit that guy, but I didn't.'"

He couldn't have, he explained, because Berra and Kucks were holding on to his arms. And, besides, the club's bouncers had assured him, *we know how to take care of this.*

"Now Dan Topping comes in. He says, 'I warned you sons of bitches.' So he fines us each a thousand dollars.

"And I went to Mickey and I says, 'Mick, when the hell did we get a warning?' He says, 'You wasn't at that party.'"

The Yankees issued a statement saying that their private investigators had satisfied general manager George Weiss that none of the players had struck anyone. The afternoon tabloids blared the news:

YANKEES' BAUER IN COPA BRAWL
—*New York Post*, May 16, 1957, front page

BAUER: I DIDN'T SOCK GUY IN KISSER
—*New York Post*, May 16, 1957

IT WASN'T A NO-HITTER
—New York *Journal-American*, May 16, 1957

The *Post* earnestly listed possible managerial sanctions, including "the silent treatment." But Stengel had plenty to say to the papers. He benched Berra and Ford and dropped Bauer to eighth in the batting order. Martin was injured and not expected to play. He didn't expect to remain a Yankee either. "I'm gone," he told Mantle, who batted third as usual.

"I'm mad at him, too," Stengel assured reporters. "But I'm not mad enough to take a chance on losing a ball game and possibly the pennant."

2.

In January 1957, Mantle took a break from rubber chicken to negotiate a new contract with George Weiss. He asked for $65,000, twice his 1956 salary. The general manager replied with a threat, slamming a fat file full of incriminating evidence on his desk. To wit: "Billy Martin and Mickey Mantle left the St. Moritz at 6 P.M. Came in at 3:47 A.M."

He had brandished the same damning (albeit thinner) dossier after the 1953 check-signing caper at the Latin Quarter. The GM's paid gumshoes were kept very busy keeping up with Mantle—though their attempts to do so could be comically inept. One evening that spring in Detroit, lobby sitters in the team hotel watched in amusement as Weiss's hired help tailed the wrong guys, following Tony Kubek and Bobby Richardson to a YMCA for an evening of Ping-Pong. A taxicab ferrying Mantle and Ford circled the block and returned them to the hotel unnoticed.

Stengel's method of surveillance was simpler: he'd send the elevator operator upstairs after midnight to get autographs, thus documenting via absent signatures the curfew breakers. Stengel's biographer, Robert Creamer, cited the Ol' Perfessor's adage about tomcatting ballplayers: "It ain't getting it that hurts them, it's staying up all night looking for it. They gotta learn that if you don't get it by midnight, you ain't gonna get it, and if you do, it ain't worth it."

There was nothing funny or subtle about Weiss's attempted blackmail, recounted by Mantle in *The Mick.* "He pats the folder, leans back in his chair, and twiddles his thumbs. . . . With slow deliberation he checks through a batch of papers and suddenly slaps them down on the desk. 'Here, take a look,' he says, the venom returning to his voice. 'I wouldn't want this to get into Merlyn's hands.' "

He also mentioned a potential trade to the Cleveland

Indians for Rocky Colavito and Herb Score. In an interview years later posted on Tim McCarver's Web site years later, Mantle compared the general manager's dictatorial ways to those of Adolf Hitler. "He felt like he was the Führer," Mantle said.

Mantle headed for Rochester, New York, to accept the Hickok Belt, annually presented to the year's best professional athlete, and went back to Commerce without a contract. One morning, he showed up at his friend Jack Meier's house as he often did, having had too much to drink. "He pulls up in a Lincoln, honks the horn, Dad goes out," said Jack's son, Mike, who was in third grade then. "It's ten, eleven A.M. I wanted to go with my Dad. He said, 'Come along, get in the backseat.' The backseat had trash, magazines, beer cans. I see something real shiny, a big, shiny, gold thing on the floorboard. I pulled it out. I said, 'Hey, Mickey, what's this?'"

It was the sports world's gaudiest piece of jewelry, an alligator strap encrusted with a four-carat diamond and twenty-six gem chips valued at more than $10,000. "He said, 'Let me have that a minute.' He handed it to my Dad. He said, 'Here, Jack, this is what you get, a prize belt, instead of money.'"

Del Webb and Dan Topping intervened with Weiss and got Mantle his desired raise.

* * *

He was the king of New York. Everybody loved Mickey. "Mickey, who?" the singer Teresa Brewer chirped. "The fella with the celebrated swing." Men wanted to be him. Women wanted to be with him. His dominion was vast, and his subjects were ardent (one fan asked Lenox Hill Hospital for the tonsils he had removed following the 1956 season). Mantle accepted his due with that great drawbridge of a smile that yanked the right-hand corner of his mouth upward to reveal a set of all-American choppers. "When he laughed, he just laughed all over," his teammate Jerry Lumpe said.

Why wouldn't he? Wherever he went—Danny's Hideaway, the Latin Quarter, the "21" Club, the Stork Club, El Morocco, and Toots Shor's—his preferred drink was poured when he walked through the door. Reporters waited at his locker for monosyllabic bon mots. Boys clustered by the players' gate, hoping to touch him. It wasn't enough to gawk at his impossibly broad shoulders and fire-hydrant neck. They wanted tactile reassurance that he was for real. They scratched his arms, his face, and stabbed him with ballpoint pens. When his little brother Larry got lost in the crowd, and a cop hollered, "Who's the brother?" A hundred boys answered, "I am."

Women—none more beautiful than he was—staked

out hotel lobbies. When Elston Howard's wife, Arlene, met him for the first time, she thought, "My God, who is that? Just the physical body, I'd never seen anything like that. There was something about his presence that was just absolutely stunning."

"He was adorable," said Gil McDougald's wife, Lucille. "We used to joke about it. Who wouldn't hop into bed with him, given the opportunity, just for the fun of it?"

His teammates were equally impressed. "He filled out that uniform like you wish you could have filled it out," said pitcher Roland Sheldon.

In the locker room, they tried not to stare: *What's he talking about? What's he doing?* Later, they would laugh, embarrassed and relieved to find out everyone else was doing the same thing. "It was the ungodliest feeling in the world," Clete Boyer said.

Not quite everyone in New York was in his thrall. At the Stadium, he shared his locker with Frank Gifford, the splendid flanker for the New York football Giants. "Excuse me, he shared a locker with me," Gifford corrected.

They both had Hollywood good looks and a 1956 championship ring. The Yankees called Gifford "Sweetness." Gifford called Mantle a "total asshole. Not a nice person. I didn't know him, but I didn't want to know him. The little bit I was around him I didn't want to be. We deified somebody who hit the ball a long way, 'cause he's got a bad knee. Other than that, what did he do? What did he really do to help society in any way?"

Gifford, whose extramarital athletics were videotaped in a 1997 tabloid sting, has survived his own public shaming, but he has hardly softened his stance on The Mick. "I'm just not a Mickey Mantle fan. I never was. He hit a ball a long way, and he was a sexist. He was not my kind of person. We were MVPs that same year. I would hate to think I was even close to what he was."

In his 1949 essay "Here Is New York," E. B. White offered a caution to restless American souls who come east to conquer the big city: "It can destroy an individual, or it can fulfill him, depending a good deal on luck. No one should come to New York to live unless he is willing to be lucky."

Mantle was willing. And he got lucky. The "Copa Caper," as it became known, might not have become public had it occurred on the watch of the baseball beat writers whose unstated code of honor was to look the other way. News-side reporters had no vested interest in maintaining cordial relations in the Bronx. When they broke the story, it was the first major breach in the Fifties seawall between "on-the-field" and "off-the-field" sports reporting.

Baseball writers ate, drank, and traveled with the

team. Their tab was often paid by the team. "You couldn't write one word of it, the debauchery," said Jack Lang, the longtime executive secretary of the Baseball Writers Association of America. "It wasn't just liquor. It was the women."

One of the more egregious examples of sports desk omertà was recalled by Bert Sugar, a Hall of Fame boxing writer who learned his trade at the rail at Toots Shor's. One day, John Drebinger, the *Times* man who, over forty years, wrote the lead story on 203 consecutive World Series games, regaled him with a tale from an overnight train on a western trip with Babe Ruth's Yankees. "The writers had their own car, and dinner had been served," Sugar said. "They'd cleared the tables, and they'd just dealt out a hand of bridge when the door to the back of the car flew open and Babe Ruth ran down the aisle naked. And about ten feet behind him a woman, equally naked, with a knife in her hand, comes running out! And Drebby says one of the guys looks up from the table and says, 'Well, that's another story we won't cover!' "

The relationship between ballplayers and writers was no different when Mantle became the face of the Yankees. "They would be making $30,000 or $40,000, and we'd be making $10,000 to $20,000," Sugar said. "They *talked to* each other. They would tell you things and—you kept 'em to yourself."

In 1957, the sports department definition of controversy was: "Casey-Weiss Feud Flares on Brawl Penalty." Stop the presses: Stengel calls Yankee higher-ups "those fellows."

So when a blind item appeared in Dorothy Kilgallen's column in the *Journal-American* on May 19, no one followed it up or wondered who had leaked it. Kilgallen wrote, "The prelude to the Hank Bauer night club fracas has been worrying the Yankee brass for some time. Ask anyone who lives in the neighborhood of the colorful Stage Delicatessen where Hank and Mickey Mantle used to be quite famous—and not just for playing ball."

The locker room code of honor was inviolable: What happens in the clubhouse stays in the clubhouse—even if it doesn't take place in the clubhouse. Which explains why Mantle refused to tell his sons what really went down at the Copa. "He wouldn't say," David Mantle said. "We asked."

Fifty years after the fact, Carmen Berra was still reluctant to say anything that might contradict her husband's syntactically brilliant party line: *Nobody did nuthin' to nobody.*

"The bouncer did it," Yogi said.

"The bouncer did it," Whitey said.

"The bouncer really screwed up," Bauer told Gil McDougald.

"That's probably right," Carmen said. "The bouncers weren't going to let anything happen to the New York Yankees."

Anthony Zingales, attorney for the pummeled bowler, pointedly noted that his client had "*also* been slugged by a bouncer as he was going down."

Mantle gave varying accounts over the years: "Somebody said, 'Meet us around the corner.' You don't have to tell Billy but once and Hank also. Next thing I know, the cloakroom is full of people and everybody's swinging and throwing punches."

"Maybe Hank," Merlyn said.

"Probably Billy," Arlene Howard said.

As far as the Yankee higher-ups were concerned, Martin was the chief suspect because he always was. Just who hit whom is far less important than the precedent set by the morning's 72-point headlines. His birthday party was prima facie evidence that things weren't exactly as they appeared in the Wheaties ads featuring The Mick's all-American mug. The Copa kerfuffle was the first public intimation of Mantle's off-field embrace of *la vida loca*.

3.

Whitey, Mickey, and Billy: grown men with little boy's names. Stengel called them his Three Musketeers. In

1956, under the full-time influence of Billy the Kid, who was released from the Army just in time for the 1955 World Series, Mantle had one of the best years anyone has ever had in a baseball uniform. "I wish somebody would influence me like that," Martin said later.

Martin was Mantle's closest friend and polar opposite. Mantle was tongue-tied, country-fed, and shy; Martin was swaggering, street-smart, and volatile. "Billy was a fighter, not a lover," said Irv Noren, who roomed with Mantle while Martin was in the Army. "Mickey was a lover, not a fighter."

When Martin rejoined the Yankees, they resumed their prank-playing partnership. Water guns, water balloons, whoopee cushions, and other simulations of flatulence were standard props. They water-bombed Eddie Robinson and his bride from the balcony of the Royal Hawaiian Hotel.

They staged water-gun battles in the locker room and took aim at unsuspecting female noncombatants standing on the ticket line from the safety of the clubhouse. "The Yankee clubhouse was, like, below street level," Mantle told me. "We had windows, like, where people are walking along. Girls used to come stand there, and we used to shoot water guns up in their puss. We could see 'em kind of flinch. They'd be looking around trying to figure out where the fuck that water is coming from."

They were gleeful peeping Toms. One night, at the team hotel in Detroit, they crawled onto the window ledge—dead drunk—hoping to see a teammate get lucky. Twenty-two stories above the street, acrophobia kicked in—and there was no going back. They had to crawl all the way around the building through decades of pigeon shit to get back to their room. To their great regret they didn't cop a glimpse of anything but macadam.

One year during the World Series, they made early-morning crank calls to their teammates. Lucille McDougald always took the phone out of the bedroom the night before a big game so her husband could sleep early and deeply. When the phone rang, she ran to answer it, worried that it might be bad news—why else would anyone call at that hour? " 'Oooh, Lucille, it's Billy. Mickey and I have just been in a dreadful accident. He's hurt bad. We need Gil to come and help Mick.'

"I said, 'Billy, call the cops. Where are you? I'll call them.'

" 'No, we don't want the Yankees to know.'

" 'Well, if he's hurt bad, they're going to know about it. Where are you?'

"I finally knuckled under and woke up Gil, and he comes out to the kitchen to the phone and says, 'What's going on?'

"Billy says, 'Oh, Gil, Mickey's hurt bad.'

"Gil said, 'Yeah, he's hurt bad. Go home. Go to bed.'

"He knew right away it was a joke. And they did this to every ballplayer on the club they could rouse that night. Next time I saw Billy, I put my hands around his neck. And they thought that was hysterical. Laughter like you never heard. Two boys, two little boys, playing pranks. They thought they were Babe Ruth, could drink all night and play all day."

Martin wasn't Mantle's only ally in misadventure; Ford also did his bit. One day, after missing a train from Baltimore to Washington, they hired a cab to drive them to D.C. A fifth of bourbon eased their way on down the road. Stopping at a fireworks stand, Ford helped himself to a haul of Roman candles, which they proceeded to shoot at each other upon reaching the Shoreham Hotel, destroying their room and Mantle's brand-new suits. Ryne Duren saw the wreckage and thought, "My God, they're killing the franchise."

They told the Old Man, and Stengel said, "Fine, pay the bill." The hotel manager took care of the mess, and Stengel did damage control. "Stengel never said anything bad about his ballplayers," said Virgil Trucks, to whom Ford confessed the next day. "He never under-rated them or overrated them. And he protected them."

Like many Yankee families, the Mantles rented in New Jersey, on the other side of the George Wash-ington Bridge, an easy commute to the Stadium. The only drawback was the damage inflicted on their cars by Mantle's adoring multitudes. Charlie Silvera's car looked like it had been through a war after one season of carpooling with Mantle. So one year a bunch of guys, including Mantle and Tom Sturdivant, chipped in $500 each and bought a 1940 Packard. "It used more oil than it did gas," said Jerry Lumpe.

"It looked like a tank," said Art Ditmar. "I sat in the front seat until I saw the way Sturdivant drove. He'd go up the one-way streets the wrong way. We must've got stopped six or seven times. Every time the cop would come up to the car, Sturdivant says, 'Hi, officer! Have you met Mickey Mantle?'

"We never got a ticket."

It was a giddy, high-octane time. They lived over the speed limit, and Mantle was a get-out-of-jail-free card. There were no rules—until Mantle was taught some manners by Miss Marjorie Bolding. He met her at Manhattan's Harwyn Club, a swell joint where Grace Kelly announced her engagement to Prince Rainier of Monaco. Bolding was a southern belle from Birmingham, Alabama, and an aspiring actress and writer. She recalled, "The maître d' came over and said, 'Mr. Mantle wants to meet you.'

"And I said, 'Well, I don't know who Mr. Mantle is.'

"He wanted me to come over to his table. I said, 'I'm sorry, I don't go to anybody's table. If he wants to meet me, he'll have to come to my table.'

"I remembered that I'd seen him on television, advertising Chesterfield or Camel cigarettes. So I get up to go to the little girls' room, which was very small, just one little booth. I came out of the stall, and Mickey was standin' in the little girl's room! He had his hand across the door and he said, 'You've got the prettiest blue eyes I ever saw.'

"I said, 'Do you really smoke those cigarettes?'

"He said, 'Nah, I don't smoke.'"

It was the beginning of what she called "a unique and personal" relationship. The first time they went out for a drink he asked her up to his hotel room. He said he had something he wanted to show her. She ignored her mama's warning about etchings and went. "We'd become friends by then. He showed me what he wanted to show me and started the procedure to kiss me. But when he got a little too amorous, I said, 'Oh, Mick, if you think I came up to your room because I'm gonna go to bed with you—let me tell you right now that's *not* gonna happen.'

"And I got up and walked out of his room and he came runnin' up the hall. He said, 'I'm sorry, Marjorie, I'm sorry.' I walked out of the hotel, and he came down and got me a cab. But he realized right then that I was totally different—I was a southern lady. He liked me because I *didn't* go to bed with him."

What a time it was, to be a young girl from Alabama barhopping with *the* man. "There were days I don't think we got out of the limousine for two days! He was the most fun. Nobody could play ball like Mickey, and nobody could *play* like Mickey."

Some nights they stayed in, hanging out at her sister Bonnie's apartment where the Bolding sisters composed lyrics in his honor—"Men in general we can handle but I have a hard time with Mickey Mantle"— and challenged him in games of Truth. "You *had* to tell the truth if you were called on," Bolding said. "You learn a lot about a person when you're playin' Truth."

One thing she learned: "The number of women that he had gone to bed with." Fifty years later, the exact figure was beyond her powers of recall. "But it was outrageous," she said.

His teammates assumed she was one of them because they cleared out of his room whenever she arrived. If call girls were present, he quickly let them know she was different. And he never said a bad word about his wife. "Ever, ever, ever," Bolding said. "He had respect, if you can call it respect, for the wife and what she represented in his life."

Merlyn was forthright about his infidelity in *A Hero All His Life*, the family memoir published after his death. She had long ago come to understand that her husband regarded marriage as a "party with added attractions." She was just one of the party favors. As she put it: "He was married in a very small geographic area of his mind."

Holly Brooke resurfaced in the spring of 1957. The New York showgirl with whom he'd had a rookie-year fling authored the cover story for the March 1957 issue of *Confidential* magazine. Mickey and Merlyn were en route to Havana in Harold Youngman's private plane when the magazine hit the newsstands: "I OWN 25% OF MICKEY MANTLE . . . and there were times when he was mine—100%!"

The exposé was coy in the manner of Fifties propriety, making liberal use of double entendre to describe the "night games" they played during their summer trysts. She confided plenty of incriminating details— her pet name for him was the same as his mother's, Mighty Mouse. Their song was Rosemary Clooney's sultry, sexy hit "Come on-a my house, I'm gonna give you candy . . . I'm gonna give you everything."

It was about money, of course. This was payback for the 1951 loan she had made to Allan Savitt, the agent who had finagled Mantle into signing an exclusive contract that had given him 50 percent of all future endorsement income.

Brooke wrote: "I'd met a Broadway publicity man and personal agent and agreed to let him handle my press notices for a year. Over the drink or two we had to clinch the deal, he revealed that 'The Mick' was also one of his clients.

"He painted a gaudy enough picture of the possibilities in managing Mickey, but it was of no great interest to me until he suddenly came up with a proposition. He was short of cash to exploit his contract with Mantle and wanted me to loan him $1,500 for a few months. If he didn't pay the money back by January 10, 1952, I was then to become owner of one-quarter interest in Mickey."

One night, she recounted, Savitt called and invited her to meet him and Mantle at Danny's Hideaway. "Let's be buddies," she recalled Mantle saying.

Buddies, indeed. "Plenty of times we'd sit in my car in the wee small hours of the morning and watch the sun rise over Manhattan. Those all-night rendezvous might come after an afternoon of baseball or golf, but the Mick was never too tired for a night game. . . .

"More than once, Mickey would ask me how I felt about marrying him, owning him 100 percent . . . permanently. But he always answered his own question.

" 'No, I guess not,' he'd say. 'You'd never be happy in a little town like Commerce.' "

When he was sent down to the minors, she spent three days with him in Columbus, Ohio, she said. When he was recalled by the Yankees, she visited him in Washington. He always wanted her to stay the night. "But it had already reached the point where *Mickey* was collecting too many dividends on *my* investment."

It went on that way for a couple of seasons, she wrote, until she was "lured out to Hollywood with the promise of a Warner Brothers contract." But they stayed in touch—she had heard from him as recently as August 1956—and she was still hoping to see her dough. So when he asked her to meet him in Boston, she went. "Don't worry, the Yankees will take care of it," he said when she asked about her money. Their song was playing in the background. Her unsentimental conclusion: "It's time he went to bat—for me."

At every layover en route to Havana, Mantle and Youngman bolted from the plane and bought every copy of the offending issue of the magazine off the newsstand to prevent Merlyn from seeing it. When she got home, a stack of *Confidential*s was waiting at the front door.

Mantle was hardly the first baseball player with an expansive definition of "to have and to hold." Extramarital sex was a perk of the job. The endless hours in faraway hotels had to be filled somehow, and there was no shortage of willing playmates to help fill them. "These girls, these women, would just do *anything*," Bolding said. "They're like flies."

Green flies, ballplayers called them. They swarmed the team hotel, and buzzed every bar and lobby. Home or away, there was a plethora of opportunities.

Mantle knew New York's demimonde as well as its café society. Irv Noren introduced him to a celebrity essential: the fixer. Julius Isaacson, the president of the International Union of Doll, Toy and Novelty Workers, also known as Big Julie, a would-be pitcher who threw with such velocity he could knock down a wall—but only if he didn't aim at it. He was a pugilistic presence—six feet three inches and a couple of hundred pounds. He later managed Ernie Terrell when he became heavyweight champion of the world. Julie was a good friend to have when falsely accused of getting someone pregnant. "Mickey had a problem," Isaacson said. "A girl was trying to shake him down. It wasn't his. We had the girl come over to the Edison. We met her there. Took her to the East River and told her she had two choices: leave Mickey alone, or this."

Mantle was one of millions of Americans on whom

FBI director J. Edgar Hoover kept tabs. He never was the subject of an FBI investigation, but when his name surfaced in other probes, Hoover kept the notes—just in case. In 1969, John Ehrlichman, counsel to President Richard M. Nixon, requested a background check on Mantle, along with a group of other baseball personalities. The FBI responded, "Our files reveal that information received in June, 1956 indicated that Mickey Mantle was 'blackmailed' for $15,000 after being found in a compromising situation with a married woman. Mr. Mantle subsequently denied ever having been caught in a compromising situation. Mr. Mantle readily admitted that he had 'shacked up' with many girls in New York City, but stated that he has never been caught.

"A confidential source, who has furnished reliable information in the past, advised in June, 1957, that a very prominent Washington, D.C. area gambler and bookmaker arranged dates for members of the New York Yankees baseball club at a Washington, D.C. house of prostitution. Allegedly, Mr. Mantle was one of the members of the team who was entertained at this house of prostitution."

The names of the other players were expunged when the file was released in 1998 in response to a Freedom of Information Act request. Merlyn had no knowledge of her husband's FBI file, but she had no doubt who was leading him astray. "Why can't you get Whitey Ford to room with Billy on the road?" she asked Noren.

He told her, "Merlyn, he won't do it."

At one point, Del Webb became concerned enough to ask his friend Bayard Taylor Horton, a doctor at the Mayo Clinic, to have a talk with Mantle. Tom Horton was sitting with his father in Webb's box when the Yankees' co-owner passed the doctor a note: "See Mantle."

Horton, who later authored several Yankee biographies, accompanied his father to the clubhouse. "My father said, 'I want to see you, Mr. Mantle.' He took him in the office for twenty or twenty-five minutes. When Mantle came out, he was crying. I asked my father what he had said. He said, 'It's a medical problem.'

"There was no question he said, 'You gotta stop doing this, and you gotta start doing more of that.' I think he was saying, 'You will get a social disease if you keep this up. You gotta quit the booze.'

"I said, 'You can't talk to Mickey that way. He's making a lot of money.'"

The next day Mantle hit two home runs. Webb told Horton, "Good job, Doc."

4.

Martin was still a Yankee on June 11 when the team arrived in Chicago for a three-game series with the league-leading White Sox. Two days later, in the bottom of the first inning, Yankee starter Art Ditmar threw a pitch that buzzed Larry Doby's head. Doby took umbrage. "He said, 'If you do that again, I'm going to stick a knife in you,'" Ditmar recalled. "We both started swinging, and the umpire jumped between us. Skowron come in and pinned Doby to the ground, and then [Walt] Dropo got on Skowron and then [Enos] Slaughter got on Dropo. I walked back to the mound and watched everybody fighting. Mickey didn't get involved. Mickey was the pacifist."

"He never left his position," said Bill Fischer, who watched the free-for-all from the Chicago bullpen. "He said, 'I'm in enough trouble.'"

Things had just about quieted down when Martin asked Ditmar what Doby had said. Ditmar told him. "Martin went after him in the dugout," Ditmar said. "'Course, Billy always got the first punch in."

Twenty-eight minutes and four suspensions later, order was restored.

Mantle once said that Martin was the only guy he knew who "could hear someone give him the finger."

He negotiated life with a chip on his shoulder, and those in the know gave him a wide berth. Even the Yankees were dismayed by his brawling on the field in Chicago—an act of wanton self-destructiveness, given the trading deadline just two days away, and the Yankees heading for a weekend series with the Kansas City A's, Weiss's personal farm team. On Saturday night, one hour and nine minutes before the trading deadline, Stengel summoned Martin from the bullpen, where he'd been trying to avoid the inevitable. "You're gone," the Old Man said.

Actually, Martin was staying right where he was—he was the A's new second baseman. Stengel also passed on a message from Dan Topping that the Yankees would refund the $1,000 he had been fined for the Copa melee (they would do the same for the others at season's end).

The Three Musketeers retired to the closest bar and cried into more than one beer. Publicly, Martin said all the right things. *That's baseball.* But he refused to talk to Stengel for years until Mantle negotiated a reconciliation.

The Yankees won twenty-four of their next twenty-eight games.

"Did Mickey stop drinking when Billy left?" Ditmar said. "No. Did he stop going out? No."

On June 24, 1957, the Copa Six were summoned to appear before a grand jury at the Criminal Court Building in lower Manhattan. The Manhattan DA had declined to prosecute Bauer, so bowler Jones pursued his only remaining legal remedy: a citizen's arrest on a charge of felonious assault. He demanded $250,000 in damages. Bauer exercised his right to have the case presented to a jury of what his lawyer no doubt hoped would be Yankee-loving peers.

Mantle was the last of five players to testify. He took the stand with a mouth full of bubble gum. Admonished about his lack of courtroom decorum, Mantle obligingly removed the offending wad and stuck it to the bottom of his chair. "I was so drunk I didn't know who threw the first punch," he testified. "A body came flying out and landed at my feet. At first I thought it was Billy, so I picked him up. But when I saw it wasn't, I dropped him back down. It looked like Roy Rogers rode through the Copa on Trigger and Trigger kicked the guy in the face."

The grand jurors were still laughing when they handed down their decision.

Hank Bauer Freed by Grand Jury,
Left-Fielder Cleared of Assault Charges
—*The Sporting News*, July 3, 1957

The next day at Toots Shor's, Mantle announced the formation of the Mickey Mantle Hodgkin's Disease Research Foundation at St. Vincent's Hospital in honor of his father. On July 3, Bauer sued Edwin Jones for false arrest, seeking $150,000 in damages. New York's most famous bowler was never heard from again.

5.

It was a liquid time in America. The language, the culture, and the national pastime were suffused with booze. Raise a glass. Wet your whistle. Line 'em up. And always have one for the road. As Toots, the Prohibition bouncer turned saloonkeeper, declared: "I never felt guilty about a drink in my life. It's beautiful, legal or illegal."

Stengel divided the Yankee clubhouse into "them milk drinkers" and those he called "whiskey slick," which is how Mantle and Ford came by their shared nickname—Slick. His notion of drinking responsibly was to warn his players not to patronize the hotel bar because "that's where I do my drinking." His idea of punishment was to put transgressors in the starting lineup. After one dispiriting loss, he threatened to make nonperformers "go out and get a double Scotch and a woman."

Yankee bon vivant Don Larsen summed up the big league ethos: "You got to have a little fun, for God's sakes. You're going to play in big cities and watch TV? Or go to movies? Those weren't even invented."

No one in baseball thought Mantle's drinking was exceptional because it wasn't. He could quit when he wanted to and "handle it"—baseball's ultimate praise—when he had to. He always took Stengel's preseason admonition to heart: "Son, you're the Yankees now."

Only Merlyn's father seemed to think his drinking was problematic. One day after his Triple Crown season, Mantle volunteered to babysit while Merlyn went shopping. He took Mickey, Jr., with him to his favorite Commerce watering hole. When they didn't return as expected, Merlyn's father, the church deacon, went looking for them. When he found them at Mendenhall's, his son-in-law was on the floor wrestling with another patron. Mickey, Jr., was sitting on the bar holding his father's beer. "The first time I went to a bar and tasted my first beer," Mickey, Jr., recounted in *A Hero All His Life*. "It wouldn't be worth telling except that I was three years old."

"Grandpa took the beer out of his hand and grabbed Mickey, Jr.," Danny Mantle said. "When he walked by, he picked up Dad's head and went, 'Don't ever do this again' and dropped it back down. Dad went back to fighting, and Giles walked out with Mickey."

Merlyn told me, "Dad was furious about that."

"That night I slept at my grandparents' home," Mickey, Jr., wrote. "Dad didn't say much when he picked me up the next morning, but I know he hated to face Grandpa Johnson. He didn't think there was anything immoral about taking me to the bar. He wanted to show me off, and this was where he was most likely to bump into his friends."

He was only doing what his father and other fathers in town had done. He told his late-in-life companion Greer Johnson that his father would take him to bars and set him on a stool.

It's unlikely that any of Mantle's drinking buddies in Commerce—or New York—would have recognized the genesis of his alcoholism. They would have thought he was "irresponsible," said "Sudden" Sam McDowell, who lost his fastball and his family and came close to losing his life before getting sober in the 1990s and becoming a drug counselor. "In 1956 and 1957, unless you were a trained therapist, you would have thought nothing was wrong," said McDowell, who developed drug and alcohol programs for the Texas Rangers and Toronto Blue Jays. "He didn't go out and act like that all the time. No alcoholic does. Up until close to 1960, you would see very little progression in the disease. He was an undiagnosed highly functioning alcoholic."

In 1957, Mantle won his second consecutive Most Valuable Player Award and became only the fourth player in major league history to get on base more than half the time over the course of a season. In June, he batted over .400. On July 26, he hit his 200th career home run. He was three years younger than Babe Ruth when he reached the same landmark. He was hitting .369, with 34 home runs and 91 RBIs, when he landed in the hospital on September 6 with what the Yankees called "shin splints." He had injured himself playing golf with Sturdivant on a day off, which, unlike imbibing, was against Stengel's rules. Hurling his putter in frustration, he had broken a tree limb and sliced his leg down to the shinbone. He missed five games, his first of the season.

While he was recuperating at Lenox Hill Hospital, America was reading about the hijinks in Hank Bauer's apartment above the Stage Deli in 1951 in the September issue of *Confidential* magazine: "There was no umpire around when . . . These Yankees Had a Ball!"

Mantle, Martin, and Bauer appeared arm in arm, grin to grin beneath the cover line. The story was a follow-up to Holly Brooke's spring exposé and Dorothy Kilgallen's May column in the *Journal-American*.

It detailed the fun and games that had taken place in the apartment above the famous deli during Mantle's rookie season.

"While waiting to go to bat, the Mick knocked off a few more highballs, then suddenly got up, started to turn white and staggered to the bathroom," *Confidential* said. "Now games have been called on account of rain, cold weather or wet ground, but this is the first time one was called because it was drunk out. When Mickey collapsed, they had to call off the game in the bedroom so Mickey could be put to bed."

His girl went home alone when "he dozed off."

None of New York's family newspapers pursued the story. The sporting press confined itself to musing about who would be the American League's Most Valuable Player—Mantle or Ted Williams—and the World Series matchup between the Yankees and the Milwaukee Braves.

Yet Billy Martin's birthday party was a watershed event, and not just because it gave Weiss the occasion to trade him. It was the day sportswriting began to grow up. The era of hear no evil, see no evil, speak no evil could not withstand TV's increasingly intrusive cathode glare or the skepticism of an irreverent cohort of young sportswriters for whom questioning authority was a generational prerogative.

Weiss saw the printer's ink on the wall and he was plenty worried. "People have been looking for incidents since the Copa affair," he lamented in a 1960 interview with *The Saturday Evening Post*. "A national TV network was considering the Yankees for the same sort of inspirational show that is built around institutions like West Point and Annapolis. This might have steered some good prospects to us, and the players could have made some extra money appearing on the program, but the project was shelved after the Copa affair."

6.

The Russians launched the first Sputnik satellite on opening day of the 1957 World Series. Mantle crash-landed in game 3. Trapped off second base on a pick-off play in the top of the first inning, he slid back in beneath Red Schoendienst's legs as the ball sailed into center field. Gazing in supplication at the ump, Mantle waited for adjudication, Schoendienst draped over his shoulder like a sack of potatoes, until he saw where the ball had landed. "The ball was thrown high and away," Schoendienst recalled. "I went up for it. I landed on Mickey. I stayed on top of him."

Mantle shook him off like a bedbug, went to third, and scored the Yankees' first run; later he singled and hit a home run. But he was badly injured on the play. No one realized how badly his shoulder hurt until Stengel removed him in the top of the tenth inning the next day. The *New York Times* headline reported:

> Bombers Face Prospect of Losing Mantle for Fifth Series Contest; Shoulder Injury Handicap to Star; Mantle's Inability to Throw with Usual Strength Leads to Removal in Tenth.

Mantle sat out game 5, the ninth World Series game he had missed due to injury, and the Yankees lost 1–0. They won game 6 without him but lost game 7—and the World Series—despite his return to the lineup. His throwing arm would never be the same.

The winter of 1957 would be his last in Commerce. Mickey Charles and his hometown had grown apart. "Off-ish to the community," is how local Frank Wood described him. Half the folks in town wanted no part of him and the other half thought he didn't give back enough. What had become of the boy who never drank, never smoked, never talked back, the all-American face in the magazine ad for the Breakfast of Champions? *BillyMartinandWhiteyFord,* that's what many of the townfolks said, expectorating their names like a foul stream of chaw.

Harold Youngman, Mantle's patron, did everything he could to entice him to stay. He gave him a house in Joplin, Missouri, and a 25 percent ownership stake in a new Holiday Inn he was building there—Mickey Mantle's Holiday Inn, the only Holiday Inn in the U.S. of A. named after an actual person, Mantle bragged. The bar was called the Dugout.

Later, Youngman outfitted a heated fishing cabin built over a pond on his ranch for Mantle and his pals. It had a hole in the floor and an electrified winch that raised the lid so they could fish in winter without getting cold, wet, or dirty. The foreman, Charles Brinkley, stocked the refrigerator and the pond.

Youngman pleaded with Mantle not to quit his roots. He talked up the benefits of being a big fish in that small pond. But Mantle was gaited to the rhythm of big city life. Moving to Dallas made it easier to fly off to all those banquets, photo shoots, and beauty contests he was asked to judge. Merlyn went shopping for Texas real estate. Mantle leased a bowling alley, where he would put his brothers to work and show everyone he could manage affairs on his own.

After the Yankees avenged themselves against the Braves in the 1958 World Series, Mantle went to Dallas to see the new $59,500 home Merlyn had selected, the one important decision in their marriage he allowed her to make. The "Mantle manor," *The Sporting News* called it. Furnished in the French Provincial style, it was located in the neighborhood where George W. Bush would settle after he left the White House.

The move distanced Mantle from Youngman's well-intentioned but paternalistic control and deprived him of much-needed advice. In cutting ties with the place where he made most sense to himself, he became a celebrity nomad, a citizen of everywhere and nowhere. Mantle later said leaving Commerce was the biggest mistake of his life. "Guided differently, he could have done better for himself as a human being," Jerry Coleman said. "He would have liked himself more."

11

August 14, 1960

SEASON UNDER SIEGE

1.

In the bottom of the sixth inning of his 1,352nd major league game—the second game of a Sunday double-header against the Washington Senators—Mantle bounced a ball down the third base line. The score was 1–1. Hector Lopez was on first base. The Stadium was half full.

They witnessed the unthinkable: the fastest man in the major leagues, who ran harder than anyone else no matter how much it hurt to do so, did not run. He jogged down the line. And he didn't arrive in time. The Senators turned a potential rally into an easy, inning-ending double play. Upstairs in the press box, reporters sighed. *Mantle's hurt again.*

But he wasn't. Casey Stengel yanked him from the lineup and dressed him down with an unprecedented dugout rebuke. "Quitter!" shouted a man in the stands. Mantle kicked the water cooler en route to the clubhouse. The indignity of losing a doubleheader to the woeful Nats was compounded by the fact that the second loss dropped the Yankees into third place, where they had finished the year before. Worse, Roger Maris had been injured sliding into second base in an attempt to break up a double play on the previous play. A knee to the ribs had sent him to the hospital for X-rays. The contrast between Maris's fierce take-out slide and Mantle's nonchalant sally was glaring.

When a very long day of baseball finally ended at 10:29 P.M., reporters found Stengel in his office, naked and fuming. *Who does he think he is? Superman? Well, he's not.* Cooling his managerial prerogatives in the shower, he hollered over the spray that there would be consequences. A fine, perhaps. Maybe a suspension. "It's up to me," he declared, seizing the opportunity to remind everyone he was still in charge even if he was sixty-nine years old.

Mantle answered questions with monosyllables and shrugs. " 'Why did you leave the game?' " *Daily*

News columnist Dick Young asked. " 'Was it your idea or his?' " " 'It must have been his,' " Mick said with a smile. " 'It sure wasn't mine.' "

Whitey Ford took up for him, explaining that the error had been one of inattention, not lack of effort, a clubhouse mortal sin. Mantle thought there were two outs, he said. But anonymous gripes made their way into print for the first time. "If he's hurt, that's one thing, but I don't think he's hurt," one veteran told the *New York Post*. "He certainly would have beaten that ball out if he'd run. When you're making $60,000, you just can't do that in front of a crowd like that."

"Maybe it'll wake him up," another teammate said.

Ryne Duren, who arrived in the trade that sent Martin into exile in Kansas City, found Mantle at his locker, pouting and mad as hell—at himself. "I've never seen him angry at anybody *other* than himself," Lopez said.

It was one of the traits the Yankees appreciated most. He never showed anyone up, never called anyone out, never blamed anyone but himself. "When you're that intense, sometimes you're too hard on yourself," Eli Grba said. "You beat yourself to death. And he did."

Duren backed his car up to the police barricade as always in order to minimize the gauntlet Mantle had to run from the players' gate. He sprinted to the car. Duren gunned the engine. They often stopped for a pop or two on the way home. Duren was accustomed to hearing Mantle's muttered imprecations aimed at the manager: "Old gimpy-legged bastard, sonofabitch." This time, Duren said, "Mickey didn't say a word except 'Turn here' and 'Turn there.'"

He directed Duren to a tavern on the Jersey side of the George Washington Bridge, one of the few Duren didn't know. Duren had tried and failed to get sober the year before. He wasn't ready to take the twelve steps required by Alcoholics Anonymous. "He ordered a double shot," Duren recalled. "I think it was bourbon. He slugged that down and said, 'Give me a double.' And the guy poured him at least two more jiggers in that one, and he threw that one down. He wanted that anesthetic feeling up there in the prefrontal lobe."

Duren counted four double shots in the half hour or so they stayed at the bar. Mantle wasn't falling-down drunk when Duren dropped him off at home. "But he got the fix he needed," Duren said.

Talking about it was not an option. Talking wasn't the Mantle family way. Merlyn had long observed her husband's inability to show his feelings and his increasingly quick reflex to numb them with alcohol. She never could get accustomed to the familial reserve. "Mick's family was cold," she told me. "They didn't visit. *He* didn't visit."

And when they did pay a call, they didn't show

up emotionally. "Nobody would talk. It was weird. Nobody would say anything," Merlyn told me. "And the only way they could really talk was when everybody got bombed. Everybody had to have the beer. They could not relax and visit with each other unless they were having beers."

"All they said was hello and goodbye," Danny Mantle said.

2.

That Sunday was a low point and a turning point for Mantle. Loafing was out of character. He had vowed that 1960 would be different from the year before, when, he admitted, he had "a lousy season." The Yankees' third-place finish, fifteen games behind the White Sox—only four games over .500—was a team effort. But in the sports pages and the front office, Mantle was deemed the chief culprit. After batting .285 with 31 home runs and 75 RBI, he wondered if the Yankees would or should trade him. The phenom had become a disappointment.

"Mantle: A Problem Child," Leonard Schecter's five-part preseason series in the *New York Post*, analyzed the lack of maturity that had become a favorite subject of columnists. In the absence of sports psychologists, not yet a de rigueur retainer for major league clubs, Schecter devoted part one of the series to the Freudian musings of an anonymous baseball man who diagnosed Mantle as a self-destructive masochist, a big kid with no judgment and no self-awareness. Schecter offered as proof of his adolescent proclivities the way he badgered batboys, bullpen catchers, and unsuspecting rookies to catch the knuckleball he pleaded to throw in a game, defying the front office and common sense. "You know how he ruined his arm?" Jerry Lumpe said. "Throwing knuckleballs."

Even before the Bombers stunk up the joint in 1959, Stengel had begun referring to Mantle as his greatest disappointment. When he signed what would be his last two-year contract to manage the Yankees in 1958, he seized the opportunity to name the greatest players of his tenure. DiMaggio was number one on Casey's list. He mentioned virtually every Yankee who had ever made an All-Star team—except Mantle. "The trouble with Mantle is Mantle," Stengel explained.

His teammates were mystified by the managerial snub. "What about Mantle?" Bauer demanded after Stengel named him, Berra, and Rizzuto as his three best players. "You gave 110 percent every time you were in the lineup," Stengel replied.

With his age and his ability to enforce discipline

being questioned, Stengel could no longer afford to wink at bad-boy behavior. His own drinking had gotten heavier as well. "Casey was drunk every night, but what he was concerned about was what the writers thought of him and the team and entertaining them and being a clown with them," Duren said. "So you look at his behavior and wonder if he was out of *his* adolescence."

Writers scrambling for copy during the disconsolate off-season wondered whether Stengel had the moral authority to curtail his party boys. No, Weiss replied, Stengel wasn't the problem. The 1959 Yankees weren't hungry enough, made money too easily, thought they deserved automatic raises just for putting on uniforms, had too many outside interests, and weren't living and breathing the game twenty-four hours a day as they had when they were rookies.

"Oh, they wanted to win and they hustled, but they had an air of complacency," Weiss told Stanley Frank, the author of the "Boss of the Yankees" in *The Saturday Evening Post*. "All the key men are independently wealthy from the high salaries and the World Series shares they've been getting for a long time."

This would have been news to the average major leaguer, who earned an estimated $12,340 in the Fifties, according to a study, "The Economic History of Major League Baseball," by Michael J. Haupert, a professor at the University of Wisconsin–La Crosse. (The Major League Baseball Players Association did not begin keeping salary figures until the mid-1960s). Duren spoke for a generation of budget-conscious Yankees when he said: "I don't want you to think that George Weiss never did anything for me. He taught me how to live without money."

The problem with Mantle was less clear cut in Weiss's mind. "He's an enigma," Weiss told Frank. "We've never had a player who was the subject of as much discussion and analysis. Our entire organization has tried to discover why Mantle hasn't capitalized on his enormous potential, and we obviously haven't found the answer.

"Physically, Mantle has the attributes of a superstar, a blend of Babe Ruth and Ty Cobb. He is much faster than DiMaggio was and he has more power, with the added advantage of being a switch-hitter and getting the benefit of the short right field fence in the Stadium. DiMaggio was a better fielder, but Mantle compensates for that with a stronger arm than Joe had after he hurt his shoulder. Add up Mantle's assets, and he's superior to DiMaggio but he hasn't come close to proving it yet."

Frank asked, "Would it be correct to say the Yankees believe it's entirely up to Mantle himself whether he ever reaches the top rung?"

"Weiss nodded gravely. 'We've done everything possible to help him.' "

In January 1960, Weiss sent Mantle a contract calling for a $17,000 pay cut from his $72,000 salary, $1,000 less than the maximum decrease allowed by agreement between the players and the owners. Mantle figured it was a misprint and sent it back unsigned. When he was photographed lounging in the Dugout bar in his Joplin, Missouri, Holiday Inn—the caption read, "Who needs the dugout at Yankee Stadium?"—the brass claimed he was holding out to generate publicity for the motel and his Dallas bowling alley. When he failed to report to camp on time, he was vilified. Joe Trimble of the *Daily News* called him an ingrate and "a hillbilly in a velvet suit." The "scared high school Okie" had become "a sullen holdout," Trimble wrote, "guilty of a disgraceful exhibition of ill conduct. He isn't grateful or even gracious. The Yankees made him what he is today."

Mantle laughed off the insult—"How can he call me a hillbilly when he's the sloppiest guy around here?" And later, passing the customarily disheveled Trimble in the clubhouse, murmured, "How is Oleg Cassini today?"

The witty bravado obscured the truth. "He was really, really hurt by that," Tony Kubek said.

Daily News columnist Dick Young damned him more gently: "He is emotionally immature. He's not too bright, and he's not too friendly, but he's a pretty nice guy, all things considered."

And they wondered why he wasn't a great interview.

Mantle was used to people questioning his mental acuity. "Never did claim to be smart," he'd say. "Doesn't bother me anymore." In high school, he was the sports editor of the school newspaper. When he and the editor in chief were chosen to represent the school at a state writing contest, he assumed the teacher was making fun of him. Mantle was stung when Schecter quoted him all too accurately, not standard operating procedure on the sports page. "You wrote a bad story about me," Mantle objected. "You had me using those double negatives."

"Don't you?" Schecter replied.

Sportswriters had long improved locker room oratory, making players sound better spoken and better mannered than they, and often their interrogators, really were. Red Smith crowed with pleasure at rhetorical verisimilitude: "I know the man that that's the house of's daughter." But the selective application of the principle of real speech was cruel and reinforced the perception of Mantle's dimness. Jackie Robinson put the IQ question in perspective before a World Series game in the early Fifties: "We got plenty of guys that stupid. But we don't have anyone that good."

Mantle had a good head for numbers—pals back home said nobody counted cards better or faster. He knew the percentages were against him in 1960 contract negotiations with Weiss. The wily, obdurate general manager won out by waiting him out. Mantle capitulated on March 14, settling for $7,000 less than he had made in 1959. Still smarting from the salary cut, he reported to St. Petersburg to find a changed Yankees team: Bauer and Larsen were gone, traded to Kansas City for Roger Maris. The newly stern skipper vowed, "This time we're going to stick to sweat and toil, and maybe there'll even be a little bleeding."

Mantle missed ten days of spring training. Six days after he reported, he got hurt.

Mantle, Eager to Make Up Lost Time,
Hurts Knee by Running Too Hard

The Yankees had not played well since July 1958 and continued their swoon into the regular season. Mantle was batting .286—with only one home run—on May 12 when Stengel moved him up to second in the batting order. By the end of the seventeen-game experiment, his batting average had dropped forty points. He was in an 0-for-20 drought when he hit his second home run of the year on May 28, the same day Stengel called in sick. The Ol' Perfessor had been experiencing chest pains since spring training. After ten days in the hospital a virus was diagnosed.

On May 30 Mantle jogged in from center field with the final out of a 3–2 win safely tucked in his glove. Spectators were still allowed to exit the Stadium through the center field gates and he was accustomed to braving a current of attentive fans. This crowd wasn't friendly. In the punching, shoving, grabbing melee that followed, Mantle's cap was stolen and his jaw pummeled. Scared he'd be poked in the eye, he bulldozed and elbowed his way to the dugout. In the clubhouse, he pressed an ice pack to his jaw before leaving for the hospital for X-rays. Nothing was broken, but the bond between Mantle and the Bronx partisans was fraying. "Mantle Is Mauled," the *New York Post* exclaimed, calling him the "victim of one of the worst mob scenes in Yankee Stadium history" and describing the attackers as a "mob of young toughs."

It wasn't the first such altercation. In June 1958, a thirteen-year-old girl punched him as he got out of a cab, pulling his hair and slapping him as he plowed through the throng outside the players' gate. The *Times* dubbed him "The King Whose Homage Is Catcalls" and sent a young reporter named Gay Talese to probe the source of home crowd displeasure. "The all-American out," one fan called him. The manager of a local movie theater compared him to a blank screen. Harry Greenbaum of the Bronx cited lingering resentment over Mantle's exemption from military service.

The antipathy bewildered him. Frank Petrillo, a New Jersey car dealer who befriended Mantle in the early Fifties, kept him company one evening after a particularly trying day at the ballpark. "He was actually crying," Petrillo told his son, Frank, Jr. "He said, 'I don't know what the fans want me to do. I don't know what more I can do. My knees are killing me. I can't perform any better than I'm performing, and it's not good enough.'"

In the aftermath of the May 30 brawl, the Yankee switchboard was flooded with angry calls condemning Mantle and accusing him of accosting innocent children. "I hit quite a few," Mantle told Stan Isaacs of *Newsday*, "but tell them I got the worst of it. They weren't all kids, either."

Later, he heatedly denied having hit anyone purposefully: "Anyone who said I did is a damn liar."

The *Times* columnist Arthur Daley deplored the mob mentality and defended Mantle's right to protect himself. The Yankees formed the "Suicide Squad," a flying wedge of beefy ushers who formed up on the lip of the infield grass in the bottom of the ninth inning to escort Mantle off the field.

Six weeks later, on the afternoon of July 19, while reading fan mail in the visitors' locker room at Cleveland Memorial Stadium, Mantle opened an envelope postmarked July 16, 8:30 A.M., at Tonawanda, New York. The handwritten letter said: "I had a son that was drafted with a bad leg & bad eyes he got killed but a rotten draft dodger that could run like you gets turned down. I have a gun with micrscopic [sic] lenses and I'm going to get you thru both of your knees and its [sic] going to happen soon."

Police were alerted. The FBI was called in. A case file identifying Mantle as a potential victim of extortion was opened. Mantle went 1 for 5 with a single and a run batted in.

Agents interviewed him in the locker room the following day. He told them that it was the first "really bad or disturbing letter" he had ever received, which wasn't true. There was the 1953 Fenway Park death threat, and a female stalker who threatened to kidnap Merlyn and Mickey, Jr., and demanded he buy her a diamond ring. Merlyn took the threat seriously and began keeping a gun by the side of the bed. The letter he received in Cleveland worried him, but he thought that if an attempt were made on his life it would be in the Bronx, where he was accustomed to hostility. And where, he told the case agent, "some person could stand on the rooftop on one of the buildings surrounding Yankee Stadium and take a shot at him."

3.

Long before the letter arrived, Mantle talked about dying young. Yogi Berra tried to jolly him out of it. Whitey Ford tried to reason with him. "Look, Mick, you got doctors watching you every day," Ford told him. "The doctors are better. You eat better. You're not in the mines like your father. You're not going to die young."

"He'd listen to me for a while," Ford said, "but I don't know how much he believed it."

By 1957, when Jerry Coleman told him about the new pension plan being negotiated with ownership, Mantle took attenuated mortality for granted. Coleman was the Yankees' player representative as well as the American League representative in the talks that resulted in $100-a-month in life insurance, a huge step forward in benefits. "He said, 'I'll never get it,'" Coleman recalled. "I knew what he meant. And he knew I knew what he meant. I know he had a death wish from very early. It's like he was waiting for it to happen before it should."

He wasn't morbid, and he didn't talk about it all the time. But, Johnny Blanchard said, "If you got him in a private moment, yeah, he would talk. Not often, but he would talk different."

Blanchard lived with Mantle one season in his suite at the St. Moritz. One night they were watching *The*

Honeymooners. "It was a real quiet night, just me and him laughing. Gleason and Carney made both of us just laugh like you can't believe. And right in the middle of it, he leans over to me and says, 'Hey, Blanch, you ever think about dying?'

"Well, Chrissakes, if I ever did at that moment it was the last thing in the world that I was thinking of. I said, 'Mick, what are you thinking about dying for, we're laughing our fannies off?'

"He said, 'It just run through my mind.'"

His apprehension may have been understandable, but some people who knew him well, among them Ralph Houk and Clete Boyer thought he exaggerated and exploited the fear; Pat Summerall thought it was a ruse that he "allowed himself to buy into." Howard Cosell thought it was a triumph. His interview with Mantle on WABC in New York on August 19, 1965, was the first widely disseminated public airing of his fatalism. Cosell was at his prosecutorial best. "It's a fact that you've lived with the memory of your father and often thought about the possibility of early death yourself, isn't that so?" he demanded.

"Well, I don't worry about it, Howard," Mantle replied mildly. "Of course, I've got a good chance of it. I don't know if it's hereditary or not. I hope not."

Cosell was insistent. "Do you really think about this a lot?"

"Nah, I try not to."

"You don't feel that you're a tragic, courageous hero?" Cosell hectored in a subsequent interlocutory. Afterward, he ballyhooed his role as psychic healer. "He did a half-hour show with me and he felt like he had a cathartic," Cosell declared. "He felt cleansed."

The refrain of doom became the subtext of every profile. Mantle began to describe his life as a tragic epic as well. "The story runs I'm the first Mantle—the first Mantle boy—to ever make it past forty," he told Cosell, who didn't challenge Mantle's recall (he'd forgotten Grandpa Charlie) or his logic. Apparently it had never occurred to him that an equally lethal genetic inheritance from his mother's side of the family might prevail.

Ultimately the fear became a public embarrassment. Who wants to go through life known as the doomed Yankee slugger? "That was overplayed," he told me. "People wrote about it, but I don't think I ever thought about it that much. Maybe I thought about it, but it didn't bug me. I didn't sit around saying, 'Oh, fuck, I'm going to die' all the time."

Nonetheless, it became convenient for others to excuse his self-indulgent behavior as a fatalistic entitlement. Merlyn told Gil McDougald's wife, Lucille, "I see him wanting to enjoy himself while he is alive because he fully expects not to live. And I let him do it because if he goes like his dad and his uncle, he won't be alive by the time he's forty."

Frank Scott, the Yankees' onetime traveling secretary and Mantle's agent, shared Stengel's perspective with the author Gerald Astor, who quoted him in an unpublished book, *Eyes Open, Mouth Shut*. Exasperated, Stengel said, "This kid still has everything and he has a lot of career ahead of him, that is if he don't screw up and he comes close to screwing up every goddamned day of his life. In the first place he has little sense about people, and I'm always worried about whether he is going to buy the Brooklyn Bridge.

"It's not Martin and it's not Whitey it's Mantle himself with all that shit about how he's going to die soon because everybody in his family it seems dies of that Hopkid or Hopskid whatever it is disease so he may as well kill himself with fun shit."

4.

The "yanked Yankee," as the *Post* called him, was booed when he returned to the Stadium on Monday, August 15, for his 1,353rd major league game. Hoots followed him as he ran out a ground ball to shortstop in the first inning against the league-leading Baltimore

Orioles. Stengel had declined to say what other sanctions Mantle might face.

Historically, Mantle visited his rage on inanimate objects—balls, water coolers, cement posts—abusing them when he was most angry with himself. He kicked the dugout posts so often his toes turned black and blue. Sometimes he restrained himself. Noren remembered one night when Mantle granted the beleaguered water cooler a reprieve. "Just as he's about to kick it, the motor went on and Mickey said, 'Okay, you sonofabitch, I won't kick you this time.'"

Jack Reed recalled another night in Kansas City when Mantle made a beeline for his preferred target after popping up. "He looked at the water cooler. I started to kind of shy away from him. And he looked at me and kind of laughed, and I relaxed. And all of a sudden, he kicked it! Knocked the side off it. The next two times at bat, he hit two of the longest home runs I ever saw."

With Maris swathed in bandages from waist to shoulder, the Yankees needed Mantle's furious clout against the Orioles. No doubt Maris's unavailability influenced Stengel's refusal to divulge what punishment he would exact for that dawdle down the first base line. In the fourth inning, with a man on first and the Yankees down by two runs, Mantle beamed his frustrations on a fastball for his twenty-ninth home run. As he touched the plate with the tying run, he was bathed in a sweet cascade of applause. He tipped his cap in grateful appreciation. The gesture is accepted baseball etiquette but one Mantle eschewed, wishing not to inflict further distress on an already wounded pitcher. This time, he acknowledged the fans and their disappointment in him the day before. "Figured I better get on their good side while I could," he told reporters after the game.

In the eighth inning, with the Yankees again trailing, 3–2, and a runner on first base, Mantle faced the old knuckleballer Hoyt Wilhelm. Catcher Clint Courtney had traded his usual glove for one of those puffed-up jobs the size of a fur hand muff flaunted on a night out on the town. With two strikes, Mantle lifted a pop fly near the screen behind home plate. Courtney dropped it, keeping the at-bat alive. "Giving Mantle a second chance under the circumstances was like fondling a fer-de-lance," Harold Rosenthal wrote in the *Herald Tribune.*

Never had he wanted to have a good day so badly, Mantle said later. He swung at the next pitch, hitting his second home run of the night, and belting the Yanks back into first place. "Even his ears were grinning," Dick Young wrote.

"He didn't have to run those two out, did he?" Stengel crowed.

He said he accepted Mantle's explanation for his lapse the day before. *It was a bonehead play.* Indeed. "This did him good, and it'll do the others good," Stengel declared. "It'll show them they got to run it out, or else when they get their paycheck they'll think they have taken a cut."

Making an example of Mantle may have been tactically useful—the Yankees were hardly tearing up the American League. But it was also a frontal attack on Mantle's sense of himself and he took up the challenge. The one thing he always did was try. "Mickey really never, ever stopped doing the best he could," Coleman said. "There might have been lapses. There were lapses from all of us."

5.

Stengel had celebrated his seventieth birthday two weeks earlier. The mileage was evident in his face, his gait, and his internal organs. "Some of them are quite remarkable," he said upon his release from the hospital. "Others are not so good. A lot of museums are bidding for them."

Players joked about his flatulence. "Like a damn goat fart," Mantle said. "When he was alone on the other side of the dugout, he wasn't just thinking."

Sometimes he was asleep. "Don't wake him up, we're ahead," Frank Crosetti, the venerable third base coach, was known to say.

"Casey was sure he ran the team," Duren said. "He made out the lineup cards. But to those regulars on the team, he was pretty much ignored."

They had grown tired of his syntactically effusive charm, an act that besotted the U.S. Senate when Stengel testified at baseball antitrust hearings in 1958. It was a forty-five-minute excursion through appositives and subjunctive clauses that left stenographers exhausted. Mantle was the next witness. "My views are about the same as Casey's," he testified, under oath.

The players knew a less amusing Stengel, one who rarely came out of his office and rarely remembered their names. Veteran players hated his inflexible platoon system, his rigid adherence to playing the percentages with lefty-righty matchups. Clete Boyer was speechless when Stengel pinch-hit for him in his first World Series at-bat—in the top of the second inning. "Everybody hated him," Boyer said. "When he come out of his mother, the doctor slapped her."

"See, Casey, he would do anything to win," Ditmar said. "*Anything* to win. If I had to pitch nine days in a row, I would. If he had to play Mickey with a bad leg, he would play him, because he knew he was the best and he wanted the best in there. I think Casey really had a deep love for the guy."

If so, it was a complicated kind of love, infused with equal measures of affection and bemusement, disenchantment and disillusion. In the beginning, Mantle was the indulged teacher's pet. By 1960, Schecter wrote, the relationship had become one of "mutual respect and mutual disappointment" with "practically no communication."

Mantle often said that Stengel was "almost like another father," a truth that went deeper than perhaps he knew or wanted to acknowledge. Like Mutt, Stengel never praised him to his face. Every day when Mantle emerged from the dugout, Stengel would say, "Gentlemen, there's the ball game," Tom Sturdivant recalled. "He complimented him everywhere but in the clubhouse."

In Schecter's preseason analysis of the inner Mick, he quoted at length "a baseball man here in St. Petersburg who has read more than just record books and has watched Mantle and the Yankees and with great interest for years now." Schecter conceded that many readers would dismiss the jargon of his amateur baseball shrink as the "Freudian prattlings of an unqualified layman." The unnamed analyst speculated that Mantle's propensity for striking out left-handed was an unconscious retort to the constant demands his father had placed on him to go against his nature and "to be better than he could be." Striking out was a way of striking out against the absent, internalized father. "Showing the old man what a mistake it was to force him to bat left-handed," the analyst concluded. "It's a form of masochism. I think he wants the fans to boo him. It's like having his father in the stands again."

Overwrought as that might have been, Mantle was hard put to please either man. Like Mutt, Stengel saw in him the player he would never become. And, like Mutt, Stengel regarded his success as a monument to his own ambition and desire. The relationships with his father and his second father were predicated on demand and expectation: you *will be* somebody. Which is a whole lot different than saying: you *are* somebody.

David Mantle likes to tell a favorite family story about Mutt receiving congratulations on his son's stellar performance for the Whiz Kids one night in 1948. "Mutt said, 'He coulda done better,'" David recalled. This on a night when, local lore has it, Mantle had hit two home runs into the Spring River.

"Just think how great he'd have been if he'd had confidence," said Mantle's teammate, Tom Tresh. "Knees and confidence."

6.

On the morning of September 16, the Yankees awoke in a first-place tie with the upstart Orioles atop the American League. The defending American League champion White Sox were lurking just two games behind. The Yankees won that night and the next fourteen straight games to claim Stengel's tenth pennant—equaling McGraw's record. Mantle led the league in runs and home runs. Maris led the league in slugging percentage and RBI and was named the American League's Most Valuable Player. Mantle was first on more ballots, but three writers left him off completely. He finished second in the voting, two points behind Maris. Now Stengel was lavish with praise, noting how hard Mantle had worked and how much he had hustled—impressive, Stengel said, considering he was a cripple playing on one leg.

The World Series opened in Pittsburgh on October 5, the day the G-man assigned to investigate the July death threat informed his superiors that he had "failed to develop a suspect in the matter."

Yankee pitchers, particularly Whitey Ford, had long chafed at Stengel's idiosyncratic pitching decisions, none more fateful than choosing to start Ditmar instead of Ford in game 1 of the Series. True, Ditmar had

had his best season. Yes, he threw a lot of ground-ball outs. But he was Art Ditmar and Whitey Ford was the Chairman of the Board, the pitcher who would break Babe Ruth's record for consecutive scoreless World Series innings.

"Casey talked to me about it," Ditmar said. "He said, 'Lookit, I'm not going to start Whitey because I want Whitey to start in the Stadium because of the short right field and to nullify left-handed hitters in the Stadium.'

"And he says, 'You've pitched great for me the last two years, so I think you deserve the opportunity.'

"And he says, 'Whitey has had some arm problems. We don't figure he can pitch three games.'"

In game 1, Ditmar failed to make it out of the first inning. Mantle went hitless. Batting left-handed had been a struggle since Red Schoendienst had gotten tangled up with his shoulder in the 1957 World Series. Watching film from his rookie year in a darkened clubhouse one day in 1959, he saw the problem—he wasn't holding his bat high enough. "I can't because of my shoulder," he told the Yankee coaches. "It hurts."

Although there was almost no change in his left-handed power numbers throughout his career, his batting average dropped 19 points after 1956. He battled left-handed 65.9 percent of the time and hit 69.4 percent of his home runs from that side; but he also struck out twice as often left-handed. After being caught looking twice in game 1, he told reporters, "Up there lefty, I just can't pull the trigger." The Pirates won 6–4.

Batting right-handed the next day, he hit a home run that Pirates' shortstop Dick Groat swore tore seven seats out of the right field stands. Then he hit another. It soared just to the left of the iron gate in right center field, fifty feet over center fielder Bill Virdon's head, by his reckoning, and disappeared into the trees, a precinct previously reached only by left-handed batters. A city cop on patrol behind the fence estimated the distance at 478 feet. Virdon respectfully disagreed with the law. After the 16–3 Yankee deluge, in which Mantle accounted for half the Yankees' runs, Virdon found Groat soaking his broken wrist in the whirlpool: "Roomie, you missed the granddaddy of all time. Without a doubt, nobody ever hit a ball further or harder."

"It had to go six hundred feet," Virdon told me, "and that's when balls weren't very live."

The Series relocated to New York, where Toots Shor, the bon vivant saloonkeeper, temporarily without a saloon, erected a beachhead in a tent on the site of his new joint on East 52nd Street. He invited some pals to help break ground on the off day between games 2 and 3—Chief Justice Earl Warren, and Mantle, Berra, and Ford. "Everybody have a booze!" Toots cried. The drinks in the tent were on the house.

The Yankee onslaught—and Mantle's slugging—continued unabated in game 3, a 10–0 shutout. But the Pirates laughed off the two laughers and stunned the lordly Yankees by winning the next two games at the Stadium. In game 4, again Mantle was held hitless left-handed. In game 5, Ditmar improved on his game 1 performance—he lasted 1⅓ innings. In game 6, in Pittsburgh, Ford threw another shutout, setting the stage for what Yankee partisans regarded as a foregone conclusion.

Vern Law, the 1960 Cy Young Award winner, held them scoreless through the first four innings. By the bottom of the second inning, the Yanks were trailing 4–0. Stengel summoned relief pitcher Bobby Shantz to save the season. Shantz stood five feet six and weighed 139 pounds. "Soaking wet," he said. "I didn't throw hard at all. I threw a lot of changeups and curveballs. I changed up on my fastball and changed up on my curve."

He held off the Pirates for five innings while the Yankees scored seven runs. They led 7–4 with two innings left in the season. Then came the fateful eighth. "First guy up, Gino Cimoli, blooped one into right field," Shantz said. "Then Bill Virdon hit a nice double-play ball right to Kubek and that damn thing hit a rock or something, bounced up and hit him in the throat. Oh my God, I couldn't believe it."

Kubek went to the ground, choking on his own blood. Everyone else was safe.

Stengel summoned Jim Coates from the bullpen to face the redoubtable Roberto Clemente, who hit the inning's second fateful grounder. First baseman Moose Skowron fielded the ball. But when he looked for someone to throw it to, no one was there. The lapse still provoked Mantle's drunken ire a quarter of a century later. "Fuckin' Jim Coates didn't cover first!"

Shantz sighed at Mantle's resilient memory—no lapse there. "Hal Smith come up and hit a three-run homer, and then they went ahead of us. That's the thing—we should have been out of the inning. They shouldn't have scored even more than one run. We'd have still been ahead if he covers first base."

Instead, the Pirates scored five runs—three charged to Shantz—and took a 9–7 lead to the top of the ninth inning. Kubek was taken to Pittsburgh's Eye and Ear Hospital, with a suspected fracture of the larynx.

The Yankees did not choke. Mantle wouldn't allow it. He drove in the Yankees' eighth run in the top of the ninth with a single that sent the tying run to third.

There was still just one out. Berra slashed a serrated grounder to Rocky Nelson at first base, who snared it on one hop, stepped on the bag, and straightened to throw to second base to complete the double play. But Mantle wasn't where he was expected to be. He stood his ground on the infield dirt, waiting for Nelson to make his move. They faced each other paces apart, like Hollywood gunslingers. It was high noon for the Yankees. Mantle drew first, diving back into the bag, and eluding Nelson's frantic, perplexed tag.

Without Mantle's reflexively balletic maneuver, Gil McDougald would not have scored the tying run. The Series would have been lost. There would have been no bottom of the ninth. There would have been no goddamned Bill Mazeroski. No kicker in Red Smith's column in the morning *Tribune*: "Mazeroski is up first for Pittsburgh."

Ralph Terry had relieved Coates in the eighth. His second pitch to Mazeroski sailed over Berra's head, over the ivy-draped left field wall, and into civic bedlam. For the second year in a row the Yankees were left looking up.

The clubhouse door was still locked when batboy Frank Prudenti finished his chores and went downstairs to change. The Yankees needed time to compose themselves before receiving the gentlemen of the press.

"It was the quietest locker room I can ever remember," Prudenti said. "Everybody was just shocked, devastated. I looked over by Mickey's locker, and he had a towel over his head. I didn't want to look too hard and make it obvious I was looking. I didn't go over; I didn't wanna see it. But it wasn't only Mickey. A lotta players had tears in their eyes."

Clete Boyer thought: "Shit, maybe I should cry, too."

When the writers were finally admitted, those who dared go to Mantle's locker saw tears running down his cheeks. "This was the first time that we lost a Series when I know we should have won," he said.

As Berra put it, "We made too many wrong mistakes."

Terry was in Stengel's office, trying to apologize and explain. Stengel was taking off his Yankee uniform for what would be the last time, his pants down around his shoes, his shirt unbuttoned. Terry had warmed up five times before entering the game, pitching off a steep bullpen mound that bore little relationship to the one on the field. His footing was off, making the fateful pitch rise instead of breaking low and away as he and the scouting report intended.

"Forget it, kid," the manager said. "Come back and have a good year next year."

Mantle was still at his locker, shrouded in a towel, when Terry emerged from Stengel's office. He had batted .400 (10 for 25), scored 8 runs, driven in 11, hit 3 home runs (for a slugging percentage of .800), and walked 8 times, finishing what was perhaps his best World Series with an astonishing On Base + Slugging percentage of 1.345.

It wasn't his fault that Stengel had let Shantz bat for himself with two on in the eighth inning. Or that, in a sacrifice situation, he chose Coates—known to his teammates as "Rock" for his less-than-sure hands—to replace a pitcher who had won seven gold gloves. Or that he had started Ditmar instead of Ford.

"How's Tony?" Mantle asked.

Kubek was under doctor's orders not to talk—especially not to the enterprising reporters who showed up at his bedside hoping for a quote, but he took Mantle's call.

That evening, as the Mantles flew home to Dallas, John F. Kennedy and Richard M. Nixon wrangled over the defense of the islands of Quemoy and Matsu off the coast of China in the third of the great presidential debates. Somewhere high above America's midsection, Mantle was still crying. "Mickey, it's only a game," Merlyn said.

But it wasn't. In a week he would be twenty-nine,

and he felt he had let them all down—Mutt, Casey, himself.

The Stengel era ended five days later, just past high noon, when he arrived at Le Salon Bleu of New York's Savoy Hilton Hotel for a formal execution orchestrated by the front office. The Yankee bosses declined to use the distasteful word "fired" but pointed out that Stengel was the highest-paid manager in baseball. "Quit, fired, write whatever you want," Stengel told his writers. "I'll never make the mistake of being seventy again."

Thirteen years later, Stengel admitted to an interviewer that he was wrong to start Art Ditmar. Ditmar sued Anheuser-Busch for libel when a 1985 World Series commercial recycled Chuck Thomson's erroneous call of the bottom of the ninth, attributing Terry's gopher ball to him. The suit alleged that Ditmar had been held up to "undeserved ridicule, humiliation and contempt" and went all the way to the United States Supreme Court before being dismissed in 1988 as ridiculous.

PART THREE

Nightcap

ATLANTIC CITY, APRIL 1983

By 11 P.M., the candle at Mickey's table in the Bamboo Lounge was guttering, and so was he. It was the end of a long, liquid day: cocktails in the limo, followed by cocktails before dinner, followed by toasts with dinner—"To the greatest center fielder of all time, behind Joe DiMaggio," one of his playing partners had said.

The affair was hosted by the sportscaster Dick Schaap. The menu was elegant—"Boston Bibb lettuce with Hearts of Palm, Coquille St. Jacques, Filet Mignon Béarnaise, Coupe Baccarat"—none of which Mickey ate. Now his fist was wrapped around a snifter of Grand Marnier. I ordered a Perrier. Mickey cackled. Some sportswriter.

Bill Brubaker of the New York Daily News *was keeping him company in the second-floor lounge, a cozy little watering hole done up in ersatz Raj decor with fan-shaped rattan chairs and fake potted palms. I joined the conversation with trepidation and annoyance. Mickey was "my account," sports department shorthand for proprietary ownership of a story. He was mine. Except that everything in my notebook was also in everyone else's. Everyone else had had a "one-on-one." Mickey had done a series of stand-ups with local TV guys and a long sit-down with Bob Lipsyte for CBS Sunday Morning. I was bringing up the reportorial rear.*

I had scribbled the questions I wanted—and dreaded—to ask on the inside cover of a reporter's pad and consulted the list reflexively throughout the day, a nervous tic that worsened as the prospects for my scheduled interview dimmed. But there was no forgetting what I needed to ask him. Over the phone, he had told me that his son Billy had been diagnosed with non-Hodgkin's lymphoma, Mutt's disease. The treatments were expensive, and he needed the house money to pay the medical bills. He knew that Bowie Kuhn would banish him for taking the job. He didn't know that the commissioner acted with full knowledge of the reason he signed on with the casino—a decision Kuhn later told me he neither regretted nor second-guessed.

This was a good story, a scoop even—"Aging Slugger Banned, Needs $ for Dying Son." But now that it was finally my turn to get a shot at The Mick, he was tanked up and talked out. The drunker he got, the closer his chair got to mine, until his famously gelatinous right knee was touching mine.

I wanted Brubaker to get lost before Mickey became thoroughly insensible almost as much as I wanted him to stay. The fact is, I was scared to be alone with my hero. When, a round of drinks later, Bill closed his notebook, I pleaded softly, "Don't go."

On the table before Mickey was a stack of oversized baseball cards commissioned by the hotel. On the front was a painting of the slugger in his pinstriped youth. His face was rendered in a deep metallic hue that made him look as if he'd already been bronzed. In the foreground were three action-figure Micks, only one of which resembled him at all. "Get rid of them three li'l sumbitches and give me a place to sign muh name," he told the new PR guy when the print run ran out.

On the back there was a brief bio and his career stats, major and minor league. "You know what they left offa here?"

I could guess. But it was a different social disease from the one he had bragged on that morning. "'Bout leadin' the league in the clap six straight years. A major league record. It'll never be broken."

Four Alabama boys stopped by the table for an autograph. The biggest and blondest of them looked a bit like The Mick of memory. "I'm a thirty-year-old man," he said, holding out a pen. "Would you mind? It's the price of fame."

Mickey shrugged. Like this yahoo knew the price of fame. "I get paid to do this."

"I'm a lot nicer than I was," he confided when the southern boys departed. "I care more about what people think than I used to. When I was playing ball, I didn't give a damn. Now I try to make people like me. Maybe more 'n I should."

"Why?"

"'Cause in public relations it matters more. It didn't matter before. When I'm around a golf course or out with my wife, I never jump on anybody's ass."

Mickey paused. "Y'know, even though it's my job to be nice, it isn't hard." He sounded almost surprised. "I run into guys who say I pushed them aside. Guys who say, 'Hey, I was at Yankee Stadium in 1958. You shoved me out of your way.'

"Hell, if I strike out four straight times, I'm not going to sign any damn autographs. I don't feel bad. I didn't shove him out of the way because I hated him. I wanted to get to the ballpark. It was mostly the same kids every day, been there for twenty years. If I saw a

little kid with crutches, I did stop. If I thought it was the same one who got one yesterday, I wouldn't."

He turned to Brubaker, who was gathering his belongings. "You wanna autograph?"

"No, but my wife would probably like one."

"What's your wife's name?"

"Freddi. F-r-e-d-d-i."

"To Freddie," he wrote. "Mickey Mantle."

"Thanks, but you misspelled my wife's name."

Mickey ripped the offending card in half and took another from the stack. "Well, how do you spell it?"

"F-r-e-d-d-i."

Again, Mickey misspelled her name; again, Bill corrected him. "Here," Mickey said, submitting a new inscription for Brubaker's approval. "To Freddi, you have a hard name to spell, asshole."

"Mickey, I can't give this to my wife."

"Gimme that."

He snatched the card, crossed out "asshole," and scribbled "sweetheart" above the offending appellation. "To Freddi, you have a hard name to spell, asshole sweetheart."

When Bill stood up to say good night, I thought about going with him. "Stay," Mickey said, reaching for my arm.

It was two in the morning. Finally, I had The Mick

to myself. He suggested we'd be more comfortable in one of the modular love seats on the other side of the room. I put my notebook on the glass table and set my tape recorder to spinning. My intentions were clear. So were his.

"C'mon," he said, brushing the notebook aside. "Lesss do thisss at brefffasss."

Just as I was about to ask about his son Billy, I felt his hand on my knee, then on the inside of my thigh. A knee is open to interpretation; a thigh means business.

His hand was thick, sure, and entitled, casually asserting its prerogative the way it would over a coffee mug. And that hand was moving inexorably upward when Mickey listed to his left and passed out dead drunk in my lap.

I closed my eyes; the room was spinning like lemons on a one-armed bandit, but probably that was just my weight shifting beneath the heft of number 7. I was pinned in a modular love seat beneath two hundred pounds of Grade A American Hero.

In Casino World it is always night but never dark. Just below, in the Hi-Ho Casino, bells clanged and sirens blared—someone was getting lucky. Briefly, I considered the man facedown in my lap, the thinning hair no longer bleached by the outfield sun. A road map of tiny broken blood vessels spread down the slope

of his nose and over the ridge of his cheek, small breaks
etched by age and hard living.

I considered my options. Moving wasn't one of
them. That would have required a crane. He was dead-
weight, and he was out cold. I lacked the fortitude to
dump Mickey Mantle on the smoke-stale carpet amid
shards of half-eaten pretzels and twisted swizzle sticks.
I told myself to look on the bright side: at least he'd
passed out before he hit the jackpot.

So I waited, assuming someone would notice. Evidence of life throbbed through the floorboards and
walls. But the lounge was empty. The TV above the
bar was on mute.

The cocktail waitress—a weary AARP candidate
with a lacquered beehive that paid homage to the hotel's tumescent tower—had last been seen disappearing
through a swinging door off the kitchen. When she finally reappeared, she was intent on inflating a gigantic
orb of bubble gum. Then she peered at the odd configuration on the love seat. "Oh, fuck, not again," she
said, popping a humongous bubble.

Sighing, she pulled Mickey off me and got him upright. Together we helped him to the elevator. Mickey
steadied himself against the gilt-edged mirror surrounding the elevator bank. Forehead and lips pressed
to the glass, he was out on his feet.

"He's dead," I thought, a surreal and fleeting moment of insight interrupted by his labored breathing.

"Okay, so he isn't dead." I just wished he was.

Each time he exhaled, the mirror fogged over, his features dissolving in a warm breathy mist. The whole day had been like that: moments of intense clarity followed by hours of reflexive obfuscation.

When the elevator door finally opened, I pushed Mickey into the car. Bracing himself against the polished brass rail, he flashed that famous, toothsome smile.

"You comin' upstairs with me tonight, Jane?"

"Not tonight, Mick."

He seemed neither surprised nor disappointed, as if the rebuff was as expected as the offer. "Oh, well," he said, cheerfully, "y' know what they call me, dontcha?"

"No, Mick, I don't."

"Well," he drawled, "they call me Mighty Mouse. 'Cause I'm hung like him."

He was still vertical when the door slid shut on his grin.

I went up to my room and cried.

Outside the window the moon rolled in on the tide. Shafts of indirect light filtered through panes of salty glass and splayed across the carpet where I'd kicked off my clothes—Loehmann's bargain suede and a 100 per-cent Virgin Orlon Acrylic sweater.

What I had seen of Mickey wasn't what I had ex-pected or hoped. But he had promised to meet me for breakfast—again. I wondered which guy would show up—if he showed up. One thing I knew for sure: to see Mickey clearly I needed to see him in direct light.

12

September 25, 1961

DR. FEELGOOD*

1.

In late September 1961 Mantle was feeling poorly. Sportswriters following the team followed his condition closely, diagnosing a cold, a head cold, a heavy cold, a virus, an eye infection, and an upper respiratory infection that lingered through a long home stand, Cleveland, Chicago, and Detroit. When he was still under the weather on September 24, Mel Allen offered help: "I have a doctor. He'll give you a shot that'll fix you right up."

The voice of the Yankees made an appointment for him to see Dr. Max Jacobson on Monday, September 25, an off day for the American League champs.

All summer Mantle and Maris had chased each other and Babe Ruth's unassailable home-run record across America. "Sixty, count 'em, sixty!" The Babe had crowed that September afternoon in 1927. "I'd like to see some other sonofabitch do that!"

For much of the season, it appeared that both of the M & M boys might just do that. The pursuit played out against the backdrop of new administrations in Washington and in the Bronx. Ike and the Ol' Perfessor had ceded center stage to JFK and Ralph Houk, the new Yankee skipper. The country was charged with energy and change. Mattel had given Barbie a boyfriend named Ken. Yuri Gagarin and Alan Shepard had defied the gravitational pull of the Earth. Audrey Hepburn's soignée gamine Holly Golightly had *Breakfast at Tiffany's*. A struggling young comic named Vaugh Meader was perfecting his long Boston A's for an impersonation of JFK and the First Family.

Vigor—make that *viggah*—was the watchword of the day. Youth would be served by youth. No one suspected how much Dr. Max Jacobson contributed to the vitality of the young American president. Nor did anyone suspect how much Mantle was flagging until he confided in Mel Allen on the flight back from Boston on Sunday, September 24. The day before he had hit his fifty-fourth home run, his first since September 10.

Like President Kennedy, Mantle had a secret that required discreet medical intervention. When he arrived at Jacobson's Upper East Side Manhattan office, Dr. Max told Mantle to pull down his pants and filled a syringe with what Mantle later described as a smoky liquid. He squirted some into the air and plunged the needle deep into Mantle's hip. Too high, Mantle said later. It hit bone and raised the question: how many demons can dance on the head of a hypodermic needle?

2.

Dr. Max Jacobson already had a cabinet full of files on famous patients, many with secret and special needs. He had flown to Europe with Kennedy aboard Air Force One for the Vienna summit with Soviet Premier Nikita Khrushchev in June 1961. Although his name was carefully omitted from the official presidential traveling party, he accompanied Kennedy to the residence of the American ambassador, treating the President forty-five minutes before Khrushchev was scheduled to arrive. The chairman, however, was running late. So Jacobson sat on a windowsill in a vestibule outside the music room where the leaders of the world's two superpowers met in case the President needed a pick-me-up.

Jacobson first treated JFK in the fall of 1960, when Kennedy's back acted up under the rigors of campaigning. He continued to treat the President at the White House, in Palm Beach, and at the Kennedy compound in Hyannis Port until three weeks before his assassination. In the summer of 1962, Attorney General Robert F. Kennedy attempted to have Jacobson's magic potion analyzed at a government laboratory, but the test sample was too small to yield results. "You don't know what's in that," Bobby Kennedy told his brother. Replied the President, "I don't care if it's horse piss. It works."

In New York, Jacobson was known as Dr. Feelgood to the jet-setters, celebrities, and pols who visited his office day and night for injections of amphetamines laced with vitamins, human placenta, and eel cells. Among them: Eddie Fisher and Johnny Mathis, Cecil B. DeMille and Otto Preminger, Anthony Quinn, Emilio Pucci, Tennessee Williams, and Truman Capote. In time, he would treat JFK's girlfriend Judith Exner as well as Jacqueline Kennedy, whose depression and headaches following the birth of John F. Kennedy, Jr., in November 1960 concerned the President enough to summon Dr. Max to Palm Beach in May 1961.

Mel Allen, whose livelihood depended on his baronial vocalization, trumpeted Jacobson's virtues, raving to friends about the medications "Miracle Max" pre-scribed. "Man, what he can do!" Allen exclaimed. "Those pills, they work."

On the first day of spring training—the last time the Yankees trained in St. Petersburg—Ralph Houk called Mantle into his office to give him new marching orders. Houk was known as "The Major" because of his valor under fire at the Battle of the Bulge in World War II. "Have you seen my helmet," he asked when I visited his retirement home in Florida, and led the way to a back room where the helmet that saved his life sat on a bookshelf. He fingered the hole where the bullet went in and the exit hole just above the brow. "You see, we had a liner in there," he said. "Otherwise we wouldn't be talking about Mickey Mantle."

Houk didn't believe in team captains but he believed in Mantle. "I told him, 'You should be the leader of our club because everybody respects you and you don't like to lose,'" Houk said. "He didn't think he could do it. I just remember him saying, 'Ralph, if that's what you want me to be, I guess I gotta be it.'"

"I said, 'You just do it in your own way. But I want you to know that I'm gonna tell the press.'"

Mantle figured Houk had made the same speech to Berra and Ford; he hadn't. Houk understood the value of Mantle's understated example, how his tolerance for pain lent perspective to an everyday charley horse. "I knew he was what we needed," he said. "Mickey, if he struck out three times and the team won, he was a happy guy in the clubhouse. But he could have a great day and nobody'd know it."

It may have been the first time in Mantle's adult life that he was charged with responsibility rather than absolved of it. "It changed me over in my thinking," he told Howard Cosell in 1965. "When Ralph came here, I didn't feel at ease playing the game. People were still booing me, and I didn't know how to take it. Ralph come over, and he says, 'This is our leader,' which I really wasn't. He kept saying all this—I was the leader and I got a lot of guts because I play on bad legs. I started thinking, 'Maybe I am a little better than I thought I was.'"

A baseball lifer, bullpen catcher, and perennial third-stringer—Houk had the affection of his players, many of whom he had managed in the minor leagues. He also had the support of the front office, where he was regarded as a company man. He did not consciously define himself in opposition to Stengel, but he did do some things differently. Gone were the clubhouse sandwiches, replaced by a post-game buffet of hard-boiled eggs, cottage cheese, and soup. Gone were the erratic starting assignments. Ford would pitch every fourth day and win twenty-five games. Gone was Stengel's hated platoon system. Houk stuck with a set lineup. Elston Howard was installed as the regular catcher; Berra established a new residence in left field.

But the biggest difference between the regimes was perhaps also the subtlest: Houk's invitation for Mantle to rethink himself. "I was close to Mickey, him being an old country boy like he was and me being one too," Houk said. "Mickey liked me. He knew that I fought for the players even as a coach before I became manager. I knew he would do anything I asked him to do."

Unlike Houk, Mantle was a follower, not a leader—everyone said so, including himself. In coaxing him to take on a new role, Houk appealed to the better angels of Mantle's nature, his conscientiousness, and sense of responsibility. "He was so modest I had to let him know what he meant to the ball club and what he meant to me," Houk said.

He got Mantle the salary he wanted, got him to report to spring training three days early, and got—ordered—Moose Skowron to go out on the town one night in St. Pete with Mantle and Ford. Mantle took it personally when teammates turned down an invitation. "They rented a car, they give me a chauffeur's cap, and we go to these bars in St. Pete–Tampa," Skowron said. "I'd meet the maître d', I'd say, 'I got Mickey Mantle and Whitey Ford,' and they didn't believe me 'til they come outside and open up the door. They'd go inside, freeload food and booze, and then the next day, my name's in the lineup. I said, 'Ralph, I can't play.' And Mickey and Whitey were along the fence, laughing like hell when I was taking infield."

Skowron sighed. "I guess Ralph just wanted to make him happy."

"Ralph is the best thing that ever happened to me in my life," Mantle later said.

3.

When the season opened, Mantle was living alone in a suite at the St. Moritz Hotel on Central Park South. Merlyn had stayed home in Dallas with the boys. Spring rains on the East Coast had played havoc with the first week of the season, scrambling schedules and pitching rotations. More wet weather was predicted.

When the April 19 game against the Angels was postponed, Mantle invited his former teammate Eli Grba up to the suite for a party. Grba, who'd gone to the expansion franchise in the off-season draft, gladly accepted. As soon as he arrived, Grba said, "Mickey took off. So here we are. Two guys and four of the most beautiful girls I've ever seen in my life. As he's leavin', he says, 'Take your pick.' And they're not prostitutes. Lord have mercy!

"It wasn't like he was pimpin'. They were friends. He went someplace with Bobby Layne. Left us there. 'Here, party!'"

Grba got home at 4 A.M. and awoke six hours later to find a cloudless sky and unwelcome news in the morning paper: "Doubleheader: Yankees vs. Angels, Game 1, starting pitchers, Ditmar and Grba. "Oh, shit," he said.

When he got to the ballpark, a teammate pulled him aside and said, "What did ya do to Mickey last night? He doesn't feel too good. Sonofabitch isn't gonna play."

"But, see, I know better," Grba said. "Because when Mickey wasn't as strong, he hit the ball further. He didn't swing as hard. I get out there, and I'm pitching a pretty good ball game. First inning. I get a man on base. He comes up. The sonofabitch hits a home run. I hung a slider. What the hell—he's Mickey Mantle."

In the bottom of the fifth inning, with two on and two out, the score 2–2 and Mantle due up next, Angels' manager Bill Rigney walked to the mound for a conversation. "Do you wanna pitch around him?"

Grba considered his options. His control was never that refined. He decided to waste a pitch inside anyway. But which pitch to waste? "I have a hard time hitting your fastball," Mantle had told him the night before. "Your ball is like a metal ball."

He's setting me up, Grba reasoned. *He's gonna look for the fastball. And he's gonna hit that sonofabitch nine miles.*

So he threw Mantle a hard slider inside. "Well, that's what he was lookin' for all the time," he said. "He hit that sonofabitch nine miles."

As Mantle circled the bases, Grba circled the mound, calling him every name he could think of, beginning with Okie.

Mantle drove in five of the Yankees' seven runs and was personally responsible for five of their first seven wins of the season. When he hit his fourth homer of the year the next day, the *New York Times* took the measure of his auspicious start: " . . . he's eight games ahead of the pace set by Babe Ruth when he hit sixty homers in 1927 . . ."

The drumbeat of historical imperative sounded often and early. By the end of the second week of the season, he had seven home runs; had driven in the winning runs on April 17, 21, and 26 and had saved two games with his glove. Cartoonists puffed out his chest. West Point cadets saluted him, awarding him an adjutant's pin with three stars.

Mantle Finally Meeting Destiny.
Mickey Taking Charge of Yanks with Stengel Gone
—the *Boston Globe*

On May 2, he hit a tenth-inning grand slam to beat Camilo Pascual in Minnesota. "Never felt better in my life," he said. Two days later, he hit his ninth home run and embarked on a 16-game hitting streak. "Just like 1956," he observed.

Maris was batting .200.

4.

All baseball players lead bifurcated lives—home and away, season and off-season. The sport's immutable schedule causes stress fractures in even the strongest marriages. Tom Tresh, who made his major league debut in 1961, once calculated that he saw his family perhaps six weeks between February and September. "I was an absentee father," said former Detroit Tiger Denny McLain. "Mickey was an absentee father. That's what we did for a goddamned living."

Mrs. Mickey Mantle, envied for the presumed benefits of being married to a baseball demigod, was often miserable. Her life was equal parts glamour and loneliness, comfort and emotional deprivation.

The Yankee wives, as they are collectively known, mostly to team broadcasters, were an entity in name only. Yes, there were bus rides through the city streets at World Series time, when New York's finest escorted them from one borough to the next, sirens wailing, half of them pregnant and counting their blessings that they didn't deliver on the Brooklyn Bridge. Yes, there were picnics and barbecues and birthday parties, but the quotidian life was one of isolation. "We did things as wives together very seldom because we all had children," said Lucille McDougald. "We would get together maybe once a season when they were on a long road trip. But there was not a whole lot of socializing going on."

Like many former tenants of the Concourse Plaza Hotel, the Mantles moved across the Hudson River to the New Jersey suburbs when they had the means to do so and the families to house. Merlyn hated the rentals, especially the one with a painting of a couple in flagrante delicto in the master bedroom, and cat poop behind the furniture. "My sons got boils," she told me. "They were sick all summer."

By 1961, she had four boys under the age of ten—not including her husband—and she was overmatched. "When we all moved over to New Jersey, Merlyn just stayed with the kids," Lucille McDougald said. "She rarely came to the ballpark. She more or less faded into the background."

The Mantles, Berras, and McDougalds occasionally shared a babysitter named Martha Helen Kostyra, a young grammar school student who was embarking on a career of entrepreneurial domesticity by organizing birthday parties for neighborhood kids. "They behaved for Martha," declared the empress of style herself, Martha Stewart. (Yogi and Carmen didn't remember her.)

Mantle didn't behave and Merlyn absorbed the worst of it. Back in Oklahoma, before the move to Dallas, she had the help and support of her sister and her parents. They could do only so much. One particularly liquid night in the winter of 1954, after Mantle had been publicly upbraided by Stengel, and challenged to grow up, he came home to find the door locked and the house empty; he cut his hand on the glass while breaking in. Merlyn had left, taking Mickey Jr. and all her belongings, Mantle recounted in *The Mick*. He was still drunk when he showed up at her parents' house, hand stitched and bandaged, demanding the keys to the car. Merlyn's father told his son-in-law he was too drunk to drive. Mantle threw Merlyn's clothes and the family groceries on the lawn and got behind the wheel. He took out a telephone pole, ripped the car door off its hinges, and landed in a ditch. She and Mickey, Jr., returned home the next day.

"Merlyn took a terrible mental beating that winter," Mantle confessed in *The Mick*.

The following year on Christmas Eve, when she was nine months pregnant with their second child, she came home after opening gifts with her parents to find a crew of drunken revelers on her front lawn. One of Mantle's pals had made off with his house keys. He blamed her. "He grabbed me by the arm and pushed me aside," she wrote in *A Hero All His Life*. "If I had fallen, I probably would have had the baby then and there."

Lovell gave him what for when the twins told her what had happened.David Mantle was born two days later.

His father was away on a hunting trip with Billy Martin in 1958 when the Mantles' third son was born. Billy Giles was named for his daddy's hunting partner and for his maternal grandfather, who had driven his daughter to the hospital.

Sometimes she tried keeping him company in the fast lane, but that had consequences, too. Like the time they both had too much to drink and she almost ran him over, on purpose; and the dinner a couple of weeks later with the Berras, when Yogi told her not to let him drive. She cracked her head against the windshield at seventy miles per hour. The investigating officer was their next-door neighbor. No charges were filed.

By the summer of 1961, he was leading the life of

a leading-man bachelor. He knew women in every American League city or where to meet them. In Baltimore, ballplayers favored a joint called the Club Troc, short for Trocadero, where a dancer known as Fern "The Flower Woman" reigned. Baltimore native Frank Deford, a Princeton student who would become one of sportswriting's most literate and graceful writers, recalled Fern's determination to show The Mick a good time one night when the game was rained out. "She says, 'I go home with Mantle. He gives me a hundred dollars.' That was a lot of money in those days.

"She probably was a ten-dollar hooker. She says, 'Man, I gave him everything for that. I'm up with him all night. And I know he'd been drinking. So I take the one hundred dollars and bet it against the Yankees. I figure anybody spends the night with me ain't gonna play baseball the next day.' "

The way she told it, his game-winning home run the next day cost her a bundle.

Stengel fretted about Mantle's sexual profligacy, once telling writer Gerald Astor: "You can't tell me he ain't getting some of them all the time. He's got enough ailments, so's he don't need to get the clap, too. His taste with broads isn't great, except for that one he's married to and hasn't been together with for a million years, so you can see what I mean about his taste."

Some Yankees were direct beneficiaries of Mantle's little black book. Others derived pleasure from basking in the aura of his prowess. "Everybody got their rocks off on Mantle," said *New York Times* columnist George Vecsey, who covered his first Yankee game that year. "One player told me once he couldn't get anywhere with some blonde. Mantle said, 'Let me give it a shot.' They hear gasps of pleasure. They were this far away. It was as if they all participated through Mantle."

His reputation was hardly a guarded secret. Among the Yankee wives there was empathy for Merlyn. Lucille McDougald wondered: "Oh, my God, how can Merlyn stand it? I think that's why she stopped coming up after a while."

She had become as adept as her husband at hiding pain. "I would know stuff but turn the other cheek," she told me. "It can be very devastating to the wife. My pride hurt me too."

Left alone to his vices and devices, Mantle was a danger to himself. "We all told him, 'Mickey, go home, don't stop at another bar,'" Carmen Berra said. "He hated that—that he didn't go home. He loathed himself."

Finally, she and Yogi lured Mantle and a handful of other Yankees to their house in Montclair for a homestand retreat. "Yogi said, 'We'll make Mickey come out and we'll keep him out there,' " recalled Joe DeMaestri, one of the guests. "We were trying to sober him up, so to speak. I guess it was '61, because Houk was involved with this, too. He said, 'Yeah, good idea, take him out there.' We were going to stay for like four or five days. Mickey was there for, I think, two days, and he says, 'I gotta get out of here. I gotta go back to town.' "

5.

One day, Big Julie Isaacson got a call from a friend who owned a bar down the street from the St. Moritz. "Mickey was in the bar the night before," Isaacson said. "Two girls tried to roll him. Mickey always walked around with a lot of money. My friend took the money. Got him back to the St. Moritz. Called me. Twelve or thirteen hundred bucks. Gave it to me. Gave it to Mickey. Never knew it was missing. Roger said, 'Jules, we got to take Mickey out of the St. Moritz.' "

Isaacson had rented a two-bedroom apartment in Queens on the Van Wyck Expressway for Maris and his roommate Bob Cerv. The building was popular with stewardesses who flew in and out of nearby Idlewild Airport. "We go to the St. Moritz and talk to Mickey. Roger says, 'Mickey, you got your own bedroom, fur-nished apartment.' Mickey objected a little at first. 'I'm not going to go out to God's world.' Queens is God's world to Mickey."

In Cerv's recollection, it was Mantle's idea to come to Queens, and he wasn't sure it was a good one. "I was skeptical about it," Cerv said. "I didn't know if I wanted him or not, 'cause I knew what he did. Roger and I talked it over and he said, 'Oh, hell, let him come.'

"So I said okay. But I said, 'These are the rules. If you break them, you're outta here—no partying, no girls.' We talked to him pretty heavy. He said, 'I'd like to have a summer like that.'"

Mantle and Maris appeared so opposite that people assumed they couldn't possibly be friends. But they had much in common. Both were miner's sons. Both were three-sport, high school athletes (both halfbacks recruited by the University of Oklahoma). Both married their high school sweethearts, who stayed home to raise their families in the shadow of the big time. Together they had the best summer of their baseball lives.

For fun, they Frisbee'd Big Julie's collection of Yiddish LPs across the Van Wyck from the balcony of their apartment. Over breakfast, they read Dick Young's account of Mantle/Maris acrimony in the *Daily News* and then read it aloud in the clubhouse again for everyone's benefit. "We all laughed that summer—'Mantle/Maris feuding'—and we were living together!" Cerv said.

Isaacson recalled one Sunday-morning SOS from Whitey Ford. The M & M boys had banged up their car and were sleeping it off at a gas station. He went to ransom them and pay off the gas station attendant. "Gave him a hundred dollars," Isaacson said. "I told him, 'Anybody finds this car, you're a dead pork chop.'

"It was Roger's car, license KC-9. Pushed the car into the garage. I took Mickey, Roger in my car. We went to the ballpark. Captain Kelly, police captain, is in charge, a good guy. I said, 'Captain, can I see you a minute?' He looks inside. 'Captain, you got plenty of cops. You put Roger between two of them. I'll take Mickey.'"

They deposited the heart of the order in the clubhouse in care of trainer Gus Mauch. It was a Sunday to remember, but Cerv doesn't.

6.

Maris had one home run at the end of April, twelve at the end of May, and twenty-seven at the end of June. Joe Trimble of the *Daily News* was the first to ask Maris if he thought he could break The Babe's record. "How should I know?" Maris replied with impolitic honesty. Between May 17 and June 22, he hit twenty-four home runs in thirty-eight games, an astonishing accomplishment that went largely unremarked in Mickey thrall. When Maris leaped ahead by six home runs on June 20, *The Sporting News* declared, "Time to sit up and take notice."

A week later they appeared together on the cover with Ruth and Gehrig:

> Dial Double M for Murder and Mayhem
> —*The Sporting News*, June 28, 1961

At the All-Star break, Maris had thirty-three home runs; Mantle had twenty-nine; and baseball commissioner Ford Frick had a problem. Before he was named commissioner in 1951, before he became the PR man and then president of the National League, Frick was known primarily as The Babe's Ghost. He was "quite aroused," as Dan Daniel put it, at the assault on The Babe by a pretender feasting on diluted American League pitching.

On July 17, Frick issued his convoluted and controversial "asterisk" decision, which never actually mentioned the word. What it boiled down to was this: Mantle and Maris had to break Ruth's record in 154 games—the number he played in 1927—or would have some "distinctive mark in the record books" attached to the accomplishment. *The Sporting News* later included Frick's edict among the thirty most shameful acts in baseball history.

As the summer progressed and the home runs mounted, and the Yankees advanced on their 154th game, the pressure doubled and redoubled. Maris lost gobs of hair; the circles under his eyes appeared etched in charcoal. Mantle basked; he became the beloved.

Mantle Thrilled by Fans' Cheers—AP

"It's a new feeling and it's nice," he told reporters. "Those fans, they've changed."

The Yankees had to revive the postgame suicide squad, this time to safeguard him from adulation. It was an honor to be selected, to jog beside him, elbows high, scanning the crowd for potential, if unintended, danger. They formed up in the aisles beside the home dugout in the top of the ninth and greeted Mantle where the infield dirt met the outfield grass. Tony Morante was a second-generation Yankee usher and a frequent member of the squad. "Some of the people were nice, wanted to shake his hand," Morante said. "I didn't try to knock people out of the way until people came close. I did have to push a few out of the way to give him enough room to run. We were out there in a flash and back in a flash."

Later Morante realized he had adopted Mantle's loping, limping gait.

Maris was presumed to be a sacrificial rabbit in the home run race, his role to encourage and then defer to The Mick. But he declined to play the part. He hit four home runs in a July 25 doubleheader, the last of which, his fortieth of the season, surpassed his career best in 1960. Mantle hit his thirty-eighth the same day and regained his rightful place as home-run leader with three on August 6.

> Blasé Broadway Buzzing Over Maris, Mantle HRs
> —*The Sporting News*, August 23, 1957

> Broadway Busting Buttons Over Bomber Thrill Show
> —*The Sporting News*, September 13, 1957

Fans from Florida sent them a thirty-foot-long telegram. Their agent Frank Scott predicted $500,000 in endorsement income for the winner of the grand sweepstakes and finalized clothing deals worth $14,000 each. A photo-op 44 jersey prepared by the PR department on August 13 was old news by the end of the day when they both had 45 home runs, and both were sixteen games ahead of Ruth's pace.

Newspapers across the country began running daily

tout sheets; *Newsday* printed "Race with Ruth" numbers below photos of the three protagonists. The pictures of Ruth and Maris looked like mug shots. Mantle looked like a choir boy.

When Maris didn't fade, the righteous guardians of baseball history demanded to be heard. *Who does he think he is? What right does he have?* "It would be a shame if Ruth's record got broken by a .270 hitter," Rogers Hornsby intoned.

The New York Times asked an IBM 1401 computer named Casey to predict if either one could match The Babe. Casey gave Hornsby little reason to cheer. The verdict: Maris, yes, Mantle, no.

Maris never trailed Mantle after August 15.

The Detroit Tigers arrived in New York for a decisive three-game series over Labor Day weekend. The race had been close most of the year. The Yankees led by 3½ games when the Tigers arrived on Friday, September 1—The Yankees won 1–0, with a run in the ninth. On Saturday, Maris hit his fifty-second and fifty-third home runs after scoring the tying run on Mantle's sacrifice bunt. On Sunday, Mantle hit his forty-ninth and fiftieth home runs. On Monday, Labor Day, they rested. Both played; neither homered. But the Yankees swept a doubleheader from the Senators. Four days later the Yankees led the Tigers by 10 games—and their lead would never be less. Frank Scott announced that Mantle and Maris would appear on *The Perry Como Show* during the World Series. Their fee was $7,000 each.

7.

On Labor Day, Mantle heard the siren call of the Great White Way and returned to the city. All summer he had played by the house rules, except once, Cerv said, "when he brought a gal up there. We ran 'em both out, and pretty soon, about a couple of weeks later, that's when he left. Labor Day, he said he had enough of this life. Went back to Times Square."

On September 9, Maris hit his fifty-sixth home run; he wouldn't hit another for a week. The next day, Mantle hit his fifty-third home run; he wouldn't hit another for almost two weeks. On September 15 they both went 1 for 9 in a doubleheader in Detroit. "Maris is so tight up there at the plate that he can hardly breathe," umpire Hank Soar said.

Maris sequestered himself in the trainer's room with his brother for forty minutes, precipitating a rhubarb with the gentlemen of the press who had been stood up at his locker. When he finally emerged, Mantle whispered damage control advice. Reporters who remembered him as a shy, sullen rookie were stunned. He waved the white flag before the Yankees left Detroit. "I can't make it, not even in 162 games," he said. He had already sent a telegram of concession to Mrs. Babe Ruth, leaving Maris to soldier on alone, stalking a record no one wanted him to break.

Hurricane Esther headed for the East Coast and the Yankees headed for Baltimore, followed by a throng of fifty reporters and unkind headlines: "Maris Begins the Big Sulk." It was make or break in The Babe's hometown.

Mantle wasn't in the starting lineup on September 19—he pinch-hit in the ninth inning of the first game of a doubleheader and struck out. He didn't leave the bench in game two, the 153rd game of the season. Maris went 1 for 9 again. He had fifty-eight home runs and one more day to upstage the Babe.

On the morning of September 20, before the Yankees' 154th game of the year, Maris went to Johns Hopkins University Hospital to visit the son of a former teammate who was dying of cancer, stiffing Milton Gross of the *Post* on a scheduled interview. He refused to explain his absence even to Big Julie, who had arranged the appointment. Gross ripped him in the next day's paper. The boy died two days later.

That night, Houk was forced to realign his outfield

when Mantle declared himself unfit to play. "Eye infection," Houk explained.

Maris hit his fifty-ninth home run and barely missed a sixtieth. The Yankees clinched their eleventh pennant in thirteen years, celebrating in the usual manner by pouring bad champagne all over each other. Maris exhaled. Chest heaving with emotion and exhaustion, he told Leonard Schecter of the *Post*: "I tried. I really tried."

The next morning, Mantle stayed in bed. His "recovery from the sniffles" had been "complicated by a penicillin rash," the *Tribune* reported. "Nothing bad," Houk assured the reportorial scrum.

Mantle was with the team when they left for Boston but was said to be "peaked" and unlikely to play. In the first inning Saturday afternoon, he hit his fifty-fourth home run, the most he would ever hit in a season. He was replaced by Tom Tresh after singling in the seventh inning. On Sunday, he went 0 for 3 and left the game in the bottom of the sixth. On the plane back from Boston, Mel Allen mentioned he knew a doctor named Max Jacobson.

8.

Most patients left Jacobson's New York office feeling energized. Billy Crystal's grandmother would come home after her "vitamin" treatments and "make nine pot roasts in an hour," her grandson said. No wonder. Jacobson was injecting patients with up to 30 to 50 milligrams of amphetamines—speed—a highly addictive stimulant that made them feel as if they could run forever, sing forever, or cook forever. But Mantle left the doctor's office in excruciating pain. The needle felt like a red-hot poker, he wrote in *The Mick*. Jacobson advised him to play hurt. Walk it off. "Don't take a cab. You'll be fine."

He wasn't fine. An elderly Good Samaritan offered to call a doctor when she saw him staggering down the street. "No, just a cab," he replied.

"Mick drug his leg all the way back to the hotel," Merlyn told me.

The next morning he was burning up with fever. She was due to arrive that afternoon. The hotel sent someone to meet her train. He awoke to find his wife at the foot of his bed, asking, "What happened to you?"

"I just got sucked dry by a vampire."

"Mick told me, 'I think the guy wanted to hurt me,'" Merlyn said. "He said the place was filthy and he had blood on his coat."

Jacobson's son, Thomas, a cardiologist practicing in Arizona, said his father never kept any of his records. They spoke about Mantle only once, a brief conversation in which Dr. Max acknowledged treating him but offered no details about his care. "He'd taken care of so many well-known people that it was just matter-of-fact," Thomas Jacobson said. "I don't think he had any of these special amphetamines or anything that I know of."

The *New York Times* exposed Jacobson's practices in a series of investigative reports in 1972. His medical license was revoked three years later, after the New York State Board of Regents found him guilty of forty-eight counts of unprofessional conduct and one count of fraud or deceit. Unable to account to the federal Bureau of Narcotics and Dangerous Drugs for quantities of amphetamines in his possession in July 1968 and March 1969, Jacobson was also found guilty of manufacturing and combining "adulterated drugs consisting in whole or in part of filthy, putrid and/or decomposed substances."

Mantle said he never knew what was in Jacobson's syringe, and he never paid the bill, either. Mark Shaw, the Kennedy family photographer, paid with his life, dying of amphetamine poisoning in 1969. Tennessee Williams's brother told the *Times* that the playwright had spent three months in a mental hospital that year as a result of taking drugs prescribed by Jacobson. Truman Capote collapsed after a series of injections and had to be hospitalized with symptoms of withdrawal. When Mel Allen was fired by the Yankees after the 1964 season, the infamous medical referral was widely cited as cause.

Mantle played only two complete games after September 17 and started only two after seeing Jacobson on September 25. The morning after, New York papers were filled with medical bulletins: Cerv was in Lenox Hill Hospital, having surgery on his right knee; Mantle showed up at the Stadium looking as if he hadn't seen daylight in a year. "He's limping as heavily as he ever has in his life," the *Post* reported. "He says he's stiff. Some of the injections he's been taking to cure the cold have proved painful. It's possible that Mantle is hurting more than either he or the Yankees will admit."

On Tuesday evening, September 26, Merlyn kept Pat Maris company in a box beside the Yankee dugout. Mantle left the game after walking in the first inning. He watched Maris hit his sixtieth home run on the clubhouse TV. Mrs. Babe Ruth cried.

When Mantle did not report to the Stadium the next day, the Yankees sent a doctor to examine him. He was admitted to Lenox Hill Hospital with a 101-degree fever. Sidney Gaynor, the team physician, operated that night, incising and packing an abscess in the area of the right hip. It was "like a boil," Gaynor explained, only under the surface of the muscle. In the *Post*, Schecter came up with an explanation befitting a fretful grandma: Mantle had contracted the infection "possibly as a result of playing on a wet field."

Sunday, October 1, was a beautiful day to make history. Maris did not feel up to it. The 35-inch, 33-ounce bat with which he had hit number sixty felt empty. The crowd was as sparse as his hair—paid attendance 23,154—despite the $5,000 reward a California restaurateur had offered for the sixty-first home-run ball. The right field stands, however, were packed. "He did not want to play the last game of the season," said Whitey Herzog, who became close to Maris during their tenure with the St. Louis Cardinals. "His teammates talked him into it. He really didn't want to break that record."

It was 2:42 P.M. when he came to the plate in the bottom of the fourth inning. The game was scoreless, the bases empty, the count 2–0. Tracy Stallard, the young Red Sox pitcher, threw the ball—"a strike," he said later, "knee high on the outside corner of the plate"—and Maris hit into the right field stands, six or seven or fifteen rows deep and just to the right of the Yankee bullpen, where it was caught on the fly by a nineteen-year-old Coney Island boy named Sal Durante. As the scoreboard flashed the official news—MARIS 61 HOMERS BREAK RUTHS 1927 RECORD FOR A SEASON—his teammates forced him to take a second curtain call.

In the clubhouse, a radio guy asked, "As you were running around the bases, were you thinking about Mickey Mantle?"

It would always be about The Mick.

"Nobody knows how tired I am," Maris replied.

All they had to do was look at him, slumped on a stool, drinking a beer, inhaling another cigarette. He was smoking two or three packs of Camels a day. It was close to 8 P.M. when he and Pat left the Stadium with Julie and Selma Isaacson, heading downtown for dinner at Joe Marsh's Spindletop. There was time for a late Mass at the Catholic church across the street from the restaurant before dinner. "Two minutes later, here comes Rog and Pat," Isaacson said. "I said, 'That's a quick Mass.' He said they spotted him in the church. Priest started talking about 'Roger Maris is here.' So he walked out."

They made another stop at Lenox Hill to visit Mantle and Cerv. Mantle sat up in bed when Maris came in. "Rog went over, and they hugged one another," Isaacson said. "He said, 'It should have been you, Mick.'

"His eyes started to tear up, Mick."

Maris was too exhausted to eat. Big Julie had invited Milton Gross along to dinner to make up for the snub in Baltimore and he re-created the scene in the *Post* the next day. A little girl approached their table to ask Maris for an autograph. "Would you put the date on it, too, please?" the she asked.

"The date?" Maris said. "What is today's date?"

"The date is the one you did what nobody else ever did," Big Julie replied.

Maris would never be that good or that healthy again. Although he played five more years in New York, it was almost as if he ceased to exist after '61.* "Six years of hell," he called them. The accomplishment of a lifetime became a source of pride and torment. The mosaic tile he commissioned for the bathroom floor in his Kansas City home showing him mid-home-run stride attested to the former. The sportswriters he had long and loudly disdained contributed to the latter, seeing to it he never got his rightful place in the Hall of Fame, an omission that appears more glaring with each passing year and every steroid-soaked revelation. His bat is in Cooperstown, his jersey, too, and the ball Sal Durante gave him without asking for anything in return.

The unasked or unreported question was: why had Mantle sought treatment from someone other than the Yankees' team physician? "It was just prior to the World Series," said Johnny Blanchard. "He didn't want anybody to think that he wasn't healthy. 'Cause any-time you go to the team doctor, immediately it goes to the front office."

Blanchard didn't say what "it" was. Clete Boyer finally copped to the obvious. "I can't believe you god-damn media people are so dumb," he said. "Nobody ever figured it out. Ever think why Mickey Mantle went to another doctor other than the Yankee doctor? Ever think about it? Why would he have gone to an-other doctor other than the Yankee doctor? How 'bout the clap? C-L-A-P."

Boyer had a good laugh at the credulousness of sportswriters who solemnly reported the progress of a virus that had somehow "lodged in his buttock," as Schecter wrote in the *Post*. "The twenty-four-hour virus, and it got *infected*?" Boyer said. "C'mon."

Mantle certainly wasn't the first or only Yankee to find himself in such a predicament. In the press box, writers spoke knowingly of "a rash of injuries."

Mantle never publicly acknowledged the indiscretions that hobbled him. "I think we just did a helluva job and he had a helluva summer until September," said Cerv. "Then he said, 'I've had enough of this. I gotta have some good times.' In two weeks, he was so screwed up he didn't even play in the World Series."

His condition wasn't news to the Cincinnati Reds'

pitching staff. Jim O'Toole heard it from Darrell Johnson, the former Yankee catcher as they watched Maris hit his sixtieth home run. "Mantle's got a little problem." Jim Brosnan heard it from the author George Plimpton, who said, "Mantle was not going to be stealing any bases, because if he had to slide, he'd be bleeding all over the ballpark. I asked what the hell all that was about, and then he told me. What the hell else would you get a shot in the butt for?"

9.

The Series was over before it began. It ended the moment Jim Turner, the longtime Yankee coach now with the Reds, decided to take a contingent of pitchers and catchers on a tour of the Cathedral of Baseball. "He took us out with all the monuments like we were supposed to worship all those guys," said Johnny Edwards, an impressionable young catcher. "I think it got us a little uptight about where we were and who we were playing."

Their only hope was Mantle.

Mick Mends—*Daily News*

Mick to Start—*Daily News*

Mickey Doubtful—*Daily News*

When Mantle was released from the hospital on Monday morning, October 2, he went straight to the Stadium, where reporters duly noted his pallor and his weight loss. He did not work out. Gus Mauch sent the batboy Frank Prudenti on an urgent mission up the Grand Concourse to procure a magic salve said to speed the healing process. Mantle summoned Moose Skowron, who had taken biology in college, to watch Mauch change the dressing on the wound, which had been left open in order to allow it to drain. "I never forgot that," Skowron said, "'cause the blood and the pus was coming out of it."

The size of the hole was generously and variously described. "Big enough to put a baseball in," Mantle said. "'Bout the size of a golf hole," Joe DeMaestri said.

And it was deeper than it was wide. "They cut a circle about the size of, oh, I'd say a small platter," De-Maestri said. "They cut it in four sections, like an X, and pulled back the layer of skin, of the meat, and they had to then put stuff in there. They had to fold back the outer layer of his skin to drain it.

"He'd come in and he'd lay down on the table on his side and he'd move his toes and you could see the tendons move. In the sore you could actually see the cords in there. And he'd laugh. Mickey had a funny way about himself."

The day before the Series opened, he took five batting-practice swings and told Houk he couldn't play. That evening he took Merlyn to dinner at the Harwyn Club but couldn't sit long enough to eat. He played in neither of the games in New York, and the Yankees and the Reds headed for Cincinnati, tied at one game apiece.

A crowd ten deep waited behind the batting cage at Crosley Field when Mantle took his swings the day before game 3. Bob Addie described the scene for *The Sporting News*:

> His mouth flew open, his face contorted with pain. He then bent over the plate and pounded his bat for a few seconds, trying to regain his composure.
>
> The 2,000 fans who had come out to see the workout seemed to sense the drama. They were quiet. So were the players on the field. Time seemed suspended and caught in a still photograph.
>
> Mantle straightened up and knocked the next pitch into the right field seats. He hit five more "homers," some over the distant, center field wall.

Having "paid his dues to pain," Addie wrote, Mantle "received an ovation from the crowd. Then he told Houk, *I'm playing.*"

Houk hadn't seen the wound. "My God, you'd better not play," The Major said after assessing the damage. But the doctors had assured him that playing wasn't going to hurt Mantle further. It was just going to hurt. Mauch bandaged and padded and cushioned the area as best he could. "They taped a big rubber doughnut as a protection so that if he bumped it, it would give him a cushion on the sore," DeMaestri said. "If you just touched it, my God, such pain you couldn't believe it."

When the Yankees took the field, Maris was jeered; Mantle was cheered. Lovell Mantle, who had not seen her son play in eleven years, was at the game. Asked if she thought she would bring the Yankees luck, she replied, "I don't think Mickey is superstitious." Mantle went hitless, striking out twice. But his wise counsel proved decisive when Blanchard consulted him before pinch-hitting against Bob Purkey in the eighth inning. "He's going to throw you a slider first pitch for a strike," Mantle said. "Then he's gonna come back with knuckleballs."

Blanchard homered on a first-pitch slider to tie the score. When he returned to the dugout, Mantle said, "Hey, Blanch, you owe me a six-pack."

"I bought him a case," Blanchard said.

Maris won the game with a ninth-inning home run.

Game 4 was a rematch of the Series opener: Whitey Ford versus Jim O'Toole.

Mantle led off in the top of the second inning with a hard ground ball to third. Halfway down the first base line, he pulled up. The effort had yanked the unstitched wound apart. Blood began to seep through the layers of protective padding. Mantle tried to disguise the spreading stain with his glove when he returned to the dugout after the bottom of the third inning. "Clete Boyer went to Ralph Houk and says, 'Do you know Mickey's bleeding?'" Prudenti said. "Ralph went over and says, 'Let me see that leg. Move the glove.'

"And Mick was trying to hide it. And Ralph Houk says, 'Come on Mickey, I'm takin' you out.'

"He says, 'Aw, that ain't bad. It's nothing.'

"You're bleeding like a stuck hog," Houk said, summoning the trainer.

"Mickey didn't believe him," Blanchard said. "He had to drop his pants to look at the red spot, the blood. He couldn't see down the back of his pants. He looked down and there it was. He said, 'Okay,' and he left. Boy, talk about a competitor."

After the game Mantle gave a somewhat different account to Joe Reichler, the baseball writer for the Associated Press. Naked except for the bandage on his hip, he told Reichler he knew he would have to quit after the third inning. "At first I could only feel the blood running down my leg," Mantle said. "Then I could see it through my ball suit. When I came in at the end of the inning, I told Ralph I couldn't run and he said he wanted me to bat and if I got a hit, he would take me out for a pinch runner."

Maris led off the fourth inning with a walk. Mantle walked gingerly to the plate, took two pitches for balls, and fouled off the third. He sent Jim O'Toole's fourth pitch on a low line drive to left center field. Any other time, it would have been a double. Mantle had to stop at first; Maris went to third; he would score the game's first run. From the mound, O'Toole could see "the bloody mess on his leg." O'Toole saluted his courage fifty years later: "He may have caused it all himself, but God bless him."

Mantle left the game for a pinch runner and received an ovation from his teammates when he reached the Yankee bench.

The entrance to the visitors' clubhouse at Crosley Field was through the home dugout. Everyone in the ballpark understood the significance when Mantle made his way across the field, down the steps, and into the shadows.

13

May 18, 1962

HIS BEST SELF

1.

A mean May chill dampened attendance for the Friday-evening game at the Stadium against the Minnesota Twins, the first of a long home stand. A forgotten tributary of the Harlem River, on whose banks the ballpark was built, ran on a diagonal from left field through the hole at short. The ancient waterway, Cromwell's Creek, buried deep beneath layers of Manhattan schist and the sedimentary rock of urbanization, deepened the chill. A mist enveloped the scalloped copper frieze that ringed the upper deck of the Stadium.

There were not quite 21,000 hardy souls in attendance, every one of them thinking the same damn thing: if he hits one out tonight, nobody will ever see the ball again. That was the thing about Mantle. You never knew what might happen when he stepped to the plate—or what might happen to him.

It was supposed to be his year. God owed him, didn't He? Mantle had greeted the spring with rookie enthusiasm, as if granted a reprieve. He seemed different, more patient. Cornered at batting practice one day by an unctuous Texas disk jockey, who insisted upon reciting Mantle's yearly home run totals and poking him in the shoulder for emphasis, Mantle let him yammer away, rolling his eyes heavenward when the soliloquy ended. "I care much more about other people's feelings now," he told the jabbering deejay. "I used to think what the hell and not waste any time with them. Now I realize they have feelings too."

That winter the M & M boys, new members of the Screen Actors Guild, made a cameo appearance with Rock Hudson and Doris Day in *That Touch of Mink* and starred in a comic caper called *Safe at Home!*, for which they each earned $25,000. Whitey Ford, Ralph Houk, and Big Julie had bit parts. The day they inked the deal, Mantle admired himself in the mirror, and offered a disarming bit of quotable hubris: "I can't understand how it could have taken them so long to discover me."

Mantle got standing ovations wherever he went, including the men's room; Maris got mugged in the press. Rogers Hornsby called him "a punk baseball player." Jimmy Cannon called him "a whiner."

Mantle understood that the terrible onus Maris carried to the finish line conferred on him—the loser—a certain grace. "I became an American hero in 1961 because he beat me," he told me. "He was an ass, and I was a nice guy. He beat Babe Ruth and he beat me, so they hated him. Everywhere we'd go, I got a standing ovation. All I had to do was walk out of the dugout."

Maris wasn't in the lineup on May 18, having injured his groin two days earlier; nor was Yogi Berra, who was making his last trip around the bases as an everyday player. Mantle was the Yankees offense.

He was batting .326 with 7 home runs and 17 RBIs when the evening began. In his first three visits to the plate, he walked and scored twice. But in the top of the seventh, Harmon Killebrew hit a two-run home run off Ford and the Twins took a 4–3 lead, which they nurtured into the bottom of the ninth inning. Mantle was due up fourth.

The murmuring began when Berra pinch-hit to lead off the ninth—Mantle was at the bat rack. Berra popped out but Tom Tresh singled. It was a measly infield hit, to be sure, but enough to get Mantle another at-bat.

The buzz turned into a thrum when he emerged from the dugout, swinging a bat with that graceful torque of possibility. *Mickey's on deck!*

Joe Pepitone—*Pepi! Pepi!*—lifted a fly ball to deep center field, where it died a predictable death, but not before moving Tresh into scoring position. Mantle walked to the plate, and Minnesota manager Sam Mele marched to the mound, summoning lefty Dick Stigman from the bullpen. Amid the frenzied crescendo—*Mickey! Mickey!*—Mele ordered Stigman to throw only curveballs, low curveballs. "I won't be mad if you walk him," Mele promised.

The first curve hung high, a tantalizing offering, and Mantle mauled it, sending a ferocious one-hopper at Minnesota shortstop Zoilo Versalles. It looked like a sure out, an easy out. But the ball was hit so hard, as Jack Reed noted on the Yankee bench, "it almost knocked him into left field."

Twins second baseman Bernie Allen remembered, "The ball came up on Zoilo and hit him in his shoulder and popped up in the air."

Mantle saw the momentary glitch and reached for a remembered burst of speed.

He was tired. The week had been grueling: a doubleheader in Cleveland, followed by a three-hour bus ride to Pittsburgh for an exhibition game; an abortive flight to Boston that returned the Yankees to New York at 5 A.M. because Logan Airport was fogged in; then, after three, maybe four hours of sleep, they flew back to Boston, played two night games, a day game, and flew back to the Bronx to face the Twins.

With two outs, Tresh was running with the pitch. "I'm on second," he said, slipping into the present tense, as ballplayers do when recalling the past. "By the time he makes contact, I should be between third and home. He's behind me. I can't see him. I see Versalles bobble the ball."

Just steps from the bag—some observers said five, others ten, maybe twelve—Mantle's body betrayed him. "The legs wouldn't go as fast as his mind made him go," Houk said.

Mantle wasn't surprised. He could always tell when something bad was going to happen. He collapsed in midstride, his legs extended beyond reach or reason. He hung there for an instant, or so it seemed, before the force of gravity sucked him to the ground, splayed in the base path, his cheek pressed to the dirt. His feet churned, his hand reached for the bag.

The Twins' first baseman, Vic Power, heard him moan. "It's my legs. It's my legs."

Allen heard the muscle pop from second base. "I thought, 'Oh, my God.'"

From where Houk sat on the bench, it looked just like October 1951, when the world went out from under Mantle's feet. Watching him writhe in the dirt, Houk thought of the rabbits he had hunted as a boy. They rolled like that when they were shot.

The big Longines clock in right center field read: 10:23 P.M. Mantle lay in a fetal position, inert with pain. Everyone else in the ballpark was standing. "I've never heard a place that big get that quiet," Mantle told me, reflexively massaging his left leg. "I thought I broke this leg then. It wouldn't come back down. It just stuck up, and when I fell, I tore this knee up."

Tresh never quit running. He cut across the infield and reached Mantle before the trainers arrived. Don Seger, the assistant trainer, was one of those attending to him. "He really, really hurt himself," he said. "He tore his groin. It didn't detach but it was strained enough that he lost function in the groin complex. I think he came out with a little meniscus on that one, too."

Mantle refused the stretcher and was helped from the field, arms draped over supportive shoulders. Bob Cerv cleared a path to the dugout, murmuring, "It's bad."

The Yankees stayed away from the trainer's room out of fear and respect. "All we could think about was

'Wow, there goes the pennant,'" said relief pitcher Roland Sheldon. "'How are we going to go on without Mickey?'"

Houk briefed the press with wishful thinking: "Maybe a charley horse."

The team doctor was on the telephone making arrangements for Mantle's admission to Lenox Hill Hospital.

He showered on crutches, telling himself, "I'll be out a week." He was gone for five. He hobbled up the steps to the players' gate, where seventy-five mournful fans, a phalanx of news photographers, and Dan Topping's chauffeur waited. His smile was tight. The buttons on his snazzy cardigan strained beneath his crutches. "See y'all," he said.

Houk stayed up until 3 or 4 A.M., waiting for word from the hospital. Mantle had torn the adductor muscle in his right hip. Landing hard on his left knee—the good knee—he had strained the ligaments behind it. Gaynor called the injury "reasonably severe."

Houk already knew what no X-ray could detect: it was the beginning of the end. "Start of it, anyway," he said.

2.

A locker room is as much an idea as it is a place, a state of mind requiring no fixed address. In major league lexicon, "locker" is also a verb, as in: *I lockered next to Mickey Mantle.*

The Yankee locker room circa 1962 wasn't plush. Amenities were minimal: wire hangers, cubicles separated by chain link, four-legged wooden stools. The concrete floor was painted a deep terra-cotta red and paved with corrugated rubber mats that stamped serrated edges into bare feet. Pepitone was lucky to find an electrical outlet when he showed up with his much-scorned hair dryer.

But for Mantle, the Yankee locker room was a sanctuary, a safe haven where he was understood, accepted, and, when necessary, exonerated. "I think the happiest time in his life was probably when he was in the locker room with the guys that he cared for," said Tresh, the American League Rookie of the Year in 1962.

The locker room was ruled by the Petes: Big Pete (Sheehy) and his assistant, Little Pete (Previte), who autographed a thousand baseballs with a convincing facsimile of Mantle's signature and kept his locker tiptop. "He looked out for Mickey better than a mother would look out for their little kid," said batboy Frank

Prudenti. "If Mickey Mantle went around the corner into the trainer's room wantin' to get a Coke or went into the lounge, he would run over to that locker, straighten out the stool, straighten out his clothes, his shoes. Mickey walked by one day, he grabbed a shoe brush and says, 'Come here, Mickey, your shoes are dusty.' Right there, he took the dust off him. Once he was shining his street shoes. He didn't like how they were shined."

Mantle didn't want to stick out, but he did; he didn't want to be treated as special, but he was. He didn't want to be the center of attention, but he was the center fielder for the most visible sports franchise in the world. In this, he was quite unlike the arrogant, demanding star he replaced. DiMaggio demanded: "Attention must be paid." Mantle deflected it. "Go talk to Moose. Go talk to Hank."

"He didn't want to be exempted as one of the great ballplayers," said Tony Kubek. "He just wanted to be with his boys."

In 1962, Sheehy stationed Mantle between two rookie infielders, Tresh and Phil Linz. Tresh was the son of a major league catcher, "Iron Mike," who had taught him to switch-hit like The Mick. For as long as he could handle a glove, Tresh aped everything Mantle did. As a boy in Detroit, he stationed himself in the upper deck of center field in Tiger Stadium when the Yankees came to town "just to watch him run out to the field." When they became teammates, Tresh said, "I actually called him 'Idol.'"

The first time he saw Mantle up close, Tresh was in an elevator during a rookie camp in St. Pete—the doors opened and there *he* was. Though they stood eye to eye, Mantle seemed a thousand steps higher—a feeling Tresh never got over. "I never could see him as just one step higher than I was."

He didn't expect sweetness from the icon. Mantle gave him four pairs of spikes, and the shirt off his back, and when the Yankees moved Tresh to the outfield, he gave him the glove he was breaking in to replace his gamer. Mantle had made the same transition from shortstop to outfield his rookie year.

When Tresh's wife got pregnant, he told Mantle, "I'm naming my son after you." He wasn't the only teammate to do so. Mantle dismissed such tributes with laconic profanities. *What the fuck did you do that for?* "He was always embarrassed about anybody paying him a compliment," Tresh said. "I think he felt very honored but he couldn't show it."

The baby was indeed a boy. And Tresh named him Mickey. "When he was born, he weighed seven pounds, seven ounces. He's blond-haired, blue-eyed. I told Mickey, 'Did you miss a road trip? If that kid has a limp, you're in trouble.' "

Linz took a more distanced approach. He watched how people talked to Mantle and was careful how he spoke to him. "I wanted to be able to say the right thing. I was always in awe, but I never let him know it. The first sign that he would think he was special, he would want to be away from that guy."

Mantle's "aw, shucks" modesty was genuine. "*I'm* in the encyclopedia?" he'd say, and gasp—"Damn!"—when shown the entry under his name. He stared with incredulity when Tim McCarver introduced him to his insurance man, an old University of Oklahoma running back named Buddy Leake. "*You're* Buddy Leake? I can't believe I'm meeting Buddy Leake!"

He was a great storyteller and the fall guy of his best tales. A favorite was set during a game in Detroit. "I make this really good catch with the bases loaded," he told me. "I used to be bashful. Head up, I'm not. In front of a lot of people, I am. Everyone's applauding and I think it's the last out. I got my head ducked. I'm running back in, and I got the ball in my glove. All of a sudden Billy Martin is going, 'Give me the fuckin' ball.'

"I'm going, 'What the fuck?'

"He says, 'Goddamn, there's one out.'

"I looked up, and there's fucking people sliding all over Yogi. Two guys scored on a fly ball."

That was the thing other players, teammates and opponents, admired most about him.

"No ego," said Gil McDougald.

"Great control of his ego," said Reggie Jackson.

"Wasn't no individual," said Jim Coates.

In the locker room, where everything is exposed, he was seen as "the best teammate ever." They saw how hard he worked at it. Kubek saw how "he didn't phony up anything." Ford was chairman of the board, who "commanded the respect of all the guys" said Eli Grba, and "held meetings without the training staff if we were playing terrible."

But Mantle was the official greeter. When Bob Turley reported to the Yankees in the spring of 1955 after an infamous nineteen-player off-season trade, he found a "Greenie" (lime soda) and a flower waiting at his locker. "What the hell is this?" he said.

"Welcome," Mantle replied.

What was most surprising was his empathy. "If a person had a problem, he could just feel that person's problem," said Gene Michael, the utility infielder who later became Yankee manager and general manager.

His people skills—emotional intelligence in modern parlance—made a lasting impression on Mark Freeman during the brief time they played together in 1959. "I think he's basically one of the most decent guys I've ever met," he told Leonard Schecter of the *Post*. "He's learned to swap places with rookies, so he understands them, sympathizes with their aspirations. He has a way of making you feel he's really interested in what you're saying. Every day, Mickey would go by DiMaggio's locker, just aching for some word of encouragement from this great man, this hero of his. But DiMaggio never said a word. It crushed Mickey. He told me he vowed right then that if he ever got to be a star, this never would be said of him."

Michael and Kubek insist he would have been that way no matter what.

When Big Pete assigned Kubek a uniform number commensurate with a quick trip back to the bush leagues his rookie year, Mantle intervened and got him the one he would wear throughout his career. "You know he's going to make it."

It was Mantle's calming voice Kubek heard that season calling from the dugout during a tough at-bat. And it was Mantle's voice he heard on the other end of the line expressing condolences when Kubek's father died. He was the only teammate who called.

Few outsiders ever saw that Mantle. Imagine the surprise when he showed up at a Chicago hospital to visit

Marjorie Bolding's mother with a raft of teammates in tow and a suitable bouquet of flowers. "People don't know what kind of man he was," said Steve Kraly, his teammate in major and minor league baseball. "If you sat down and talked with him at the breakfast table, he came across as pleasant and gentle, but sportswriters wrote the opposite."

Beat writers suspected there was another Mantle, one who didn't begin every conversation with "Fuck you" or "Go fuck yourself" or, when he was pressed for time, just plain "Fuck." But Mantle turned on his stool when they pried and probed. "For me, the frustration when I was covering him was, you knew he was a nice guy and you knew he was terrific with his teammates," said Stan Isaacs, who covered the Yankees for *Newsday.* "You'd try to talk to him, and he'd look right through you. This was his defense."

Writers were relieved when he was in a good mood and didn't mind being twitted for their bad haircuts and shiny suits. "When he was happy, as a young reporter, I was happy because he wasn't tormented at that moment," said George Vecsey, who began traveling with the team in 1962. "Nobody should be that miserable."

He was a guy's guy who called everyone "bud" or "pard." But he cried easily. He wept at mournful country-western tunes, and at the morning headlines. "Somebody got killed or something, he'd get tears in his eyes," Irv Noren said.

He cried when a dying child was placed in his arms outside the clubhouse at Griffith Stadium in Washington and he cried when he failed. "One day he strikes out four times and goes back to the clubhouse, and he's crying," Hank Bauer said. "Moose says, 'Mick, what's wrong?'

"He says, 'I let down my teammates and my fans.'

"Moose says, 'You know, there is tomorrow.'"

Locker room newcomers learned fast—"If he went up and struck out, nobody said a word until he broke the silence," said Stan Williams, who arrived in 1963. "He was a hero in the clubhouse because of the respect the other players had for the way he played the game—not just his ability but the intensity he played it with."

One day Coates made the mistake of welcoming him back to the bench with a consoling spank after a poor at-bat. "He smacked my arm off of his butt so fast I didn't know what hit me," Coates said. "You don't say nothin' to him. He wants to do it so bad and so good."

"Was he nasty sometimes?" Kubek said. "Oh, yeah. Could he lose his temper sometimes? Oh, yeah. Did he snap his bats sometimes? Oh, yeah. But he just wanted to be among his people."

No one was spared The Mick's unrepentant teen-age humor. He was the guy who froze a plastic snake and stuck it down Marshall Bridges' pants—the relief pitcher had a serious case of ophidiophobia. He left a dead fish in the whirlpool for Johnny Blanchard one spring training morning because he knew Blanch would be needing to sweat off a hard night. He put the talcum powder in Pepitone's hair dryer. Pepitone retaliated by putting Joy dishwashing liquid into Mantle's whirlpool. Nothing remarkable about that, except that he was a rookie.

One day after Al Downing, the Yankees' first African American pitcher, joined the club, Mantle and Ford joined him in the shower room. Taking the measure of their new teammate, Mantle told Ford, "Hey, Slick, he's one of us."

Downing wasn't amused hearing the story for the first time fifty years after the fact. "*Who* gave you that information?" he demanded. Upon learning that the source was Mantle, he laughed.

"He was an equal-opportunity offender," Jim Bouton said.

In the clubhouse, Mantle shed the residue of bias he had brought with him from Commerce. He laid a carpet of white towels at Elston Howard's locker after he hit his first major league home run, easing the path for the Yankees' first African American player. Traveling secretary Frank Scott told reporters that Mantle had refused to attend a cocktail party hosted by Cardinals' owner Augie Busch when Howard was excluded.

He may have left prejudice behind, but Kubek says he remained a good old boy from Oklahoma. When Bobby Cox arrived in 1968 and Mantle learned that he was born an Okie (albeit one raised in California), he gave the rookie his first-class seat on the first plane flight of the year.

Cox was what Stengel used to call "one of them milk shake drinkers." Mantle knew the best soda fountain in every American League town. "He'd ask us to go out and have milk shakes because he knew what kind of guys we were," said Cox's roommate Andy Kosco. "He'd say, C'mon, there's a great place across the street.' One of his favorites was having Coke and milk together. After a while it really didn't taste too bad. He loved that drink. He'd have it even on airplanes. It's like a root beer float."

There was some self-interest involved—nobody went looking for The Mick at a soda fountain counter. "He knew if he went with us he wouldn't be bothered," Kosco said.

Rookies or veterans, slumping teammates got invitations to dinner. *You come with me tonight.* "He'd take you downtown for a steak and a talk but not a word about baseball," Blanchard once told a reporter. "If you were a farm boy, he'd talk about farming. If you were a city boy, he'd ask where you went to school."

He took in homeless teammates like strays. "When I got divorced, he saw I was depressed," Pepitone said. "He says, 'I got two rooms at the St. Moritz. You come stay there.' So I stayed for a year."

One season he gave Jerry Lumpe the keys to his house in New Jersey. Merlyn and the boys had gone back to Oklahoma. "He'd already paid the rent, and he said, 'Just go out,'" Lumpe said. "We stayed for two months or something. He never asked for a dime."

Lumpe also inherited the sports jackets Mantle left behind in the closet. "Take 'em," Mantle said. They fit. Lumpe would have worn them even if they hadn't.

As with many who grow up poor, money didn't stay in his pocket long unless he had mislaid it—like the paycheck that Sheehy found stuck between the cracks of the trunk he used to store players' valuables. Or the uncashed $1,000 check from Wheaties he dug out of his pocket one day on a Manhattan street corner while searching for a piece of paper on which to write his home address for a visiting Oklahoma fireman.

He was a soft touch, always good for a loan. He picked up every tab (that wasn't picked up for him) and overtipped on every one of them. "Mickey would tell you before you go out, 'Don't even try it,'" Clete Boyer said. "If you tried to pay your share, he'd say, 'I won't go out with you no more.' And he meant it."

Pepitone saw him leave $50 on a 50-cent cup of coffee. "He'd say, 'You workin'? You deserve whatever I'm gonna give you,'" Pepitone recalled. "A lot of these waiters, they'd come up to Mickey, 'Can you autograph this?' Well, he didn't like that. But you didn't say nothing, you served him, shit, you'd get a two-hundred-dollar tip for a dollar-and-a-half sandwich."

Once on a spring training bus ride from Tampa to St. Pete, he saw a man lying on the railroad tracks. *Hey, bussie, stop!* The driver pulled over. "Mantle got off and gave him a hundred dollars," Kubek said.

Years later, Kubek was greeted by a homeless man as he left his New York hotel: "Hey, Tony, Mickey just walked by an hour ago and gave me a hundred dollars."

"He always did it in the dark so no one knew," Kubek said. "I gave him twenty dollars."

Mantle gave away so much—including his first uniform, at a birthday party—Merlyn finally bronzed one of his gloves and a pair of spikes before they disappeared from the house. "I used to get mad at him because we weren't exactly in the bucks," she told me.

The quarter horses he received on Mickey Mantle

Day in 1965 were shipped to Harold Youngman's ranch as a gift for the ranch foreman's children. A matched set of holsters and replica six-shooters? Handed over to a lost boy being minded by a cop at the players' gate.

Every winter just before spring training, he hosted an elaborate banquet at "Mickey Mantle's Holiday Inn" in Joplin, inviting all the major-leaguers who lived within a 200-mile radius. All expenses paid. "The first year it was a very formal dinner," said Shirley Virdon, Bill's wife. "Everybody was all dressed up in white gloves and tuxes. We had all this service, wine for every course, silverware all lined up. None of the ballplayers knew how to use it. I can still hear him giggling."

His generosity was reflexive and dutiful. After Mutt died, Lovell never worked or wanted; she received regular checks from the Social Security Administration and the New York Yankees. "Every time the Yankees paid him," his sister, Barbara, said.

He provided in a way that Mutt never could—far more than the necessities. He made sure Barbara had the fashion accessories the popular girls flaunted; she never forgot that pair of beaded Indian moccasins. His childhood friend LeRoy Bennett thought his largesse did not always benefit his twin brothers. "Mickey brought back big cars and big hats and big money to

Commerce. Mickey kind of spoiled them rotten before they got to the point where they could be men of enough character to play baseball."

Faye Davis, his half-brother's widow, thought some family members, including his mother, treated him like an ATM machine. Whenever Ted's friends brought up the money he fronted for Merlyn's engagement ring, her husband replied, "Yeah, but he gave it back to me a hundredfold, many times over."

"What's more," she said, "Ted was always getting arrested or something, and he'd get him out of jail. "

His teammates saw the guilt he felt about his absence from his own family. "Really, really, really guilty," Tresh said—and how, out of the spotlight, he doted on other people's children. Some days when he was car-pooling with Ryne Duren, Merlyn would call and say he didn't need a ride to the park. Duren figured the sonofabitch had gotten drunk without him the night before. "I found out later that Mickey Mantle visited an awful lot of kids in a hospital," Duren said. "He didn't allow her to say where he was."

A *Sports Illustrated* reporter traveling with Mantle for a story about "life on the road with The Mick" was in Mantle's hotel suite in Kansas City when a man called from the lobby identifying himself as a friend of Mantle's brother Frank and asked to bring his kids up to say hello. Whitey Ford was puzzled, pointing out that Mantle didn't have a brother by that name. "I know I don't," Mantle replied. "I figured the fella'd be embarrassed in front of the two kids and all."

One year at spring training, he paid a cabana boy to climb a royal palm and fetch fronds to make hats for all the Yankee children, recalled one of the recipients, Tony Kubek's son, Tony III. He heard from his parents how after home games, while waiting for the crowd to disperse, Mantle would go to the family room to say hello to the wives and children. He was the only player who did that, his mother told him. When arrangements needed to be made for the inevitable World Series trips, Kubek said, "He always made sure the players' wives knew where they had to be."

His father, the Yankee shortstop, wondered whether it was because Merlyn and the boys weren't there.

3.

Mantle remained in the hospital for five days after collapsing in the base path on May 18. He was released at noon on May 23 and appeared in the clubhouse an hour and fifteen minutes later. He was still on crutches. He told reporters he could do without them if he dragged his leg but not if he planned on bending it. The Yankee brass was counting on the whirlpool to heal him sufficiently to make the West Coast trip the first week of June.

When that didn't work, he went home to Dallas, where the Cowboys' trainer, Wayne Rudy, supervised his rehabilitation. Running back Don Perkins said the Cowboys didn't know who he was, but they knew he was somebody when he took off his shirt.

In his absence, the Yankees won fourteen of twenty-eight games. Ford sent him a bouquet of eight tired daisies, because that's what the batting order was without its heart. General manager Roy Hamey summoned him to join the team in Los Angeles to provide moral support. "I won't say the players are brooding about him, but maybe they'll feel better if he's around," Hamey said.

The Yankees won two out of three games from the Angels and moved back into first place. Mantle cracked wise to his boys: "I got you into first place. Now you're on your own."

For five weeks, he was team mascot and comic. Elmer—his locker room nickname—appeared in the dugout during batting practice one day decked out like Mortimer Snerd, Edgar Bergen's dim-witted puppet, with a shred of chewing tobacco pasted over a prominent front tooth, his cap pulled low on his brow, and his ears protruding. Mantle rendered the Orioles' batting-practice pitcher incapable of throwing the ball over the plate.

When he began taking batting practice, *Sports Illustrated* reported that his legs "trembled like those of an old card table." His teammates gave him a new nickname—"B & G," for blood and guts. He replied to the incessant inquiries for medical updates by pinning a sign to his chest: "Slight improvement. Back in two weeks. Don't ask."

He came off the disabled list on June 16, summoned to pinch-hit against Gary Bell in Cleveland. Fifty thousand witnesses saw him blast a three-run homer in the top of the eighth inning to give the Yankees the lead. That proved fleeting; the memory endured. It was his first—albeit unacknowledged—hangover home run. "Pinch-hit, eighth or ninth inning, when he was too drunk to play," Linz said, one of perhaps three or four times he saw Mantle play in that condition.

Mantle returned to the starting lineup on June 22. *Sports Illustrated* excoriated Yankee management for taking "a cold and ruthless gamble" with his health. "Faced with a losing streak and the distasteful prospect of not winning the pennant for a change, the New York Yankees rushed the most valuable property in baseball back into action last week and ran the risk of losing him forever."

The Yankees insisted that it had been his decision. Ten days later, they were back in first place and he was back in a groove. He celebrated his independence from the disabled list by going 6 for 10 with 5 home runs, 6 walks, and 7 RBI during a July Fourth series against the Kansas City A's. He hit .340 for the month with twenty-five RBIs. On August 5, he had a slugging percentage of .635. "If Mantle doesn't get the MVP award this year—and get it unceremoniously—there should be an investigation," Bill Veeck, baseball's beloved renegade declared.

He finished the year with 89 RBI and 30 home runs, in 123 games. He led the American League in on-base percentage (.486) and slugging percentage (.605) and won the only Gold Glove of his career despite playing twenty-three games in right field. He finished second in the league in batting with an average of .323, despite the help of Jim Ogle of the Newark *Star-Ledger*, who generously awarded him a base hit one day late in the season, when he was doubling as official scorer. Seeing his benefactor in the locker room later, Mantle brandished a stack of World Series tickets and an impish grin. "How many do you want?"

Mantle's sense of humor was severely tested during the Series against the Giants in which he hit .120 and drove in exactly one run. At one postgame news conference a reporter solicited his opinion on the quality of the baseballs in use. "Are the balls livelier?" The timing was odd, given his dead bat. "No, but the players are," Mantle replied.

In the end, the Yankees prevailed because Ralph Terry won two games, exorcising the ghost of Bill Mazeroski, and because Bobby Richardson stepped in front of Willie McCovey's line drive with the tying run at third base in the bottom of the ninth of game 7.

Mantle was at a banquet in Terry's honor in Kansas when word came that he had won his third Most Valuable Player Award. He said Bobby Richardson deserved it more.

Mantle was a teammate for life in an era when teammates remained teammates long after their playing days ended. They were his most successful and most enduring relationships. When the memorabilia industry crystallized around him in the Eighties, Mantle brought his teammates along to autograph shows and hired them as counselors at his fantasy camp, where they congregated in old clubhouses, filling out newly issued pinstripes that fit more snugly than they might have liked. Tom Tresh was with him at a card show in Memphis, Tennessee, when his wife called to say that they had become grandparents. The boy Tresh had named after Mantle had named his new son, Tommy, for his dad. "I'll get a bottle of champagne," Mantle said. "We'll give him a toast."

"So he calls down and gets a bottle of Dom Perignon, about a hundred and a half at that time," Tresh recalled. "I call my son and congratulate him. Mickey gets on the phone and congratulates him. After we hung up, Mickey took that bottle—it was a green bottle with a gold shield on it—and took a gold marker and wrote, 'Happy Birthday, Tommy, we had one for you. Mickey Mantle.' And we tied the cork on it."

The cork stayed in the family. The outfielder's glove Mantle gave Tresh in 1962 was sold. It brought enough for a down payment on a cottage in the country.

14

June 5, 1963

THE BREAKING POINT

1.

On the morning of May 22, New York City threw a ticker-tape parade for Gordon Cooper, the last of the Mercury Seven astronauts to fly into space. Four million New Yorkers jammed the Canyon of Heroes to welcome him home after twenty-two orbits of the earth. Bill Fischer, a pitcher for the visiting Kansas City A's, was invited to the fiftieth floor of a Manhattan skyscraper to watch the parade for Cooper, another Oklahoma boy with the right stuff.

Fischer was having a very good year; his record was 6–0. Mantle was thriving. The Yankees had celebrated the Most Valuable Player Award with a $100,000 contract, making him only the fifth player in major league history to reach that financial pinnacle. The bosses were so proud of their largesse that they called a press conference to witness his signature. Red Smith wrote that Mantle looked "healthy and expensive."

The night before he had smashed two home runs, driving in five of the Yankees' seven runs, and smashed his bat in frustration at not doing more. His batting average was .303.

The A's manager, Eddie Lopat, was an old Yankee—a junkballer about whom Stengel had once said, "He looks like he's throwing wads of tissue paper." He had a "phobia about Mantle beating him," Fischer said. "He had a rule that he wasn't going to let Mantle tie or win a game."

There was a mandatory $200 fine for any pitcher who allowed him to do so.

As cleanup crews were contending with 2,900 tons of ticker tape on lower Broadway, Fischer was preparing for another evening in the bullpen, where he had spent most of his career. He was nearing the end; Tony LaRussa, an eighteen-year-old bonus baby, was preparing for another night on the bench. He appeared in only thirty-four games in 1963—and this wasn't one of them—but rules governing the amateur draft meant that the A's had to keep him on the major league roster. He watched as three A's pitchers hewed to their manager's admonition, and walked Mantle three times. "For some reason Eddie got it into his head he was going to bench jockey, agitating Mickey, and that it was going to work to our benefit," LaRussa said. "The dugout was close to the plate. He yelled something like 'This is not your day. You don't have a chance.'

"Mantle just looked over and laughed."

Why not? The Yankees were leading 7–0 in the top of the eighth inning. Bored spectators began to chant— "Let's go, Mets!"—inspiring the historically feckless A's to score six runs. Mantle was no longer amused by Lopat's remarks.

You're washed up. We got your number. You stink.

"You could tell, by the third at-bat, Mickey had enough," LaRussa said. "Eddie crossed the line. It was personal."

Fischer was summoned from the bullpen in the bottom of the eighth and retired the Yankees in order. New York was one out from victory when Ed Charles came to the plate in the top of the ninth inning. Charles was a poet of the infield; he coped with racism by writing verse during eight long years as a minor leaguer in the segregated South. Teammates called him "The Glider" because of his graceful footwork. Mantle called him "Muffie," baseball talk for a less-than-comely female. Muffie tied the game at 7–7 with his second home run of the year.

Fischer walked Mantle leading off in the bottom of the ninth. "I pitched around him," he said. "They bunted him over." But he didn't score, and the game went into extra innings. Fischer stranded another runner in scoring position in the bottom of the tenth.

It was a school night, a Wednesday. There were 10,312 people in the Stadium when the game began and notably fewer than that when Mantle led off again in the bottom of the eleventh inning. The upper deck in right field was almost empty.

The A's defense shifted toward right field, expecting Mantle to pull the ball. Center fielder Bobby Del Greco was stationed almost in front of the Yankee bullpen. Right fielder George Alusik was playing twenty to twenty-five feet in front of the outfield wall and about that far from the foul line. Fischer was not supposed to give Mantle anything to hit—to walk him intentionally if he had to. "I threw him four straight curveballs," he said. "He hit one of them a little, a weak grounder foul by inches. Haywood Sullivan, my catcher, called for another one. I shook him off. I figured he would know it was coming."

The count was 2–2. Alusik assumed baseball's age-old posture of patience: hands on knees, he awaited the pitch. From the visiting dugout, Lopat continued his verbal assault.

You got no bat speed. You can't get it out of the out-field. You were up too late last night.

"Eddie's really having a good time," LaRussa said. "He thinks he's doing something good. Mickey literally stepped out, gave him a look. You know that phrase, if looks could kill? That vein in the side of his neck when you get angry—that vein was *out* there and the adrenaline was pumping, he was so pissed off. When he cranked up to stride into that ball, he unleashed every bit of that anger and frustration. He made a perfect swing, everything working."

The ball creased the night air, heading straight for the copper frieze atop the third deck in right field. It soared over Norm Siebern's head at first base. "Ten, twelve, fifteen feet over my head," he said. "I couldn't leap up and get it—it was too high for that. So I just turned right around and watched. It just kept going up and up and up and up."

No one had ever hit a fair ball out of Yankee Stadium. Mantle had come closest in May 1956 against Pedro Ramos. The faithful had been waiting ever since for the next assault on the Stadium facade. "My God, that's it!" Yogi Berra yelped, leading a charge up the dugout steps.

"As Mantle came around, he looks around to Eddie and he just looks in," LaRussa said. "Eddie didn't say a word. He just looked at him like 'You know what, that one was for you. You should have kept your mouth shut.'"

The flight path took the ball directly over Alusik's head in right field, maybe a smidge to his left. He never moved, except to straighten with the pitch. His reflexes were no match for the speed of the ball. He leaned back, glancing over his shoulder just as the ball met the frieze. "I knew where she was headed," he said.

Joe Pepitone, who was at the bat rack in the Yankee dugout, swore, "It hit so hard, you could hear *boom*!"

Alusik didn't linger in the outfield. He returned to the bench disgusted because he thought Fischer had "taken it upon himself to pitch to the guy."

LaRussa said he also returned hard evidence of the force of the collision on his person. "He told us that it was hit so hard—it shook so much stuff—that he got showered with pigeon shit," LaRussa said."

In the locker room, Lopat loudly excoriated Fischer and the fates. "It was like he was screaming into the wind because everybody else was laughing," Fischer said. "It was so funny. Every time Ed would look at a guy, the guy would bust out laughing."

He fined Fischer $200 for allowing Mantle to beat them.

Mantle always said that it was the hardest ball he'd ever hit. "The one in Washington I had a 50-mile-per-hour tailwind," he said. "The one off Ramos was coming down more."

This was the only time the bat actually bent in his hands.

Sportswriters exhausted their arsenal of military clichés. It was unanimously ballistic—a rocket, a bullet, a blast, a shot.

Again, the fictive tape measures unwound. The *New York Times* said the ball hit the facade "374 feet away from the batter's box and 108 ft and 1 inch above the playing field."

The *New York Post* said it traveled "at least 475 feet," sailing "over the 367 marker, about 100 feet high, and was moving in a 45-degree trajectory."

Newsday said it soared "380 feet and was five feet from the top of the facade."

James E. McDonald, a professor at the Institute of Aeronautical Research at the University of Arizona, told the Associated Press the ball was traveling 230 miles per hour and would have gone 620 feet uninterrupted. Clete Boyer offered a more accessible calculus: "That's a three-dollar cab ride up there."

The speculation still excites a frenzied legion of baseball paleontologists: Was the ball still rising when it collided with the frieze? And if so, how far would it have traveled if uninterrupted by the Stadium's infrastructure? Where would it have come to rest?

Interested parties on the field, in the dugouts, and in the bullpen testified with vehement unanimity. "It *wasn't* coming down," Fischer said. "It was going up, like a jet taking off."

Only two of the next-day stories addressed what became of the beleagured ball. The *Daily News* said, "It dropped down into the upper deck." The *Post* said, "It bounced against a seat and came back down on the field." Pepitone said, "It bounced back about halfway between the right fielder and the second baseman. The trajectory coming back, it was like a line drive back to the field. It was, like, boom, boom."

He saw second baseman Jerry Lumpe trot over to retrieve the ball and saw it when it was delivered to the dugout. "The ball was scraped," Pepitone said. "Really scraped it bad. It looked like the stitching was up."

Lumpe has no such recollection. "I know I saw it bounce back in. I don't remember picking it up. To me it wasn't that big of a deal. He hit a lot of 'em a long ways."

He sighed. "Probably Alusik picked it up."

For nearly four decades, Alusik remained largely silent on the matter, perhaps because his phone number was often unlisted. He was not happy, when found, to hear that Tony LaRussa, a future Hall of Fame manager, was telling people that George Alusik had come back to the dugout covered in bird shit. "I'm not calling him a liar," Alusik said. "I never came in the clubhouse with any bird crap on me."

But what about the ball? Did he see it? Was *it* covered in bird shit? "The ball landed by my left side," he said. "I picked it up and put it in my pocket. I probably have it someplace. I'll have to look around."

The next day he called back to apologize. His lady friend, a churchgoing woman, had told him it was wrong to trifle with history. "I just laid it on a little bit," Alusik said. "Like these guys are layin' their bird stuff on me."

This is what really happened:

Alusik saw the ball smash into the frieze and heard a fluttering of wings. "There was a darkness underneath the stadium lights," he said. They cast illumination onto the field, not downward into the copper filigree. The sheltering darkness was home to something— birds, bats, he wasn't sure what. "I looked up and saw something fly out. He must have woke 'em all up."

He counted five or six winged creatures "flying for their lives and probably wondering, 'What the hell happened?' They were probably bitchin' about that one, the birds."

He didn't see the ball ricochet. He didn't see it land, and he didn't pick it up and put it in his pocket. "Matter of fact, nowhere on the field did I observe it," he said.

He, too, assumed it had dropped into the stands.

Sometime after midnight, still smiling and still in his undershirt, Mantle posed at his locker with the well-traveled ball, pressed against the sweet spot of Dale Long's borrowed bat.

A half century later, the Facade Home Run remains the subject of obsessive scientific inquiry for Mantle-ologists. Bruce Orser is among the most dogged of investigators. He sent detailed questionnaires to players on the field that night, most of whom did not reply. Hector Lopez told him he thought the ball hit the facade straight on. "But everyone else says it's going up," Orser said.

The inquiry taxes memory and patience. "The guy had all these questions: Did it do this? Did it do that?" said Lumpe. "I have no recollection of where it went or what happened to it. I wish I did. What does it matter if I picked the ball up or not?"

It matters, home run historians say, because if he picked it up—and if he could pinpoint exactly where he did so—they might be able to deduce its velocity off the bat and thus its hypothetical unimpeded distance.

In addition to measuring every major league homer, Greg "The Hit Tracker" Rybarczyk has also measured a handful of "historical home runs" for which there is sufficient data. According to his analysis, the ball struck the facade 102 feet above field level, 363 feet horizontally from home plate, reaching an apex of 108 feet. If, as he believes, the ball left Mantle's bat at a 27-degree angle, traveling at a speed of 126 miles per hour, it would have gone 509 feet, landing on the roof of Ballpark Lanes, a bowling alley occupying the block between 157th and 158th Streets on River Avenue. (A simulation of the trajectory can be viewed at www.dig italcentrality.com/Yankee_Stadium/video.html under mantle_hr_63.) Rybarczyk does not believe the ball was rising when the copper frieze interrupted its flight.

The morning papers were full of astronomical allusions and puns. They called the home run a "space shot." They called it "Gordo's day and Mickey's night." The headline Fischer remembered said, "Mantle Sends Ball into Orbit." When the A's took off for Los Angeles two days later, Bobby Del Greco was sitting in the window seat beside him. "We were, like, at 30,000 feet, and I started pointing toward the window," Del Greco said. "He said, 'What's the matter?'

"I said, 'Mickey Mantle's ball just went over the plane!'"

2.

"Is Mantle off on another batting spree?" the *New York Times* wondered. "Every so often the Yankees mighty man of muscle, finding himself in one of those infrequent periods where he manages to keep himself in one piece, goes off on a hitting binge."

The answer came two weeks later in Baltimore.

There was no room on the bench when Mantle emerged from the tunnel for the second game of the first "crucial series" of the season. The dugout was crowded with writers and players waiting out a rain delay and anticipating a cow-milking contest—Miss Udderly Fascinating lived up to her billing. "You'd think one of you writers would get up and let a $100,000 ballplayer sit down," one Yankee said, noting the breach of dugout etiquette.

In the top of the sixth, Mantle doubled and scored when Roger Maris hit a home run to give the Yankees a 3–2 lead. When, in the bottom of the inning, Mantle set off after a high fly ball hit to deep center field by Brooks Robinson, he was trying to keep the Yankees ahead in the game and put them back in first place, wrested away by the Orioles the night before.

"Mantle turned and raced back for it, running at full speed in his choppy, high-stepping sprinter's stride," Robert Creamer wrote in *Sports Illustrated*. "He looked up over his right shoulder as he ran, and as he neared the ball he lifted his glove to catch it. But the point of juxtaposition of glove and ball that Mantle had anticipated was theoretical—it lay a few feet beyond the seven-foot-high wire fence bounding the outfield . . . as the ball went over for a home run Mantle ran into the fence. His left foot hit on the downward stroke of its stride, and his spikes caught in the wire mesh. The front part of his foot was bent violently up and back."

The Orioles knew all too well the perils of the unmoored chain-link, which was exacerbated by the absence of a cinder warning track. They understood immediately what had happened. "The bottom of it caved in," said first baseman Boog Powell. "His leg went under and sprung back."

Mantle was surrounded by Yankees before Robinson rounded second base. "It's broke," he said. "I know it's broke."

This time he couldn't refuse a stretcher. He was carried off the field looking like a "warrior on his shield," one reporter wrote, and taken by ambulance to the hospital, where an X-ray detected an undisplaced, slightly oblique fracture of the third metatarsal neck, which means he broke one of the long bones in the foot that attach to the toes. His leg was placed in a cast up to his knee, and he was back at Memorial Stadium in time to hear the crowd cheer when the public address announcer confirmed the fracture.

"Isn't there some way they can strap this thing up so I can play?" he asked when trainer Joe Soares predicted, accurately, that he would miss at least six weeks.

Dan Topping dispatched his twin-engine Grumman Mallard from Southampton, where he had been fishing the waters off Long Island. Reporters met Mantle on the tarmac at LaGuardia Airport, and watched as he was helped off the plane. A photograph taken as he hobbled toward a waiting limousine was converted into an annotated medical chart with dates and arrows fixed to every part of his body except the grimace on his face: knees, tonsils, shoulders, rib cage; abscessed hip, fractured finger, fractured foot; pulled, sprained, and torn muscles; surgery, surgery, and more surgery.

In the limo, Milton Gross tried and failed to get Mantle to say he was unlucky, which furthered the apotheosis of The Mick. By the time the morning papers rolled off the presses, the "Man of Mishaps" had been transmogrified into "a tragic figure," "the champion hard-luck guy," and "the most fabulous invalid in the long history of sport."

Team physician Sidney Gaynor removed the cast on June 24 and happily predicted a return by the All-Star

Game on July 9. But on July 26, the cartilage in his left knee, torn in May 1962 and torn some more in the altercation with the cyclone fence, gave way. "It's my fault," Mantle said. "My foot hurts, so I have to run on the side of it. I think that's what makes my knee bad."

He had reached a breaking point: the cumulative injuries and mishaps, the residue of (unacknowledged) bad luck and primitive sports medicine, rendered him a part-time, one-dimensional player. He would play only sixty-five regular-season games in 1963. Of those, five would become intrinsic to the mythology of The Mick. He soared, he crashed, he persevered, he indulged, and he looked bad.

3.

Mantle was in the trainer's room when Al Downing arrived early Sunday morning, August 4, for the usual day-after ministrations on his pitching arm. He had won his seventh game the day before, having joined the starting rotation in Mantle's absence. Though it had been two years since his major league debut, Downing felt, "I had no credibility. Until you showed what you could do, nobody's gonna be around every day asking you how you feel."

Mantle called the kid over. "He said, 'You've been

doing a great job. Keep it up.' This was the first time he had a chance to say something to me. And that meant a lot, because all the guys had been telling me all along, but of course, *he* hadn't been there."

Mantle was feeling chipper. On the field before the first game of the Sunday doubleheader with Baltimore, Ralph Houk asked if he thought he could hit. Mantle said yes. Houk said he might use him if Mantle promised not to run.

The Yankees were trailing by a run in the bottom of the seventh inning of the second game when Mantle emerged from the dugout to pinch-hit. The ovation began at the bat rack and reached a crescendo when Bob Sheppard announced his name. The roar "shook the windows of the Bronx County Courthouse," one paper said. Equally shaken, Mantle dug his spikes into the dirt on the right side of the plate and told himself, "Don't just stand there and take three pitches—*swing*."

The Orioles were well aware that it was his first at-bat since the injury in Baltimore. "Big George Brunet was on the mound," Brooks Robinson said. "He just reared back and threw on every pitch."

Brunet threw one pitch, which Mantle took for a called strike, and then another. "Mantle swung his bat in anger for the first time in 61 games," the *Times* reported, redirecting the ball into the left field stands.

"The most amazing thing is, it was not a pitch that most right-handed hitters are ever gonna get airborne," Downing said. "And not only was it airborne, it was airborne about twenty feet off the ground and just hit those seats and ricocheted like a rocket!"

Mantle wasn't sure he had pulled the ball enough. When the umpire signaled home run, he thought, "Gee, I'm a lucky stiff." He broke into a cold sweat and something that resembled a trot. Later, he said he wasn't sure how he made it around the bases and, in fact, didn't remember doing so. The roar of adulation eclipsed the two-minute standing ovation that had greeted him when he hobbled to the plate. It got louder and louder as he headed toward home. The Orioles applauded silently. "Gives you chills standing over there at first base," Powell said. "Just being in the ballpark gave you chills."

As Mantle rounded third base, Brooks Robinson thought "That's why he's Mickey Mantle."

By the time he reached home plate, "there was tears runnin' all over his face," Yankee pitcher Stan Williams said. He noticed because it was one of the rare occasions when Mantle allowed the outside world to see "how much it meant to him, how much the fans meant to him, how much the moment meant to him."

After All These Years, New York Discovers
Mickey Mantle
—*New York Herald Tribune*

4.

In the days that followed, Houk made it clear that
Mantle would be limited to pinch-hitting duties for
the foreseeable future. He batted only seven times in
August and started only eleven regular-season games
in September. One night after the Yankees clinched the
pennant, Bill Guilfoile, the assistant director of public
relations, was leaving the Stadium with *Post* reporter
Maury Allen. "We saw this figure walking up the
stairs ahead of us, taking each step with both feet, like a
small child," Guilfoile said. "It was Mantle. We stayed
behind. We didn't want to embarrass him. He finally
made it to the exit and departed in a waiting taxicab."

Players around the league knew "The Brute's" reputation for playing hurt, so Mike McCormick, the Orioles' starting pitcher, was surprised that Mantle wasn't
in the starting lineup when the Yankees returned to
Baltimore on September 1. And he wondered why
Mantle wasn't in the dugout when the game began.
"Apparently he was in the training room on the table,"

McCormick said. "I guess they were tryin' to sober him up."

On the day Mantle and Ford were inducted into the Hall of Fame in 1974, they fessed up to Dick Young of the *Daily News*—"Some Tales They Dare Tell."

"Ralph told you you didn't have to play the next day, so we visited those friends of yours," Whitey said.

"That was the Cressens, good friends of mine from Dallas, who had this farm near Baltimore. So when Ralph said I wasn't playing the next day, we went out to see them."

"I had pitched the night before," said Ford, "so I knew Houk didn't need me."

"We stayed up most of the night," said Mickey. "I slept a couple of hours on a hammock.

"We decided to go straight to the ballpark, about 8 o'clock in the morning, and we stretched out on the tables in the trainer's room 'til about 10. Then we suited up and were the first ones on the field."

Their old running mate Hank Bauer was coaching third base for the Orioles. He took one look at Ford's eyes and said, "Has Ralph seen you?"

Once a comrade, always a comrade. "I sent the clubhouse boy out to get mouthwash," Bauer said. "Whitey really smelled bad. Whitey said, 'It's July, but number seven is lying on the table singing "Jingle Bells."'"

Actually, it was September, and the Yankees were losing 4–1 in the eighth inning when Houk went in search of a pinch hitter. Mantle was asleep on Ford's shoulder at the far end of the dugout. Boyer, who was on first base, thought, "He shouldn't have come out of the clubhouse."

Anyway, Bauer said, "Whitey hit Mickey in the ribs, you know, woke him up. 'Here comes Ralph.'

"He asks Mickey, 'Can you hit?'

"Mickey says, 'Yeah! But I'm on the disabled list.'

"He says, 'No, you're not. You came off today.'"

In fact, he wasn't on the disabled list, so he couldn't have come off it. Like all the best baseball stories, this one has aged very well. "So, meantime, Mickey gets up and goes to the bat rack and gets all his stuff and I go to manager Billy Hitchcock," Bauer continued. "I told him, 'He's not feelin' very good.'

"I said, 'Billy, if Mickey's going to hit, tell McCormick, "Don't throw that ball upstairs to him, it might end it."'"

It was McCormick's first year in the American League. "Billy came running out to the mound and wanted to know had I ever pitched to Mantle," McCormick said. "And I said I had pitched to him a couple of times in All-Star play. He said, 'How do you pitch to him?'

"I said, 'I try to waste my fastball and get him out with an off-speed pitch.'"

A screwball, preferably.

"Billy said, 'Okay, don't let him hit a fastball.'

"I said, 'I won't.'"

Hitchcock didn't tell him that Mantle was feeling under the weather; none of the Oriole players knew. Hitchcock couldn't have taken more than one step down into the Orioles' dugout when McCormick tried to waste a fastball up high. "Mantle hit the most god-awful tomahawk-swing line drive into the left field bleachers," McCormick said.

"Over the hedges," Bauer said.

"Honest to God, I didn't think he'd make it around the bases," Boyer said.

"He kinda sobered his way around," McCormick said.

He would hear about it for the rest of his career: "Even the drunks can hit home runs off of you."

Safely back in the dugout, Mantle said, "Those people have no idea how hard that really was."

Boyer marveled fifty years later, "Jesus Christ, he could play with a hangover."

Mantle returned to the starting lineup on September 14, the day after the Yankees clinched the pennant. He had two goals: get into shape for the World Series and finish what he started in May against Bill Fischer. The still shell-shocked A's returned to New York a week later, this time with Mantle's old friend Tom Sturdivant, who had been traded to Kansas City in July. Mantle told him there was only one right-handed pitcher in the league with good enough control to help him hit a ball out of Yankee Stadium. "He had been hitting the top of the Stadium," Sturdivant said. "He wanted to be the first one to hit it completely out of the Stadium."

The Snake agreed to help. Saturday, September 22, was a cold day in New York, almost football weather. The Yankees were ahead 4–3 when Sturdivant entered the game in the seventh inning. Bobby Richardson was on first base. Bob Sheppard invoked the familiar words: "Now batting for the Yankees . . ."

The sparse crowd was treated to an unfamiliar sight: Mantle batted right-handed against the right-hander, a transgression his father had abhorred. The departure in form was duly noted upstairs in the press box. Queried about it later, Mantle said he wanted to see if he could hit behind the runner batting from the right side.

Sturdivant had promised to throw it right down the

middle. But manager Eddie Lopat had ordered him to throw a curve, and there was a big fine for violating one of Lopat's rules. So, Sturdivant said, "I threw a crossfire, sidearm curve. He didn't swing at it."

Opportunity rode in with the next pitch. "I let Mickey know I'm going to throw it real quick and it's going to be a fastball," Sturdivant said. "Doc Edwards, my catcher, told him, 'You better get ready.'

"The second pitch was the fastball, just a hair over letter high and inside. He hit a line drive into the monuments, and Bobby Del Greco caught it. We just had three then. It was between Miller Huggins and Babe Ruth. I told him, 'You can't hit me.'

"I got my butt chewed pretty bad. I said, 'Eddie, what difference does it make?'

"We played baseball for fun."

5.

The first World Series between the Los Angeles Dodgers and the New York Yankees convened in the Bronx on October 2 before 69,000 witnesses. Teachers throughout the metropolitan area suspended class. Black-and-white TVs were wheeled into thousands of school gymnasiums for NBC's broadcast with Vin Scully and Mel Allen. Five hundred forty radio stations across the country carried the voices of Joe Garagiola and Ernie Harwell. The Vegas line made the Yankees 8-to-5 favorites to win the Series and 6-to-5 favorites to win the opener. Bookmakers posted 25-to-1 odds against a Dodger sweep.

No one expected the hitting-challenged Dodgers to break up the Yankees. Not even after Sandy Koufax struck out the first three batters on twelve pitches—the first of fifteen strikeouts, a new World Series record.

By the time Mantle led off in the bottom of the second, the Dodgers were ahead 4–0, thanks in large measure to John Roseboro's schnapps-fortified home run. This at-bat would set the tone for all the innings to come. It was one thing for Koufax to have his way with Kubek, Richardson, and Tresh, but this was The Mick.

Swing and a miss. Strike one.

High with a fastball. Ball one.

Foul back on the screen. Strike two.

"Then he struck him out with a fastball around the letters," Roseboro said. "Mantle looked back at me and said, 'How in the fuck are you supposed to hit that shit?'"

There was nothing else to say. Except to pray for more Jewish holidays, as Boyer pointed out on the team bus when the Yankees left for Los Angeles, trailing

2–0 in games. "You mean like Yom *Koufax*?" Mantle replied.

Tom Tresh bravely and brashly asserted, "Now that we've had a chance to look at him we'll get to him easier the next time."

Koufax said, "I was feeling a little weak."

In Los Angeles, Johnny Podres and Don Drysdale also had their way with the Yankees and Koufax returned to pitch game 4 on October 6, a perfect day with a perfectly blue Dodger sky. With one out in the top of the seventh inning and Koufax nurturing his customary 1–0 lead, Mantle strode to the plate. "He challenged him with a fastball in there about belt high," Ralph Terry recalled. "It's like two bulls—just power against power. Mickey fouled one of them back, and he came in there again. Mickey choked the bat just a little bit and met the Koufax fastball. That thing got about halfway up the bleachers so quick. I never seen a ball leave the ballpark that fast. Bing, *bing*."

The ball was hit so hard it made shortstop Dick Tracewski's ears ring. Mel Allen got so excited he lost his voice. It was Mantle's fifteenth World Series home run, tying him with The Babe as October's most prolific slugger.

A half inning later, Joe Pepitone—Moose Skowron's successor at first base—lost Boyer's throw in an ocean of white shirts and the Dodgers took the lead 2–1—a typical L.A. rally. "See, Joe, I told you you'd fuck up one of these games," said Whitey Ford, who had given up only two hits.

In the top of the ninth inning, Koufax faced the heart of the Yankees' order—minus Roger Maris. Bobby Richardson singled, and Mantle came to bat with a chance to put the Yankees back in the game and the Series. Quickly, he found himself down two strikes.

Koufax was a master puppeteer. He manipulated the ball and expectation with equal legerdemain. As he went into his windup, Roseboro belatedly wiggled two fingers, telling him: *Take something off the curve.*

Koufax was thinking the same thing. Mantle had made his show of strength. Now Koufax intended to use that strength against him. The best curveball in baseball punctuated Mantle's futility like a comma. He was caught looking. And he had never looked worse.

Mantle's shadow clung to him as he trudged away from greatness: head down, shoulders hunched, a solitary figure framed against combed dirt. It was a gnome-like vision of The Mick, gnarled, diminished, alone. He gripped the bat below the label, where no one wants to make contact with the ball. The shadow of his former self embraced him, neither trailing behind nor pointing the way forward.

15

September 26, 1968

LAST LICKS

1.

It came as no surprise to anyone aboard the team bus that Mantle was the instigator of the Great Harmonica Crisis one muggy Chicago afternoon in the heat of the pennant race of his last great season. The Yankees had just lost their fourth straight game to the White Sox. No one was in a good mood. Phil Linz was feeling particularly aggrieved. He thought he should have been in the starting lineup and nursed his resentment along with more than one clubhouse beer while waiting an hour and a half for the bus to arrive. "We always had beer in the clubhouse," Linz said. "I don't think I'd ever have done it if I hadn't had a couple of beers."

Earlier that week in Baltimore, Tom Tresh had bought himself a Hohner harmonica; Tony Kubek appropriated it. So when they got to Chicago, Tresh took Kubek to buy one of his own. Linz tagged along. Kubek serenaded Neiman Marcus shoppers with his rendition of "Streets of Laredo," a mournful cowboy lament about a guy who spent more time in the dram house than in the saddle.

Linz liked the sound and decided to buy a harmonica for himself. It came with instructions on how to blow in and blow out, and with the sheet music to "Mary Had a Little Lamb." Linz practiced assiduously. When he left for Comiskey Park on Thursday, August 20, getaway day, he stuck the harmonica in the pocket of his sport coat and forgot about it until after the dispiriting 5–0 loss that left the Yankees in third place, 4½ games behind the White Sox.

En route to the airport for the flight to Boston, the bus was mired in the same traffic that had caused it to be late in arriving. Linz was sitting in back with Pepitone, Bouton, Maris, Tresh, and Mantle. "All the guys who kid around sat back there," he said. "You couldn't goof around in the front."

Everything was quiet until he began to play "Mary Had a Little Lamb."

Yogi Berra, the first-time manager, was up front

with third base coach Frank Crosetti, an old-line baseball man if ever there was one. "I said, 'No joking around,' " Berra recalled.

Linz asked, "What did Yogi say?"

Mantle answered. "He said, 'Play it louder.' "

Linz knew Mantle well enough to know he was egging him on. He recognized the wicked-angel smile. But he played on anyway. "Shove that harmonica up your ass!" Berra said, charging down the aisle, throwing arm raised in anger.

"Here, take it," Linz replied, and tossed it to the manager. "I flipped it to him, a double-play flip." And then, because he was in a bad mood, he added, "What are you getting on me for? I play as hard as I can to win."

Berra cocked his arm and fired the harmonica at the bad boys in the back of the bus. "He threw it at me," Linz said. "I mean, he *threw* it. It hit Pepitone in the knee."

Pepitone swore it drew blood and left a scar. That might have been the end of it, except beat writers traveled with the team then and witnessed Crosetti's indignation at what he called "the first case of open defiance" he had seen by a Yankee player in thirty-three years.

"When we got to Boston the next day, it was all over the country," Linz said. "It was like World War III."

That afternoon at Fenway, he apologized to Berra, who was serenaded throughout the game by hundreds of harmonica-wielding Red Sox fans. "We shook hands, we hugged. He said, 'Phil, I gotta fine you. The writers are coming. How much do you think I should fine you?'

"I said, 'Whatever, Yogi. I was wrong.'

"He said, 'How 'bout $250?'

"I said, 'That's fair.'"

More boldface headlines: "Yogi Fines Linz $250."

Writers pounded Berra with invective—he was a "flip-top manager" who had "betrayed a sense of panic" in "un-Yankee-like" fashion. His team was said to be "cracking up." Berra's job was said to be "in jeopardy." Western civilization was coming to an end.

This grave insurrection was regarded by the Fourth Estate as prima facie evidence that he was incapable of managing his former teammates. In fact, what Berra lacked was not gravitas but what he might have called young youth. It was getting late early in Berra's managerial career.

The Yankees rallied for him, winning thirty of their next forty-three games to overtake the White Sox. Mantle slugged them back into first place on September 17, with a double, a single—his 2,000th major league hit—and a home run, the 450th of his career. He got a standing ovation. Linz got a $10 check from a bandleader, presumably for furthering the cause of music. He also received a barrelful of harmonicas, a $5,000 endorsement deal with Hohner, and a $250 bonus in his 1965 contract. "Houk put it in—$20,000 plus $250 for music lessons."

Mantle had a good year—a career year for anyone else. He led the American League in on-base percentage and on-base plus slugging. He batted .421 right-handed (.241 left-handed), drove in 111 runs, and hit more than 30 home runs for the last time in his career, 15 right-handed, 20 left-handed. On August 12 in a home game against the White Sox, he hit one of each for the tenth and last time in his career. The left-handed homer came off Ray Herbert in the fourth inning. Mantle hurled his bat in disgust after the swing because he thought he hadn't gotten enough wood on the sinking fastball. Herbert thought so, too, and turned to watch center fielder Gene Stephens make the catch. Back, back, back he went to the 461-foot sign in deepest center field. "Is he going to sit in the bleachers and watch the rest of the game?" Herbert wondered.

The ball landed fifteen rows into the black in the center field bleachers, where only one ball had gone before, also courtesy of Mantle. The Stadium manager said it was the longest home run ever measured in the Bronx. "Nice play," Johnny Blanchard said, greeting him in the dugout. "You broke your last bat."

Mantle was breaking down. Too often his right knee buckled beneath him when he swung left-handed. He surrendered center field to younger, more reliable legs. Reams of Conco tape and stoicism could not camouflage the deterioration of soft tissue. "They taped his legs so tight it cut off the circulation," Pepitone said. "Sittin' on the bench, they'd puff up from all the blood rushin' down to his feet."

When the tape came off at the end of the day, his legs were chalk white. Multiple surgeries and inattentive rehabilitation had left him with almost no muscle above his knees, compounding the instability in the knee. "There was no skin, no cartilage, no nothing," Pepitone said. "Everything was gone. And scars. I mean lines down the side of his leg, Xs across the top, underneath. It looked like somebody got a knife and just kept slashing at his knees. You could see the stitching marks where they cut."

To his buddy Roy Clark, the country singer, Mantle's body looked like "a statue by Michelangelo that somebody had just started chopping at." Mantle was self-conscious enough that he refused to put on a bathing suit at a family picnic when Frank Petrillo shot home movies. He played 125 games or more only three times after his thirtieth birthday.

Kids who imitated his stiff-legged arrival at second base didn't realize it was the only way he could bring himself to a stop. His peers expected him to transcend because he always had. He was like a horse who "comes out of the stable just hobbling along," said pitcher Billy Hoeft. "You figure, 'The Brute's not gonna play.' Then he's out there taking batting practice and he's in the lineup, and then he hits a ball and he's gotta run like hell and he does."

2.

The 1964 World Series between the Yankees and the St. Louis Cardinals marked a paradigm shift ably documented by David Halberstam in *October 1964*. Stan Musial had just retired. The Cardinals' lineup, with Curt Flood and Lou Brock in the outfield, show-cased the future of America—young, black, ethnic, fast. They left the old white Yankee establishment in the base-path dust. Mantle, Milton Gross wrote in the *New York Post*, was "playing on memory and nerve."

The shoulder injured in the 1957 World Series be-trayed him and would require off-season surgery to tie the tendons together. Four times he threw wildly to the infield; twice in one game he was caught off base; he was thrown out trying to stretch a single into a double and thrown out on an RBI grounder that should have been a hit. He threw his batting helmet in despair and disgust—it was his best throw of the Series. "Couldn't throw the baseball the length of the room I'm sitting in hardly," Cardinal shortstop Dick Groat said. "And yet he knew that the Yankees were a better team when he was in the lineup and the team felt better when he was in the lineup."

An egregious error in the fifth inning of game 3 at the Stadium allowed the Cardinals to tie the score at 1–1 and set up the last great unorchestrated moment of his career. "By that time I couldn't run too much anymore," he told me. "They put me in right field and Maris in center. Somebody hit me a ground ball. I nonchalanted it. It went through my legs, and the guy scored."

In the bottom of the ninth inning, the Cardinals' manager, Johnny Keane, gave him a chance at redemption by summoning Barney Schultz from the bullpen. Jim Bouton, the Yankees' exhausted starting pitcher, was at the water cooler at the end of the dugout when Mantle came to collect his bat. "He was standing there with the bat on his shoulder watching Barney Schultz. His warm-up pitches were coming in about thigh high and breaking down to the shin, to the ankles—two or three in a row. Mickey said, 'I'm gonna hit one outta here.'

"It wasn't a big announcement. He wasn't like that.

He wasn't a grandstander. He understood that Barney Schultz was the wrong guy for them to bring in. Mickey had two different batting strokes: right-handed, he would hit on top of the ball. He would tomahawk the ball. Even his home runs to left field, a lot of 'em were line drives with top spin on 'em. They get out there real quick and sink and dive down into the bleachers. Left-handed, he undercut the ball. So here's Barney Schultz throwing right into his power. Barney Schultz's ball is breaking down, and Mickey's bat goes down and . . ."

Schultz was an old knuckleball pitcher, playing in his first World Series at age thirty-seven. In the bullpen he felt "the good pop" between his right elbow and his hand that presaged a promising outing. "The ball just bounced all over," he said. "It was the probably the best stuff I had all Series."

Tim McCarver, the Cardinals' young catcher, knew Mantle was hurting. "I could even hear him groaning on some swings," McCarver said. "A swing and a miss was real bad."

Had he been privy to the dugout conversation, Mc-Carver would have walked Mantle intentionally. "He was not a man who said boastful things about himself," he said. "When somebody like that says something like that, you have a tendency to think that he's gonna do it."

Mantle stood in. Schultz wound up. McCarver knew right away: "Nothing good was gonna come of this pitch."

It didn't dance or flutter or defy expectation. It didn't do anything at all. "It wasn't thrown," McCarver said. "It was dangled like bait to a big fish. Plus it lingered in that area that was down, and Mickey was a lethal low-ball hitter left-handed. The pitch was so slow that it allowed him to turn on it and pull it."

The ball sailed over Len Melio's head in the right field third tier, a high, majestic thing—he could see the laces spinning, a pinwheel of red and white coming right at and then over him. As he turned to see where it landed, his elbow hit the beer of a Yankee hater and "spilled all over him," which, he said, "was what I was happy about."

Schultz took one quick look over his shoulder and walked off the mound. "I crossed the third base foul line as he was rounding third base," he recalled. "I didn't even watch him run the bases. I wasn't interested in that. I was interested in punching myself in the mouth."

In the locker room, a clubhouse attendant handed him a note from a secret admirer, the actress Rosalind Russell. It said, "Barney, you're still the greatest."

Schultz finished the World Series with an 18.00

ERA. Mantle hit two more homers, the last of his eighteen World Series home runs—still the major league record. But the one off Schultz mattered most because it lessened the sting of his fielding error. "It got me off the hook, being the goat," he told me. "And besides that, it broke Babe Ruth's home-run record in World Series play."

As fans poured out of the box seats, Crosetti escorted Mantle home—a departure for the self-contained third base coach. McCarver waited at the plate amid a pinstriped scrum to make sure Mantle crossed the plate.

The next day, Groat engaged Mantle in conversation at second base. The Yankees were leading 3–0 and had men on first and second with no outs. They were on the verge of putting the game out of reach. "I said, 'Mickey, you didn't have any doubts about that home run goin' out?'

"He said, 'No, why?'"

Groat described Mike Shannon's elaborate pantomime in right field. "I said, 'I thought the ball was off the end of the world, and Mike stood like he was gonna catch the ball. I thought, 'Maybe the ball is not hit as far as I thought it was.' And Mickey started to laugh."

Roger Craig went into his motion, pitching from the stretch position. Mantle took his lead off second base. "I could see in his head he was still chuckling," Groat said. "He put the weight on that right foot, and I said, 'Oh my, do I have him.'

"I went over. Roger Craig gave me an absolutely perfect throw, and Mickey is out by a mile."

Mantle dove back into second base face first and had a few choice words for Groat after he was called out. Good manners precluded him from repeating what was said, but Mantle called him "the same dirty name" from then on. "Every time I saw him at an exhibition game, he ended up saying, 'Sure, tell me a funny story and then pick me off second base.'"

The Cardinals came back to win the game, to tie the Series, and become World Champions. Later, many pointed to the pickoff play as a turning point and a portent for the future of the Yankees.

3.

The Yankees followed Mantle downhill. Berra was fired. Johnny Keane was hired. Jerry Coleman, Mantle's onetime roommate, then a Yankee broadcaster, came to believe that the Yankees' decline paralleled their behavior off the field. "They slid into a degenerate-type thing," Coleman said. "That's why they went downhill."

This is a notion that rendered even Bouton speechless. "We were as debauched in 1963 as we were in 1965," Bouton said. "We weren't as good at it in 1965. We didn't have the energy for it. Guys were older. We didn't really lose anything on the debauchery front."

They just lost. In 1965, the Yankees lost more games than they won for the first time since 1925. "I shoulda quit right then," Mantle told me.

But the Yankees needed him. "He was really the only ballplayer on the club that anybody cared anything about," said Howard Berk, a team vice president from 1967 to 1973.

Berk and company invented reasons to come to the ballpark to see him, including the first Mickey Mantle Day, held on September 17, 1965, his 2,000th major league game. The Tigers' starting pitcher, Joe Sparma, walked off the mound to shake his hand.

Mantle asked that all donations be made to the Mickey Mantle Hodgkin's Disease Research Foundation at St. Vincent's Hospital, which had been dedicated the year before. Still, he received enough loot to fill two mimeographed sheets—a car, a year's supply of gasoline and bubble gum, two rifles, two quarter horses, a mink for Merlyn, and a six-foot-long, 100-pound salami in the shape of a bat. He told the sellout crowd that he hoped to play another fifteen years.

The 1966 season was even worse. When a reporter

asked Mantle in spring training how the Yankees were going to do, he replied, "We're going to surprise some people—we're going to finish last."

They did—twenty-six games out of first place. Days before the All-Star Game in St. Louis, he called American League manager Sam Mele and said his legs were killing him. "Hell, yeah, take three days off," Mele replied, replacing him on the roster with Rookie of the Year Tommie Agee.

The response was vituperative. "You're a nigger lover, picking Agee," a caller to Mele's hotel room snarled. "And by the way, when you go on the field I'm going to shoot you and Tony Oliva."

Local police and the FBI provided security, Mele said, but no peace of mind. When he delivered the starting lineup, he ran to home plate, hid himself among the men in blue, and raced back to the dugout.

Mantle played in 661 games over the last five years of his career. If they all ran together in a blur of at-bats and road trips and getaway days, it was partly the result of the repetitive motion of eighteen years in the major leagues and partly the result of increased consumption. The longtime trainer for the Detroit Tigers regaled pitcher Mickey Lolich with a tale about the time he tried to drink Mantle and Ford under the table. There was an afternoon game the next day, and he figured he'd take one for the team. When the bar closed at 2:30 A.M., he bought a bottle of vodka from the barkeep and suggested a nightcap back at the hotel. "When he left at 6:30 A.M., Mickey and Whitey were in no condition to play baseball," Lolich said. "He had done his job as far as he was concerned. He said, the next day Whitey pitched nine innings of shutout ball and Mantle hit two home runs."

If the morning after was rough, the trainers provided a pharmacological pick-me-up. Greenies, the players called them. Everybody took them. Mantle, too? "By the handful," Linz said. "Then he'd go out and play."

Mantle's capacity was admired, envied, and imitated—to the detriment of some of the young Yankees, who were expected to become him. "You're going to be a helluva player," he told Joe Pepitone when he was living with Mantle in his hotel suite. "Don't do what I do."

Easier said than done. Steve Whitaker fell under Mantle's spell his rookie year after Mantle extended an unexpected and much-coveted invitation to dinner. Pretty soon, Whitaker says, they were going out every night—and staying out. Mantle made New York City feel small. He knew every restaurant and every maître d'. He introduced Whitaker to "the life," as Linz calls it, to joints like Dudes 'N Dolls, where Goldie Hawn danced and Joe Willie Namath partied. "I didn't care if I slept at all in New York," Whitaker said. "It's open 24/7, and, trust me, I closed it."

There was a quid pro quo, though it didn't feel like one at the time. "I was kinda like his shill, the guy to make sure nobody was hawkin' his table," Whitaker said. If there were women present, they had to be with him, not The Mick. The price didn't seem high until later, when he realized the opportunity he had squandered. "In fact," he said, "it probably was the end of my career."

By 1967, Mantle was the only reason to go to the Stadium. The weight of the listing franchise rested uneasily on his fragile pins. He became a first baseman and kept his promise to Merlyn to hit his 500th home run on Mother's Day which the Yankees turned into another occasion to honor him and boost attendance. "The gimmick was, every five hundredth fan entering the ballpark got an autographed ball," Berk said. "We asked him if he'd sign. He did it. Signed for hours and hours. I got a summons from the state of New York for violating the lottery law. A five-hundred dollar fine."

Mantle conceded that he would never catch or match Willie Mays statistically. That debate was over. But

Madison Avenue saw a different future for The Mick. George Lois, the groundbreaking art director and adman, had a certain editorial genius—his conceptual covers for *Esquire* magazine were the subject of a 2008 retrospective at New York's Museum of Modern Art. He also had a keen instinct for marrying pop-culture pitchmen to unlikely clients. He put Mantle (and Joe Louis) in TV commercials for a Wall Street brokerage house. "When I came up to the big leagues, I was a shufflin', grinnin', head-duckin' country boy," Mantle said to the camera. "But I know a man down at Edwards & Hanly. I'm learnin'. I'm learnin'." It became one of Johnny Carson's stock lines.

Lois also put Mantle to work on behalf of the baby cereal Maypo. "Mickey was visiting me on a set for some other commercial. I said, 'Mickey, do me a favor.'"

He handed Mantle a bat and a Yankee jersey and explained how they were going to transform Maypo into "the oatmeal cereal that heroes cry for." All he had to do was blubber—"I want my Maypo"—when the voice-over said, "Mickey Mantle?"

Mantle boo-hooed perfectly and on cue.

"I shot that in three or four seconds," Lois said. "The sales tripled. I hardly had to work with him at all. I gave him the shtick once or twice, and he'd give it back better. He was a natural, fluid."

A.N.C.

Aug. 1951   25c

BASEBALL
MAGAZINE

In This Issue:
MICKEY MANTLE
BONUS PLAYERS
MEXICAN BASEBALL
THE BEAN BALL!

Founded 1908

"He has it in his body to be great."
—Yankee manager Casey Stengel

WEEKLY

JUNE 18, 1956

SPORTS

ILL

25 CENTS
$7.50 A YEAR

MICKEY MANTLE
THE YEAR OF
THE SLUGGER

"Nobody could play baseball better than Mickey Mantle played it in 1956—fastest man in the league, strongest man in the league, switch-hitter, he could play the heck of center field. Nobody had the charisma; nobody looked as good in a uniform as Mickey Mantle did."
—outfielder/manager Whitey Herzog

"One day in spring training he bunted on me. I fielded the ball. He ran so hard it sounded like a bunch of wild horses running by."
—pitcher Ed Roebuck

SPORTS ILLUSTRATED

MARCH 4, 1957
a Time Inc. weekly publication

25 CENTS
$7.50 A YEAR

1957
SPRING TRAINING

THE REAL MICKEY MANTLE
BY GERALD HOLLAND

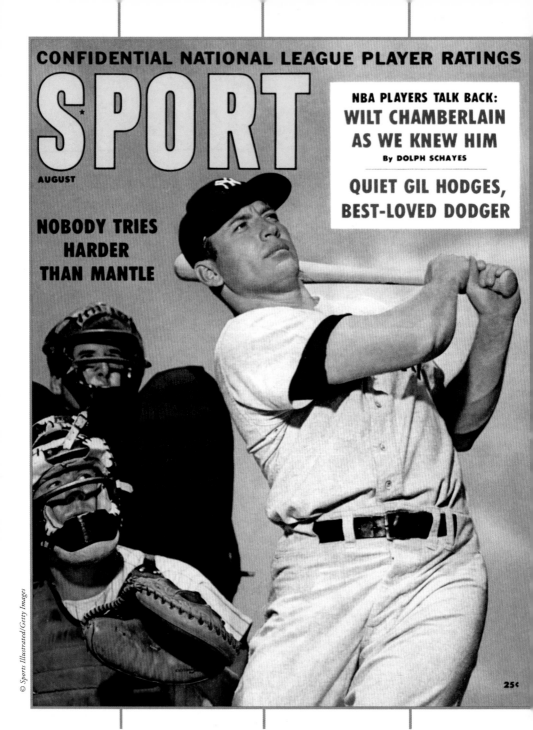

CONFIDENTIAL NATIONAL LEAGUE PLAYER RATINGS

SPORT

AUGUST

NBA PLAYERS TALK BACK:
WILT CHAMBERLAIN
AS WE KNEW HIM
By DOLPH SCHAYES

QUIET GIL HODGES,
BEST-LOVED DODGER

**NOBODY TRIES
HARDER
THAN MANTLE**

25¢

"If you were sitting out past the 457 marker in left-center field, in the last row in the bleachers, and he started to even walk to the plate, you knew who he was. You didn't need Mel Allen to tell you."
—first baseman Mike Epstein

"Sixty, count 'em, sixty! I'd like to see some other sonofabitch do that!"
—**home run king Babe Ruth**

EXCLUSIVE LIFE STORIES • OVER 200 CAREER PHOTOGRAPHS

Roger **Maris** *Mickey* **Mantle**

COMPLETE PICTORIAL COVERAGE OF HOME RUNS 60 & 61

LIFE

MANTLE'S MISERY

He faces physical pain and a fading career

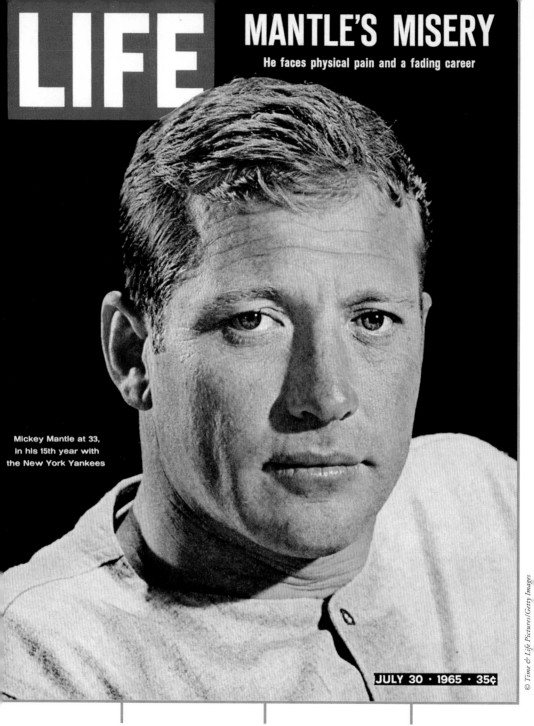

Mickey Mantle at 33,
in his 15th year with
the New York Yankees

JULY 30 · 1965 · 35¢

"One day I was standing by the players' gate when a beer truck pulled up. The beer distributor said, 'Hey, kid, you want to get into the Yankee clubhouse?' What kid doesn't want to get into the Yankee clubhouse? The door to the trainer's room was open. Mickey was lying on the table. The trainers were stretching out and measuring yards upon yards of tape. He had to be put together like a mummy."
—Yankee fan Paul Berkman

"What's the one thing you miss the most about baseball?" Mantle was asked after he retired.

"Stepping in the batter's box at Yankee Stadium and hearing that ripple of applause make its way through the crowd. It sounds farther away than it really is. Sort of like rain falling on a tin roof. When it rains now, that's what I always think of. I'd give my best year in baseball just to hear that sound again."

—author Richard Andersen, "The Dust of the Fields Behind Us"

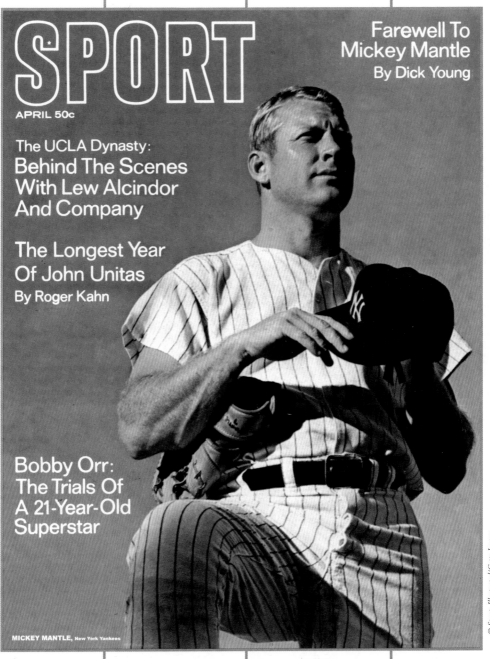

SPORT

APRIL 50c

Farewell To
Mickey Mantle
By Dick Young

The UCLA Dynasty:
**Behind The Scenes
With Lew Alcindor
And Company**

**The Longest Year
Of John Unitas**
By Roger Kahn

Bobby Orr:
The Trials Of
A 21-Year-Old
Superstar

MICKEY MANTLE, New York Yankees

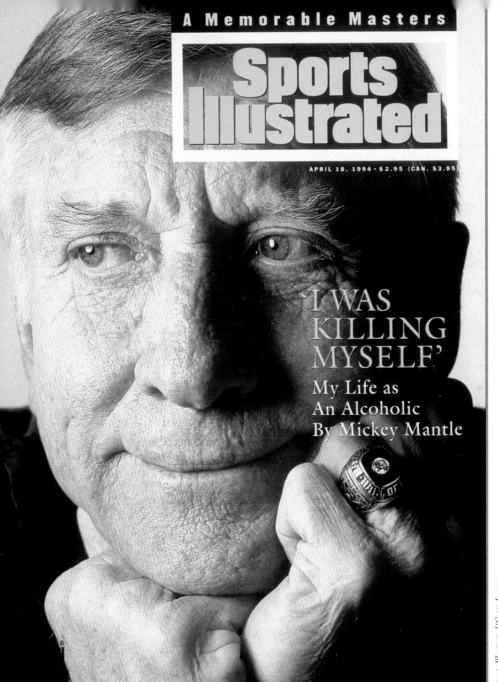

Sports Illustrated

APRIL 18, 1994 · $2.95 (CAN. $3.95)

'I WAS KILLING MYSELF'

My Life as
An Alcoholic
By Mickey Mantle

"He just wanted to have fun and feel good. Baseball was his favorite toy. Life was his favorite toy. You wanna stay nineteen the rest of your life no matter how old you get."
—Yankee teammate Joe Pepitone

Mantle was such a hit, agents for Wilt the Stilt, Dandy Don Meredith, and The Say Hey Kid called, wanting in on the action. Mantle called, too: "Hey, George, go fuck yourself." It seemed that wherever he went, he elicited a chorus of "I want my Maypo."

"He was kidding," Lois said. "He loved it. Probably got five hundred dollars and was happy to get it."

4.

Mantle's last opening day was Marty Appel's first as an assistant to PR chief Bob Fishel. Growing up in Rockland County, New York, Appel wore his Mickey Mantle T-shirt until his mother disposed of the shreds. On his first day of work, Fishel took him down to the clubhouse to meet Mantle, leaving him alone at his empty locker, where Appel discovered his hero's trove of hard-core porn—"guys in black masks and socks." All he could think was, "Oh, my God, I'm looking at Mickey Mantle's porno!"

Every day that summer, he invented a reason to consult The Mick about his fan mail, picking three or four letters or invitations from the hundred or so that arrived daily for Mantle's inspection. "He saw through the scam and would look at me with a little smile," Appel said. "Then he'd crumple them up and toss them into the trash. Once he got an invite to serve as a judge for the Miss Nude America pageant. He didn't crumple that."

When he returned from road trips, Mantle brought the unpaid "vice president of fan mail" the gift certificates he received on out-of-town postgame shows, pulling rumpled vouchers from his pockets: Brew Burger, Howard Johnson's Motor Lodge, Big Yank Slacks, Midas Muffler, Getty Gasoline, and Thom McAn shoes. Some Appel kept as souvenirs; others, like the one from Thom McAn, he used. He thought of The Mick with every step he took.

Appel visited him in the clubhouse while everyone else was on the field doing the daily chores that Mantle now eschewed—stretching, running, infield, and batting practice. "That whole season was surreal to him," Appel thought. "He just felt lost. 'Who are these guys? What am I doing batting .238 and being a teammate of Thad Tillotson?'"

Mantle's 1968 farewell tour of America was a welcome distraction in a year of national anguish and upheaval. In the wake of the assassinations of Martin Luther King and RFK, the inner-city riots and Vietnam protests, his familiar grin—and the possibility he still brought to the plate—brought out the love.

On road trips, he disguised himself in a pair of glasses and a derby hat, his collar pulled up around his ears. Sometimes he got off the Yankees' bus and walked the last block to the team hotel. Sometimes Whitey Ford, the Yankees' new pitching coach, rented a car and went back to the ballpark to collect him. In New York, he shared his suite at the St. Moritz with Mike Ferraro. In the evenings, they played cards and ordered in. Going out was too much of a hassle.

Most of the eyewitnesses to his best years were gone: Kubek and Linz left after 1965; Richardson and Maris after 1966; Elston Howard was traded midway through the 1967 season. The isolation in the locker room was compounded by his frustration on the field. "He'd strike out, and he would explode," said Ross Moschitto, who arrived in 1965. "He would come in the dugout— it didn't matter if he had gone 3 for 4 the day before. Next time he struck out, he'd come back, and *wham*, beat the locker."

On May 30, 1968, the 100th celebration of Memorial Day, the day the Beatles began recording the *White Album*, Yankee Stadium offered a reprise of the good old days. Mantle had one of the best days of his career. Frank Howard was stationed at first base for the Washington Senators. The starting pitcher was Joe Coleman, who had struck out Mantle four times in a row the week before. "I went to the mound and said, 'How are you going to pitch the Big Fella?'" Howard recalled.

"He said, 'I'm going to blow that dead red in that blind spot.'"

Nine innings and six Washington pitchers later, Mantle had two home runs, a double, two singles, five RBI, and three runs scored. It was the third time in his career that he went 5 for 5. He raised his average from .221 to .254.

First inning: two-run home run.

Third inning: singled and scored.

Fifth inning: home run.

Sixth inning: RBI double. "I'm holding the runner on first base," Howard said. "He hits a line drive between me and the bag that's no further than it is across this table. I go to stab it. If it had been hit right at me, it would have taken me to the 296-foot marker."

Eighth inning: RBI single. "I've never seen five balls scalded like that in a ball game before," Howard said. "He laid out two thousand feet of line drives."

The scoreboard flashed the news: "Mantle's finest game since his Triple Crown season of 1956."

Privately, he was wistful—the performance served to remind him of what he could no longer do. One night, he asked Ferraro, "Did you ever see me run? When I could *run*?'"

Clark Griffith was then vice president of the Minnesota Twins, and he and Mantle had become drinking buddies. "Near the end of his career, I had a marvelous evening with him," Griffith said. "He was hurting and in decline. He said, 'Man, it really hurts. I just can't do it anymore.'

"We got into a long discussion about how the game had changed. He started talking about pitchers he had to hit against in the Fifties and early Sixties, Chuck Stobbs, Spec Shea, Connie Marrero, short, round guys who didn't throw hard—and who he has to hit against now, guys who are six foot five and throw hard. His skills had not been able to keep up with the evolution of the game. The size had gotten to him. It had literally outgrown him."

In July, Mantle was named captain of the American League All-Star team, but he stayed in Anaheim only long enough for a quick how-do-ya-do. "All-Star Games were like a cocktail party to me and Whitey," he told the writer Tom Callahan years later. "I got there late, missed the team picture, hurried to get dressed. They pinch-hit me in the first inning. I got dressed, got back on the helicopter, flew back to Dallas, changed clothes to play golf. I went to Preston Trail, and the game was still on the TV. Tony Perez hit a home run in the fifteenth inning. I'm damned ashamed."

As the Yankees headed into a series at home against Cleveland on the weekend of July 19, Bill Kane, the team statistician, walked into the PR department and told Appel, "Mick's lifetime average is going to drop under .300 this weekend unless he goes 3 for 10 or better, and he's not gonna do that."

He went 2 for 8. His average was .299545 at the end of play on July 21. A week later, on July 27, Mantle ceded the title of lifetime .300 hitter for good. He went 0 for 12 in his next three games, striking out four times on July 29. As he passed pitcher Stan Bahnsen on his way back to the dugout, Mantle muttered, "This is my last year. I missed about five pitches I should have hit."

On August 4, the Yankees held yet another day in his honor, Mickey Mantle Banner Day. The contestants with the three best entries won season tickets. All Mantle had to do was shake the winner's hand. "The elderly star," the *Times* called him. He sat on the dugout steps as two thousand fans paraded past him carrying signs proclaiming I WANT TO BE JUST LIKE THE MICK!

Five days later, he was thrown out of the game for cursing the home-plate umpire, only the seventh time he'd been ejected in eighteen years. Soon after that, Appel watched him try on a new pair of spikes in the clubhouse, tossing the old ones away with a perfect peg to the trash can. "This will be my last pair of shoes," Mantle said. Fishel assured his concerned young assistant that Mantle had said the same thing the year before.

Six weeks passed between his 529th home run on

June 29th and his 530th and 531st, against the Twins on August 10. He tied Jimmie Foxx for third place on the all-time home-run list with 534 home runs twelve days later.

On September 17, the Tigers clinched the American League pennant by beating the Yankees in Detroit—a rainout the next night enabled them to recover their senses in time to play an afternoon game on September 19. When Mantle came to bat in the eighth inning, Denny McLain walked off the mound to allow him to bask in the standing ovation offered by the meager crowd. Mantle was his hero, the reason he had become a switch-hitter in high school. McLain had already won his thirtieth game of the year, becoming the first pitcher since Dizzy Dean to do so. He could afford to be magnanimous.

None of the Tigers knew what he was thinking when he called his catcher Jim Price out to the mound. "Listen, he only needs one home run," McLain said. "Let's give him a shot at it. You just go behind home plate, put your glove up, and let me see if I can hit it."

Price nodded and squatted behind the plate. Like McLain, Price idolized The Mick—"he still had that look, that Mickey Mantle look, but all he had was the look."

As Price got down in his crouch, he gave Mantle a look. McLain gave him a batting-practice fastball.

"So the first pitch it was, like, fifty miles per hour and almost on an arc," McLain said. "And dummy takes it for strike one. Mantle looks down at Price and said, 'What the fuck was that?'"

"And Price says, 'I dunno.'"

"So Mantle says, 'Is he gonna do it again?'"

"And Price says, 'I dunno.'"

"So Price starts trotting out to the mound. He gets about halfway, and I said—so loud the whole park can hear it, and there weren't a lotta people in the park—'Just tell him to be ready.'"

"Mickey could certainly hear it. The dugout could hear the whole thing. So Price turns around and goes behind home plate and I throw the next pitch almost the same way but the pitch slid a little bit and he fouled it off. I'm thinking, 'Oh, my God, Jesus, now I got him oh and two.'"

"I gotta tell you, the worst thought that went through my head is 'I'm through fucking around with this. I'm just gonna strike him out.'"

Exasperated, McLain yelled out, "Where the hell do you want it?"

Mantle pointed to where he wanted it.

"I throw one more pitch, and he hit a line drive into the right field seats," McLain said. "I think we all had tears in our eyes because Mickey Mantle represented the game of the 1960s right up to the day he retired. So he goes by first base, Norm Cash hits him on the ass with the glove. He goes by second base and short. 'Nice going.' 'Congratulations.' Now he gets between second and third, and he starts screaming at me, 'Thank you, thank you, thank you.' I'm thinking, 'Jesus Christ, I'm gonna get a letter from the commissioner.'

"So he hits third base, going down the third base line, and he's still saying, 'Thank you, thank you. I owe ya, I owe ya.'

"I says, 'Mick, that's enough.'

"So now he steps on home plate. The crowd is going crazy, and there's a standing ovation."

The entire Tigers team stood on the top step of the dugout applauding. "And he comes back out of the dugout, and don't you know he starts coming toward the mound? I almost pooped in my pants. I just did not want him to get to the mound."

Among the 9,063 spectators, there was some initial confusion about what had transpired. But Yankee announcers Phil Rizzuto and Frank Messer had no doubt about what they had seen. "Aw, you gotta give that McLain some credit, I wanna tell ya," Rizzuto said.

Pepitone waited at home plate to shake Mantle's hand. It was a tough act to follow. In the batter's box, he motioned to McLain where he'd like to have the pitch. "He threw a ball ninety miles an hour at my head," Pepitone said. "I went down. I get up. I look at the dugout. Mickey's got his hand over his mouth, laughin' his ass off."

Not everyone was amused. McLain was cross-examined by righteous scribes defending the purity of the game and questioning Mantles right to third place on the all-time him run list. "They charged down to our clubhouse," said Dick Tracewski. "McLain denied it, of course. What else could he do?"

A letter from baseball commissioner William Eckert arrived soon after. "Word had come to him that I was attacking the integrity of the game of baseball," McLain said, paraphrasing the letter. "Mickey Mantle didn't need any help with home runs, and they would start an investigation and da-da-da-da."

Red Smith offered much-needed perspective in his column, "Sportif," in *Women's Wear Daily*, his flagship paper after the demise of the *Herald Tribune*: "When a guy has bought 534 drinks in the same saloon, he's entitled to one on the house."

After the game, Mantle autographed the ball for McLain: "Denny, thanks for one of the great moments in my entire career, Mickey." It (and the commissioner's letter of opprobrium) was destroyed in a house fire in 1978. Mantle sent another ball. "Until the day he died, he kept thanking me and thanking me and thanking me," McLain said. "I said, 'You could have popped it up. You coulda hit it on the ground. You coulda fouled it off again. You could have missed it.'

"And he says, 'Nah, I wasn't gonna miss the last one.' "

There was palpable relief when Mantle hit his 536th and last major league home run the next evening at Yankee Stadium with no help whatsoever from Red Sox pitcher Jim Lonborg. He played his last home game five days later on September 25. Mantle: 1 BB, 2 SO, 1H. The single off the Indians' Luis Tiant was the Yankees' only hit of the night. Later, in the locker room, Mantle made no declaration of intent. He stuck resolutely to the facts—"I just can't hit anymore, it seems"—and packed a bag for the last road trip of the season.

5.

There are two kinds of baseball fans: those who bellow invective at the opposition no matter what and those who stand for a worthy adversary. On September 28, 1968, 25,534 Fenway Park congregants stood up for Mickey Mantle. Gail Mazur evoked the heterogeneous homogeneity peculiar to Red Sox fans in her gorgeous 1978 poem "Baseball": "the four inevitable woman-hating / drunkards, yelling, hugging / each other and moving up and down / continuously for more beer / and the young wife trying to understand / what a full count could be / to please her husband happy in / his old dreams, or the little boy / in the Yankees cap already nodding / off to sleep against his father / . . . and the old woman from Lincoln, Maine, / screaming at the Yankee slugger / with wounded knees to break his leg."

The old woman wasn't at Fenway Park when Mantle walked to the plate in the top of the first inning. Ralph Houk, who had returned to the dugout in 1967, told Mantle to expect one, maybe two, at-bats in Boston. Mantle had been taking Butazolidin (phenylbutazone alka), the anti-inflammatory drug that caused the disqualification of Kentucky Derby winner Dancer's Image that spring. By 1968, doctors knew that bute was potentially lethal to human tissue—the Dodgers' team doctor joked it worked great, except when it killed the patient. "Ralph told him he wasn't going to play, so he took himself off the stuff," Gene Michael recalled.

The Yankees were accustomed to seeing him struggle up the dugout steps. If there was a railing, he made use of it. Otherwise, Andy Kosco said, "he put his bat down, used it as a cane."

Houk wrote Mantle's name third on the lineup card, as usual. With one out in the top of the first, he walked to the plate to face Jim Lonborg. Like Mantle, Lonborg wasn't the player he had been. That winter, after pitching the Sox pennant-clinching game and winning the 1967 American League Cy Young Award, he had torn up his knee while skiing. He was too worried about his future to consider Mantle's past. "There was a sense we might not ever see him again, but I was so wrapped up in myself, you didn't care," Lonborg said. "I had no sense of history."

Going over the lineup before the game with his catcher, Haywood Sullivan, Lonborg resolved to "run hard stuff in on him," and not leave it out over the plate, as he had eight days earlier in New York. "We knew he couldn't get around on the ball," Lonborg said.

Indeed, he couldn't, and he cracked his bat trying. The tepid pop-up to shortstop Rico Petrocelli was routine, an easy out. Lonborg's friend sitting in a field box between third and home snapped a picture at that last moment of contact and gave it to him as a memento of an occasion whose significance he did not yet fully appreciate.

Baseball has rituals for everything, and its choreography of farewell has been long established. So, after the third out in the top of the first inning, Mantle went out to first base, firing the ball around the horn and accepting the practice throws sent his way. On the mound, Mel Stottlemyre threw his last warm-up pitch. The batter was announced. Then there was a pause as a defensive substitution for the Yankees was announced. Kosco trotted onto the field; Mantle limped off it. "Thank you," Kosco said. "Thank you for everything, Mick."

"He just smiled," Kosco said.

He turned to watch as Mantle crossed the baseline accompanied by a standing ovation from all of Fenway Park. It was then that Lonborg realized the import of the moment. "We had to wait a minute for the crowd to acknowledge him," he said. "Red Sox fans don't always acquit themselves well in those circumstances. That's when it dawned on a lot of us that this might be the last of Mickey."

On Monday morning, Yankee batboy Elliot Ashby presented Marty Appel with a cracked, 31-ounce, model S2 Louisville Slugger, making good on an early-season promise to deliver an ash souvenir. Neither of them realized that it was Mantle's last game-used bat or what it might be worth until Appel sold it in 1997. He tried and failed to locate Ashby to share in his good fortune. The bat paid for three semesters of his son's college tuition.

In February 1969, Mantle made his annual pilgrimage to spring training, having delayed the inevitable in deference to the Yankees' desire to use his name for season ticket sales and in deference to the leadership of the Major League Baseball Players Association, which wanted to use his name in negotiations with the owners. "We asked him not to tell anyone he was going to retire," said Marvin Miller, former director of the MLBPA.

Miller wasn't optimistic, given what he had heard about the anti-labor sentiments Mantle had absorbed as a boy in the Tri-State Mining District. "But he agreed with alacrity," Miller said. "He said if it would help other players, absolutely. You could have knocked me over with a feather."

Mickey and Merlyn checked into the Yankee Clipper in Fort Lauderdale on Friday afternoon, February 28. Sitting by the pool, she told clubhouse man Mickey Rendine, "My husband knows he can't play."

Houk tried to persuade him otherwise. The Yankees needed him. The Major would take care of him. Bob Fishel related the conversation later to Howard Berk. "He said, 'I can't do it anymore,' " Berk said. " 'My body doesn't respond.' Ralph said, 'You know I'll never let you embarrass yourself. I'll handle it.'

"He thought about it and said no."

On Saturday, March 1, the Yankees called a press conference. Mantle's expression was as somber as his plaid sports coat was loud. "I don't hit the ball when I need to. I can't score from second when I need to. I can't steal when I need to."

His uniform was hanging in his locker when he went to the clubhouse to gather his things. He gave the ribbons of unused Conco tape to Pete Previte to give to trainer Joe Soares. He took his last pair of spikes for Mickey, Jr., who had gone out for the baseball team at the military academy he was attending in Hollywood, Florida.

In the days that followed, the press box filed its many encomiums. Perhaps Jimmy Cannon said it best: "It is as though someone stole fifty floors out of the Empire State Building."

But it took a poet to capture the import of the moment for Mantle.

"The game of baseball is not a metaphor / and I know it's not really life," Gail Mazur wrote. " . . . and / the order of the ball game, / the firm structure with the mystery / of accidents always contained" is "not the wild field we wander in."

For eighteen years, that firm structure had contained Mantle's demons, structured his dread, and ordered his days. Now, at age thirty-seven, he was free—too free—to wander in wild fields.

PART FOUR

Dream On

ATLANTIC CITY, APRIL 1983

Before he passed out in my lap, Mickey promised to meet me for breakfast at 8 A.M. I wasn't sure he'd show and I wasn't sure what I'd do if he did. I had a story I couldn't write and questions I hadn't asked. I went to bed with a headache and woke up with an empty notebook.

He was waiting for me in a booth in the coffee shop above the Hi-Ho Casino. He looked up from a greasy, laminated menu, peering through a pair of dime-store reading glasses, half spectacles that magnified the broken capillaries in the once-hawkish batting eye. "I wake up most every day at 5:30 or 6:00 A.M.," he said. "All those dreams."

A medley of recurring nightmares that make the long nights longer. Missed trains, missed planes, missed curfews—missed opportunities. *"The hard part is just getting through it,"* he said.

Buses don't stop. Fly balls don't fly. Doors no longer open for The Mick. *"I always felt I should have played longer."*

He hovers above a chiseled tombstone—HERE LIES MICKEY CHARLES MANTLE: BANNED FROM BASEBALL. He's banished from heaven, too. Saint Peter meets him at the Pearly Gates and delivers the bad news. "Well, Mick, because of the way you acted on earth, you can't come in here. But before you go, could you autograph these six dozen baseballs?' "

That one started in the hospital.

Diamond dreams: can't buy a hit, can't get to first base. And if he does get in, say in the late innings—can you hit, Mick?—all he sees are fastballs. Fastballs he can't turn on. Fastballs he can't duck. Fastballs aimed at his head. There's one where he gets his bat on the ball, skies it to the right fielder, and gets thrown out at first base.

The team bus leaves without him.

In the night, he rides the pines. Sweet dreams are few. There is one.

He's in the Astrodome, where he hit the world's first

indoor home run, when the big blue 7 was still jumping off his back. He can still hit 'em where they ain't, where baseballs have never gone before. He circles the bases—and keeps circling them—thinking, "This is the way it should be."

"Then there's that one where I'm trying to get to the ballpark."

He's in a cab racing toward the Stadium—late but in uniform. He hears Bob Sheppard calling his name. "Number seven, Mickey Mantle! Center field, number seven." The attendant at the players' entrance doesn't recognize him. He's turned away at every gate. Billy and Whitey and Casey are waiting for him on the other side of the fence. He tries to crawl under but he's too fat to squeeze through. Halfway in, halfway out, he gets stuck between past and present.

"I don't need anyone to analyze that," he said, *glancing at me from beneath the half specs perched on a well-cured nose. "I got one where my golf ball is up against a wall or a fence and I can't get a back swing. And another I'm going pole vaulting . . ."*

He's gliding down the runway with the pole in his hands, and he's running as fast as he ever could. He plants the pole and thrusts himself high into the air, soaring above the bar.

"When I get to the top it's like fifty stories high, and

you look down and there ain't nothin'. And you think you're gonna die."

He's in free fall, plummeting, down, down, down. He sees the people looking up at him. But there isn't anything anyone can do except watch him fall out of the sky.

"Then I wake up," he told me, *"sit straight up, sweating."*

Home in Dallas, in the dark middle of the night, he takes refuge in the trophy room that Merlyn did up with pinstriped flannel on the walls, one whole wall plastered floor to ceiling with Mickey Mantle magazine covers. The shelves bulge with the jewels of accomplishment, the Triple Crown, the Gold Glove, the Hickok Belt, three Most Valuable Player trophies, a golf ball that traveled to the moon with Alan Shepard, and scrapbooks assembled with glue and ardor by fans who have outgrown them. He's amazed they care enough to send them. I was just a fuckin' ballplayer. He doesn't recognize himself in their avid eyes or see his likeness in the remnants of the frayed, yellow newsprint they collected.

"It's like I'm reading about somebody else," he said.

Mickey pulled his head out of the past when a man approached the table. "Hey, Mickey Mantle! I saw you strike out five times in 1950 against Mel Parnell!"

Mickey nodded and smiled, and gave the guy an au-

tograph. "There's an example of being nice," he said. "I wasn't even there in 1950, and it wasn't Mel Parnell. It was 1951, and it was Walt Masterson."

He wasn't always patient with such intrusions. Like the time in Minneapolis when the mother of little boy with a broken arm stopped him en route to the team bus and asked him to sign her son's cast. "Let me know when he breaks the other one," Mickey told her.

He was in a hurry then. "I wasn't a real nice guy, probably," he said.

I told him I was at the Stadium that raw May night in 1962 when he collapsed in the baseline like roadkill and about the Faustian agreement I had made with my mother in order to attend the game. My grandfather, a manic-depressive immigrant tailor, had fashioned two wool plaid skirts for me, one in sepia tones of beige, brown, and gold; the other in Christmas-tree red and green, perfect for Hanukkah. They were reversible and indestructible. They went with everything and nothing. Their oscillating hems attested to his ups and downs. My mother exacted an ugly price: wear one of these atrocities or forget the sleepover at my grandmother's after the game. I chose the more muted tones and an overly generous straw-colored Irish cable-knit sweater worn over a white turtleneck. I looked like a prepubescent haystack. But I did it for The Mick.

He squirmed the way he always did when people

told him he was their hero. And squirmed more when I told him about another time when I waited by the players' gate, clinging to a blue police sawhorse, hoping for an autograph. He didn't stop. He farted in my face. He was a regular MVP of flatulence—a ventriloquist, even. He could cut 'em on cue.

"What were you thinking?" I demanded. "What kind of an adult acts like this?"

If I'd been a guy he'd probably have told me to go fuck myself. Instead, a look of wounded incredulity passed over his face. "Well, hell, I'll give you an autograph now."

Signing his name had become a default mode of apology. Out came the signing pen and an eight-by-ten glossy, a headshot of the aging slugger resplendent in a fuchsia golf shirt: Mickey Mantle in the pink. "To Jane, Sorry I farted. Your friend, Mick."

He said he had a headache and ordered a beer. I gave him two aspirin. He didn't want to answer any more questions. A limo was coming to get him at 11 A.M. He said we could talk in the car on the way to the airport in Philadelphia. I decided to wait to deliver the high, hard one.

"So who's better? Mickey Mantle or Willie Mays?"

"Fuckin' Willie," he replied.

"That's what my father said."

"If you didn't know either one of us and you wanted to know who's better, Willie or Mickey, and you opened the record book, what would it say? I guess Willie. Ya have to go by the bottom line. I don't alibi. I played in a lot more World Series games. I feel when I was good I was better. Willie was probably about the best overall player I ever saw in my era. Him and Hank Aaron, neither one ever got hurt. He was a lot better fielder than I was. When I was healthy and going good, I was a better offensive player. And I played in a terrible ballpark for me. If I played in Atlanta County Fulton Stadium there's no telling, I might have hit eight hundred home runs. Baseball was easy for him. Willie loved it. You could see it in him. I did too, when I felt good."

What about the injuries? The knees, the hamstrings, the shoulder, the spike caught in the drain, the what-ifs?

"That was overplayed. A lot of times I felt great. I wasn't always a one-legged guy who looked like a mummy. I never had any problem from the waist up. I didn't hurt as bad as everybody thought I did. I still ran down to first base faster than anyone. I played twenty-four hundred games. That's three hundred more than anyone ever played for the Yankees. Out of eighteen years, probably fourteen was good. Just not the last four. Well, I don't know if I was ever a hundred percent. I was eighty percent.

"It does bug me that I'm not up in the record books more. Every time I'm introduced and someone days, 'Lifetime .298 hitter,' I think, 'You cocksucker, quit saying that!'"

He began to hum. "Have you ever heard that song 'Willie, Mickey, and The Duke'? That guy made some money."

The refrain was all he knew of Terry Cashman's 1981 anthem to Fifties baseball. *Say, hey. Say, hey. Say, hey.* "I know all the words to 'I Love Mickey.' Teresa Brewer wrote it."

They recorded it together—in counterpoint—in 1956.

"I love Mickey."

"Mickey who?"

"You know who, the fella with the celebrated swing."

It reached number 51 on the charts.

"Betcha don't know what's on the B side? 'Keep Your Cotton Pickin' Paddies offa My Heart.'"

It was the year he won the Triple Crown. He was on the way up. Life was a dream.

16

June 8, 1969

HALF-LIFE OF A STAR

1.

Mickey Mantle Day at Yankee Stadium was the third one, actually. And unlike the previous jubilees, it was intended to celebrate his retirement rather than to prevent it. The Yankees billed it as "A Day to Remember." It was also the day he joined the past tense. Merlyn said it was the worst day of her husband's life.

There were more than 60,000 fans on hand in the big ballpark in the Bronx when number 7 was retired. Yogi and Joe D., big league coaches now, detoured off road trips to attend the festivities. Harry Craft, his minor league manager, came; so did Tom Greenwade, who had given him over to Craft's care, and Green-wade's son, Bunch. Tom reminded Harry that he owed him a new felt hat because he had said Mantle was too small to make it to the major leagues. Harry demurred. But two weeks later, he sent a pair of alligator shoes that Bunch admired in a store window with a note: "To Tom, here's your goddamned hat."

The Yankees thought one of Mantle's sons should make an appearance. They flew Mickey, Jr., to New York—he was sixteen and the only one of the boys old enough to have seen him in his prime. The others stayed home to watch a backhoe dig a new pool in the backyard. Merlyn was looking forward to having a husband and father who came home and stayed home. She didn't know—as his lawyer, Roy True, did—that Mantle had reserved a room in the hotel for his current girlfriend. "It was a busy weekend," True said.

Mel Allen was hauled out of retirement to preside over the revels, the first summons since he had been summarily dismissed after the 1964 World Series. The Voice filled the Stadium again with a sugar-cured bari-tone: "Ladies and gentlemen, a magnificent Yankee, the great number seven, Mickey Mantle!" The ball-park "throbbed with love," George Vecsey wrote in the next morning's *New York Times*.

The choreographed ceremony called for DiMaggio

to present him with a plaque, which, DiMaggio noted with customary grace, "will be right along in a modest spot out there in center field." Mantle replied in kind but with actual kindness, handing Joe D. a plaque, which, he said, "has got to be hanging a little bit higher than mine."

And it did for a while until the plaques were taken down to be rebronzed and put back at equal height. Mantle's graciousness spoke volumes about the difference between them. "People have always placed Joe and Mickey on a pedestal," Tony Kubek told *Daily News* columnist Bill Madden years later. "The difference is Joe always liked being there and Mickey never felt like he belonged."

Two months later, when Mantle suited up with the Old Timers for the first time, the ovation for him so dwarfed Joltin' Joe's that the public relations department decided to reverse the order of introductions the following year, hoping to ensure that DiMaggio got his "full cheer." The Clipper was so angry he swore he would never return to the Stadium. He was wooed back with sweet nothings and lucre. But when Mantle's monument was unveiled in center field in 1996 at a ceremony emceed by Billy Crystal, DiMaggio actually punched him in the stomach because he failed to introduce Joe D. as "Baseball's Greatest Living Player."

His jealousy was palpable. "He would never look at Mickey Mantle until Mickey spoke to him—every time," Clete Boyer said. "Mickey never said a bad word to the public about Joe D. Just to us."

But he had ways of making his feelings known. One day in the dugout at yet another command performance of the Living Legends, Mantle asked his aging comrades, "Ya wanna see me piss the sonofabitch off?" Of course they did. DiMaggio did not observe professional courtesy and refused clubhouse requests for autographs. "Mickey gets one ovation at home plate," Boyer said. "Then he walks down the third base line, waving. He gets another two-minute ovation. You think he's through. He comes back, walks down the first base line. He gets another ovation. You figure, 'Okay, that's it.' Then he walks out to the mound. You could just see Joe D., thinking, 'You SOB, get off of there.' The next year Joe D. made George buy him one of those big Mercedes and Mickey says, 'Joe D. is pissed off because it only had a half a tank of gas.'"

The Clipper managed to mask his resentment on Mickey Mantle Day. When Allen intoned, "And now, Mickey Mantle, Yankee Stadium is all yours," the ovation lasted five or six or eight or ten minutes, depending on which morning paper you read. The roar of the full-throated Stadium was a sound unlike any other.

"Like an animal might make," Mantle once said.

He had watched Gary Cooper deliver Lou Gehrig's farewell address in *The Pride of the Yankees*. Now he was standing in the same spot, invoking Gehrig's parting words: "I always wondered how a man who knew he was going to die could stand here and say he was the luckiest man in the world. Now I think I know how Lou Gehrig felt."

What was lost in all the huzzahs attendant to the occasion—the last lap around the Stadium in a bullpen cart with hand-painted pinstripes—was that he cast himself as a dying man. In fact, he was already planning his funeral. His friend Roy Clark had just recorded the Charles Aznavour ode to wasted youth, "Hier Encore," recast as a country anthem, "Yesterday, When I Was Young." Mantle heard him sing it for the first time at a golf tournament that spring.

> *Yesterday, when I was young*
> *There were so many songs that waited to be sung,*
> *So many wild pleasures that lay in store for me*
> *And so much pain my dazzled eyes refused to see,*
> *I ran so fast that time and youth at last ran out and*
> *I never stopped to think what life was all about.*

When Clark returned to their table, Mantle's eyes were wet. "But he was laughing at the same time," Clark said. "He said, 'I want you to sing that song at my funeral.' " Every time they saw each other for the next twenty-four years, Mantle reiterated the request. "Don't forget," he'd say. The last time, at the golf tournament Mantle hosted in Joplin, Missouri, in 1993, he sat on the stage at Clark's feet as he sang, "tears just drippin' off the side of his face."

The moment was preserved in a photograph. Jim Abercrombie, the tournament chairman, commissioned three oversized copies. Mantle autographed Clark's: "Hang in there. I want to hear 'Yesterday' at my funeral."

2.

The summer of 1969 was one of meta-events: Ted Kennedy drove off a bridge at Chappaquiddick, and Neil Armstrong walked on the moon. Charles Manson rampaged, and Muhammad Ali was convicted of draft evasion. Woodstock and the Stonewall Inn became part of the American lexicon. The Mets conjured a miracle in the World Series.

It was the summer of love and Mantle's first season of freedom. It was like "getting out of prison," said True. "Put out on the streets to find himself."

Mantle tried fast-food franchising with Mickey Mantle's Country Kitchen. *Forty-five franchises sold!*

Mickey Mantle's Men's Shops marketed the Mantle look. *Fifty-five franchises sold!* He and Broadway Joe Namath were going to supply office temps to the greater metropolitan area through an employment agency backed by George Lois, Mantle Men and Namath Girls. From their offices in the Chrysler Building they were placing five hundred secretaries a week.

By the end of the year, Mickey Mantle's Country Kitchen had gone belly up, posting losses of $1,279,777. Mickey Mantle's Men's Shops went into reorganization. Mantle Men and Namath Girls was in bankruptcy by February 1972, followed soon thereafter by the company that had bought his share in his Dallas bowling alley. He was hauled into court to account for money he had received for endorsing bowling products.

Every time another deal went sour, Mantle would tell his brother Larry, "I dumbed this" or "I dumbed that." Driving through Harlem one day with Yankee VP Howard Berk en route to a baseball clinic in Central Park, he gazed out the limousine window at the crumbling neighborhood and said, "How come I didn't invest in this?"

True, the Dallas attorney who became a confidant and consigliere, unscrambled the financial mess. But, he said, the biggest losses were psychological, not financial. "Mickey never lost a lot because he didn't have a lot," he said. "He was going to get rich through the restaurant and clothing business. That evaporated. He didn't have any money. He always lived the first-class life. I had to loan Mickey money. He never knew how much he had or asked or cared. I paid all his bills. Mickey didn't know he didn't have any money. He had a ten- to twenty-thousand-dollar rotating line of credit with me. I didn't want to tell him. He would have been depressed."

On October 20, 1970, the day he turned forty, Mickey Mantle woke up and he wasn't dead. "I don't know what to do," he told his friend Joe Warren. "Nobody's ever lived this long."

There were two possibilities: Go home, meet your wife and children, and thank your lucky stars that you are not another doomed Mantle man. Or consign yourself to the living dead.

He treated the reprieve like a death sentence.

"He had no job, he felt useless," Merlyn told me when she and her sons, Danny and David, met with me at her home in Dallas. The Mick was everywhere—in the famous psychedelic rendering by LeRoy Neiman, in the batter's box, in center field, in an auction catalog on the coffee table. "He lost the thing he loved most, to play ball. He felt like a has-been. He was happy only if he was smashed and not coherent."

True said, "To me he was one of the saddest, loneliest people I've ever known. He had no place in this world."

Mantle had a home in Dallas but he wasn't there often enough to get to know his sons or to be a husband to his wife. Home was a hub between connecting flights. "He didn't want to go home," his friend Mike Klepfer said. "He would look for a reason for Roy to send him someplace."

His family treated each homecoming like a state visit. Dinner had to be served perfectly, plates decorated with cottage cheese and pineapple slices. The house, the kids—everything had to be scrubbed clean. Even the dog got a bath before The Mick came home. They feared that the slightest domestic misstep might set him off or send him away again. "We all walked on eggshells," Merlyn told me. "He had profound mood swings. One moment he was happy and joking. One second later, that mood turned angry. I often wondered if he was manic-depressive."

The boys—Mickey, Jr., David, Billy, and Danny— watched what they did and what they said. And there were few heart-to-hearts with Dad. "That's what was bad," David said. "It taught us how not to communicate. When Dad was at home, we just wanted to be close to him and we didn't want to do something that might upset him that he might leave the room. I was always amazed. Why were we so scared?"

" 'Cause we didn't know him," Danny replied. "That's why."

"No," David said, "I think it was respect."

"Maybe we had a feeling he might hit us," Danny said.

That never happened, though he did challenge them on occasion when he'd had one too many. He taught them not to back away from a fight, to stick up for themselves, and to stick together, which they did, bonded by their love for an AWOL dad and by the fierce resentment they encountered in toughs who thought they had it so good. The last Christmas they spent together was the first time David could recall that his father stuck around for his birthday the next day. After his father's death, David went through a box of several hundred family photos, looking for one of just him and his dad. He couldn't find one. There was always somebody else in the picture.

He never asked his father to pose with him because he was worried what he would think. When Billy Crystal told the Mantle boys he wished he'd been Mickey Mantle's son, David replied, "We wished we had been, too."

Mantle never paid attention to what his boys did with their lives, admitting later, "I didn't even make them finish high school." Nor did he object when David opted to skip his own graduation because, he told me, "Dad didn't get to go to his."

He affected Mutt's absolute control without actually asserting it. "Who's the boss?" he'd ask.

"You are, Dad," they'd reply.

"Don't you forget it."

He was as absent and laissez-faire as Mutt had been present and domineering. "He was not gonna repeat," David said. "But I wish he would have. I needed discipline."

"He never told us to do anything," Danny said.

Billy was the wild child and also the most fragile of the boys, slighter than the others, built like his grandfather, Giles. Like his father, Billy wet his bed as a child. Like his mother and his brother, David, he struggled to read, and attended a school for learning-disabled students. "He was frustrated and scared a lot," Merlyn told me. "Actually he shied away from the family, which broke our hearts."

One morning in 1977, when he was seventeen, Billy woke up with a lump in a lymph node under his ear. He got the biopsy results the same day: non-Hodgkin's lymphoma, the disease his father had been waiting for his entire life. Billy had to wait to tell his parents the diagnosis. The doctor wanted to inform them at the same time, and Mantle was out of town.

Danny, the youngest, said the only time he spent alone with his father was when he poured his beer while watching college football on TV. He was the practical son, the reliable child, always there to fetch another cold one, to mop up any mess. He was the first to clean up his own life when it spun out of control. He felt his father's absence in the family occasions they missed and the stories they didn't share. "I wish he had talked to us more about his childhood," he said.

But then, Danny said, "he didn't have time to be a child. It would have interfered."

David was the poetic son, who gave his mother a talisman she always wore around her neck, inscribed ALSO MY HERO. He was the family clown, who hid his sadness with manic energy and verbal overflow. "In 1995, I was diagnosed as having attention deficit disorder (ADD) and obsessive-compulsive behavior," he wrote in *A Hero All His Life*. "That should not have surprised anyone. As children, Billy and I were both dyslexic, and those disorders often go together. I might have been a much better student if the professionals figured this out thirtysomething years ago."

In school, he was teased about being Mickey Mantle's numbskull son. "I wasn't good at sports like Mickey, Jr.," he said. "I had heart. I played softball with Mickey, Jr., when they needed somebody. I slid into second and broke my ankle. I finished the game. I wasn't going to be a quitter.

"When Dad was at one of our things, he'd be in the car. He never said if we screwed up. One game, I struck out three times. The guy's jumping up and down on the mound, 'I struck out Mickey Mantle's son!'

"Dad, he'd never get mad or anything. He never forced baseball on us. We had to do it on our own. Like with Mickey, Jr., if he woulda worked with Mickey, Jr., shit, there'd be two Mantles in the Hall of Fame."

Mickey, Jr., the oldest, was most like his father and just as hardheaded. He floundered in school, quit the military academy his parents had hoped would instill discipline in him, earned a GED, and made a belated stab at a baseball career, in part to please his father, in part to gain his attention—but it was too late. He didn't blame his father. "It was Mickey Mantle who kept saying what an awful father he was, not his sons," Mickey, Jr., wrote in *A Hero All His Life*. "I thought he was a great dad. He wasn't what you would call a regular dad. But then, he didn't lead what you would call a regular life."

He had a hard time believing that they loved him. One day David awoke to find his father sitting on the edge of the bed in the room he shared with Billy. "He was sitting there at the foot of the bed. And I said,

'Dad, what's up?'

"He said, 'You hate me, don't ya?'

"I said, 'No, Dad, I love you.'

"When I said that, I remember him walking out of my room and saying, 'Merlyn, he said he loved me.'"

Upon seeing a homemade Father's Day card his friend Tom Molito received one year from his son, Mantle told him, "'I never got a card like that from my sons. They don't love me.' I had to reassure him that they did."

Merlyn said: "He thought no one ever loved him."

3.

On August 30, 1970, Mantle went back to doing what he knew best. He climbed back into pinstripes to coach first base for the Yankees, a team that couldn't get anyone aboard. He quit his part-time gig as part of NBC's *Game of the Week* broadcast team without bothering to tell anyone at the network. He described his new job as "patting a guy on the ass and saying, 'Nice hit.'" He never listened to the first base coach. Why would they listen to him?

For these two months of hard labor, he earned $37,500, including the $12,500 the Yankees paid him for making himself available on important occasions. He hated it. Hated the way the Yankees trotted him out to the coaching box for the middle three innings, embarrassing himself and the incumbent first base coach, Ellie Howard, who had to go sit on the bench until the bottom of the seventh. It was like being a lawn jockey.

But writers wanted to hear what he had to say, to size him up in column inches, which was the whole point. Diane Shah, the elegant sportswriting pioneer, interviewed him for the September 7 *National Observer*—a story headlined: "A Lady Fan Has a Chat with Mickey Mantle." He told her about the dream he'd had the night before—where he couldn't get into the Stadium and nobody knew who he was. The Dream became a recurring nightmare and a recurring narrative theme. Soon, every interviewer wanted to hear about The Dream. He told it funny. He told it sad. He told it to further an unstated agenda. On the golf course one day, Tom Callahan, of *Time* magazine, asked, "Did you really dream that?" "Nah," Mantle said, "I was just trying to get laid."

It was the beginning of life as performance art. He became his own ventriloquist. When reporters came around, he had time for them he never had before—hell, he said he missed some of the old beat guys—and he had answers to their questions ready to go when the red light flickered on, polishing sound bites of self until all the edge was gone.

That summer, Jim Bouton committed the ultimate

betrayal of clubhouse protectionism. He squealed—loudly. *Ball Four* was a shocking exposé of groupies and greenies and "beaver shooting" with The Mick from the balcony of a Washington hotel. It may appear positively Victorian by today's debased standards, but baseball responded with righteous indignation. The public evinced hand-wringing shock, as if it had never crossed anyone's mind that the Bombers might be Bad Boys. The Yankees excommunicated "The Bulldog." Forty years later, Arlene Howard, Elston's wife, still referred to Bouton as "a person who's such a dog, I hate to mention his name in the same breath as Mickey."

When he was in Dallas, Mantle organized his life around the clubhouse at Preston Trail, the posh all-male Dallas golf club he had joined as a charter member in 1965, trying to re-create the camaraderie of the locker room. He made new friends, among them doctors Art DeLarios and Mark Zibilich, who would later help with his medical care; Bill Hooten, a commercial real estate broker; and Lanny Wadkins, the 1977 PGA champion. "I think the places he felt comfortable were probably the Yankee clubhouse and Preston Trail," Wadkins said. "Outside of them, he couldn't be himself. And in both situations, they were men's places. His stories were laughed at and appreciated, and you never got tired of hearing them."

Mantle behaved as if the same locker room rules

applied, which explains his penchant for playing a round of golf or two in the buff and passing through the buffet line buck naked. His liberal interpretation of acceptable attire necessitated action. "You had to wear something," Zibilich said. "The Mick was fond of coming out of the hot tub, trying to ward off the effects of last night's party, and then would walk into the dining room with nothing on at all, which would not be appetizing to anybody. So we had to have a rule that said you have to be wearing at least your skivvies or a towel!"

But one day Mantle abruptly ordered True to sell his membership, telling him, "They're presidents, vice presidents, rich, I don't have anything in common with them."

Of course, all they wanted to do was play a round with The Mick to brag on later. But he didn't feel that he belonged—anywhere. When he and Merlyn were invited to a state dinner at the White House in 1976, he called True from the shop where he was getting fitted for his tuxedo and said he wasn't going. "What the hell do they want me there for?"

True reminded him that the Secret Service clearance was done; that Merlyn, who didn't get many opportunities to go places with him, was counting on it. "You can't back out now," True said.

After an hour, Mantle relented, only to call again at

5 A.M. to say, "I've been up all night. I don't have a thing to say. I won't understand what the hell they're talking about."

After another hour, he said, "Oh, fuck it, I'll go."

He sat with President Gerald Ford and talked about golf. Merlyn sat with Vice President Nelson Rockefeller and talked about children. In the afterglow, Merlyn and Mickey indulged in some rare alone time. "They had the best time in the world," True said. "So good they went to Vegas, booked a room, and had three great days together."

The second honeymoon didn't last any longer than his encore in pinstripes. By the time Mantle reported to spring training camp in 1971—three days late—his coaching career was over.

In retirement, he no longer had to genuflect before Bomber pomp and circumstance, the Yankeeography that dictated, for example, the identity of his boyhood hero. Given the opportunity, The Mick unloaded on the canon with a Bronx cheer.

In 1973, when the Yankees celebrated the fiftieth anniversary of the House That Ruth Built, the public relations department sent a questionnaire to former players. It read:

I consider the following my outstanding experience at Yankee Stadium:

In the lined space provided below, Mantle wrote:
I got a blow-job under the right field bleachers by the Yankee bull pen.

This event occurred on or about: (Give as much detail as you can)

It was about the third or fourth inning. I had a pulled groin and couldn't fuck at the time. She was a very nice girl and asked me what to do with the cum after I came in her mouth. I said, don't ask me, I'm no cock-sucker.
  Signed: *Mickey Mantle

Beneath that, he added:

*The All-American Boy

Marty Appel, who crafted the questionnaire, and received and edited the infamous reply (substituting the Barney Schultz home run for posterity), gave the original document to Barry Halper, the minority team owner and memorabilia maven who would later sell his vast collection for more than $30 million. The X-rated writing sample circulated through the baseball underground for years before emerging into the LED glare of the World Wide Web. Appel was appalled. Mantle was just trying to shock the Yankees' straitlaced PR chief, Bob Fishel, he said.

Absent Mantle's impeccable 1930s Palmer Method penmanship, the asterisks and the appellation, it's just another example of locker room crude. The self-mocking touches turn it into something altogether different and far more interesting—a send-up of Yankee grandiosity and a self-knowing appraisal. Who knew he had a sense of irony?

"That may be the best thing I've ever heard about him," said Robert Pinsky, the bard of Red Sox partisans, and the former poet laureate of the United States. "He's saying, 'I am not going to be your all-American boy.'"

The literary effort can also be read as a bad boy's cry for help. "He was looking to get caught," Bouton said. "'Stop me before I fuck up too much.'"

No one did. Mantle lurched from city to city, banquet hall to banquet hall, fashioning a life and a living from a string of absurdist events. He raced Whitey Ford at harness tracks and won praise as a "sulky sitter," piloting Big Time to victory over Ford's Anchor Boy. He accompanied Max Patkin, the Clown Prince of Baseball, on a grand tour of minor league ballparks. He made a promotional appearance at a Florida mobile home community and toured the Lone Star State on behalf of Cameron Wholesalers, hawking Mickey Mantle Grand Slam Specials.

Buy fifty doors and get one free.

He batted in a home-run-hitting contest before the second game of a doubleheader between the Rochester Redwings and the Memphis Blues; he celebrated the tenth anniversary of the Astrodome, where he had hit the first indoor home run, by hitting three more. He flacked for New York's new Baseball Instant Lottery and was signed by the Oklahoma Malt Beverage Association to appear in ads endorsing the proposed legalization of beer sales in the Sooner State. He received the Earl Smith Nostalgia Award from baseball writers in Kansas City and the Pride of the Yankees Award at the team's annual Welcome Home dinner; and he graced the dais at social functions, though not always graciously. "Craig, you've gotten fat," he said by way of greeting his former Whiz Kid teammate, bank president Ben Craig.

But he astonished Joe Torre at the 1971 MVP award banquet by recalling Torre's first home run as a major leaguer a decade earlier—in an exhibition game. To be remembered by The Mick told you something no award could about who you were. He counseled students at the Saints Philip and James School in the Bronx and Quaker Ridge Junior High in Scarsdale not to drop out. He earned his real estate license. His golf pal Bill Hooten arranged for him to take the broker's exam. Though Mantle didn't know what kind of real estate they were in, he called Hooten every year when it was time to renew the license. "It was a big deal to him," Hooten said. "It was something he got on his own."

In 1973, Alan R. Nelson Research ranked him second behind Stan Musial as America's most trusted pitchman. But there was little to pitch. After he and Willie Mays sang their off-key praises for Blue Bonnet margarine in 1980, gnawing on margarine-slathered ears of corn and wearing calico bonnets, Mays asked for five pounds of the stuff to take home. Who could blame him? It wasn't as though they were raking in big bucks. Mantle got $500 for his first post-retirement gig in 1969.

He wanted to do something that mattered, but he didn't know how or what. "Is there something I can do to really help people?" he would ask Roy True, gazing out a smoke-tinted limousine window. True took him to meet executives from the Chase Manhattan Bank and the Rockefeller Foundation at the Rainbow Room, a planning session for a youth sports commission for the state of New York. Mantle was to head up a program for inner-city children. He was uninterested and too uncertain of himself to join the conversation until it turned to golf.

That commission never materialized. Mantle settled for random and frequent acts of charity, such as the time he saw a young couple in a broken-down pickup outside a 7-Eleven on a sweltering Dallas afternoon; the daddy was buried under the hood, as the baby on its mama's lap howled in discomfort. Mantle emptied his pockets, sent a friend to deliver the contents, about $5,000 in cash, and took off.

He played a lot of golf—for money, for charity, and for fun. He hit the ball a ton but didn't always know where it was going. One of the game's great players, Tommy Bolt, the 1958 U.S. Open champion, assessed his skills this way: "He was nice. But he couldn't play golf."

One year at the Reynolds Plantation course in Georgia, Glenn Sheeley, the golf writer for the *Atlanta Journal-Constitution*, watched Mantle attack the par-4 second hole, which had a low brick wall and a muddy ditch in front of the green. Two of his first three approach shots stuck in the mud; another ricocheted into the great beyond. His fourth ball reached the green. "Instead of walking around the wall, he decides to hoist himself up on the wall and falls back into the creek," Sheeley said. "Everyone's going, 'Mickey, Mickey, are you all right?'"

Number 7 emerged muddy but unscathed. "Tough fuckin' hole," he said.

He needed the action. He learned to hustle with the best of them, thanks to his homeboy golf tutor Marshall Smith, who likes to tell about the day Wadkins, the future Hall of Famer, challenged Mantle to a round at Preston Trail for $100 a hole. "Mickey said, 'If you give me two putts and two drives, I'd play you,'" Smith said. "So when they finished, Lanny owed him $2,300."

He bet college football, too. Bill Handleman, a young reporter for the *Asbury Park Press* in New Jersey, spent a soggy Saturday October afternoon with him at a VFW hall in Fredericksburg, Virginia, where he was making a promotional appearance. The Yankees were on the TV perched above the four-stool bar. Mantle wasn't interested in the World Series game and he wasn't interested in Handleman's soliloquy about how Mantle stories in the *International Herald Tribune* were the reason he'd learned to read. "He says, 'Yeah, so? What are you telling me for?'"

When Mantle got up to use the restroom, he said, "See if you can find the Brown score," Handleman recalled. "He hands me this folded, mimeographed sheet, maybe a half an inch thick. He'd bet every game! Brown was one of his bigger games."

If Mickey Mantle Day put him in the past tense, his induction into the Hall of Fame in 1974 was an embalming. He chartered a bus to Cooperstown for family and friends—including Carl Lombardi, Joe Warren, Phil Rizzuto, Whitey Ford, the other inductee—and his mother, Lovell, who later exclaimed upon examining his ring, "Well, who in hell is in the Hall of Fame?"

Her son said, "Well, that'll putcha down, won't it?"

The bus was ambushed by autograph hounds. Ford and Rizzuto escaped. Mantle stayed and signed and was blamed when he couldn't satisfy them all, Lombardi recalled. At 1 A.M. he was still working on his speech. He wrote two versions, one clean, one unexpurgated, and shared the latter with his minor league buddy. Lombardi told him, "Don't you dare say that. We'll all be in jail."

Mantle delivered the clean version but left in the chicken joke Merlyn had begged him not to tell. It was the slogan he'd penned for Mickey Mantle's Country Kitchen—"to get a better piece of chicken, you'd have to be a rooster." He also praised her for staying married to him for twenty-two years. "That's a record where I come from," he said.

After the ceremony Mantle made a quick getaway. "I had to leave," he told Billy Crystal ten years later. "I said, 'Get the fuck outta here.' It's like a cemetery to me."

After his internment in Cooperstown, Mantle's

thirst went from steady to nonstop. No one could keep up with him. En route to a baseball clinic in Central Park one tar-melting New York summer day, the limo driver took the scenic route through Harlem. "We stop at a red light," said Mantle's companion Howard Berk. "There's an old black guy with a bottle of gin, just pouring it down. I said, 'This guy's gonna die. He's going to explode from drinking that gin.'

"Mickey says, 'Watch this. We'll stop him from drinking.'

"He rolls down the window. 'Hi, old-timer.'

"The guy says, 'Mickey Mantle?'

"He looks down at the gin bottle. 'Mickey Mantle!?!'

"He throws the bottle away.' "

But who was going to tell Mickey Mantle? One day near the end of the Me Decade, he met up with Ryne Duren at a bar in Madison, Wisconsin. After multiple stints in rehabs and hospitals, Duren had gotten sober and was training to become a drug counselor. He told Mantle about his recovery and tried to get him to confront his alcoholism. "He had the classic denial," Duren said. "He thinks, 'You've got to have a drink every day.' Mickey said, 'I'll go home and won't drink for ten days or two weeks.' "

Duren told him, " 'Laying off doesn't make any difference. It's not how much you drink. It's what you do when you do drink.'

"He never asked the big question, which is 'Do you think I'm an alcoholic?' In fact, he told me he wasn't an alcoholic. I tried to contact meaningful people in his life, doctors, lawyers, family. I had no luck with anyone who would go to bat to do the thing necessary to be part of an intervention. I told Merlyn, 'You may have to go to the point of divorcing him.'

"She said, 'I couldn't do that. He's so good to me.' "

Duren gave up. "You couldn't get Merlyn, and you couldn't get his lawyer to do it. Everybody wanted to be Mickey Mantle's friend. Then they shake their heads at the sickness."

The toll was beginning to show. Eli Grba, a Yankee teammate and recovering alcoholic, said, "It took a long time for alcohol to really mess him up badly. It took Mickey longer because Mickey was performing. He was *doing*. When you can't do shit and you're drinking just the same, now comes the slide down. Fast."

Mantle had to leave spring training camp in 1972, abandoning the gilded title of "hitting instructor," to have his gallbladder removed. Four years later, Dick Young sounded a warning in the New York *Daily News*: "Mickey Mantle has friends worried about his soaring blood pressure, which led to an unscheduled visit to a Dallas hospital. He has been told to slow down."

In the summer of 1978, he was hospitalized in critical condition with a bleeding ulcer. Six months later, he was back on the job, knocking back straight vodkas as a "vice president in charge of special marketing" for the Reserve Life Insurance Company, a gig that paid him $50,000 a year to play golf and attend cocktail parties. He dismissed the ICU episode with bravado as a publicity stunt. "I wasn't getting enough ink," he said.

Less than a year later he told a reporter from a Florida newspaper, "I can see how a guy can commit suicide."

He was living the half-life of a fading star. Dimmed as that light may have been, he was still the celestial Mick. The gravitational pull of celebrity rendered those in his orbit mute. "I just think at this point he was going to do what he wanted to do," said Wadkins. "It's a tough situation to have the skills and the talent he had, and then say, 'What am I good at?' All he was good at was destroying himself."

17

December 19, 1985

18 BELOW IN FARGO

1.

Mantle was in Fort Lauderdale shooting a commercial when Julie Isaacson reached him at his hotel. "What are you calling me for?" he demanded.

"Mick, Roger died."

The line went dead. "He slammed the phone," Isaacson said. "I called him back. I said, 'Mick, did you hear me?'

"He starts screaming at me, *screaming*, like he was right across the table. 'Damn you, damn you!'"

Mantle knew Maris was dying of cancer. He called Maris once a week, every week, for the last two years of his life and persuaded him to go to the M. D. Anderson Cancer Center in Houston, where Billy Mantle was treated. But Maris's lymphoma had too good a lead. He died there on December 14, 1985, at age fifty-one. Mantle took it hard. In fact, he took it personally.

Once again, he hadn't done enough. "Roger was a better family man," he would tell friends later. "If anyone went early, I should have been the guy."

Maris had left the Yankees on a gust of bitterness, traded to the St. Louis Cardinals after the 1966 season. He played two years for the Cards and made a new life away from the banquets and the Old Timers games, running his beer distributorship in Gainesville, Florida. He didn't return to the Stadium until Mantle escorted him onto the field on opening day in 1978 to raise the Yankees' first championship banner since M&M were the heart of the order. Maris thought he'd be booed. Applause rained down on him; Reggie Jackson, the newest Yankee slugger, was pelted with Reggie bars.

Maris's funeral was held in his hometown, Fargo, North Dakota, where he is memorialized in the only museum willing to have him, in the West Acres Shopping Center. Cooperstown's snub was as cold as Fargo on the day he was buried.

Ed Hinton, a reporter for *The Atlanta Constitution*,

arrived the night before. It was minus 18 degrees when he checked into Fargo's only Holiday Inn and followed Whitey Ford down a corridor to a private meeting room where the '61 Yankees were attempting to drown their sorrows. Clete Boyer, who knew Hinton from his years with the Atlanta Braves, waved him in, offering the writer's bona fides to skeptical teammates. "They were trying to remember the good things," Hinton said. "The only people who weren't drinking were Bobby Richardson 'cause he was a Baptist and Ryne Duren 'cause he was a recovering alcoholic. Mantle swayed, and he says, 'Before you take out that notebook, let's get one thing straight. I am not drunk!'

"I said, 'Okay, Mick, that's fine.'

"So we stand there and drink a few minutes. You hear the wind howling outside. I said, 'Did you guys ever imagine you'd end up burying Roger Maris in a blizzard in a little old town in North Dakota?'

"Mantle stood there, and little tears started to run down his cheeks, no expression on his face whatsoever. He was very collected, very serious, very emotional in the way that alcoholics can be. And he said, 'I want to go back to Commerce, Oklahoma.'"

By the next morning it had warmed up to three degrees with a wind chill of minus eight. The funeral director organized the twelve pallbearers, among them

Mantle, Ford, and Big Julie, at the bottom of the steps leading into the Cathedral of Saint Mary. "I looked at Mickey, he was half bombed, I guess with fear," Isaacson said. "So, I says, 'Mickey, don't drop this casket or I'll kill you.'"

There were only eleven steps to negotiate with Maris's coffin. But they were icy and precarious; their balance was hindered by the glare of camera flashes hitting the ice. The funeral director called out a cadence: "Step, step." "It took us maybe twenty minutes to get the casket to the top," Isaacson said.

Felt like it, anyway. Struggling up the steps with his share of the burden, Mantle muttered under crystallized breath: "Roger, you sonofabitch, I knew you'd screw me one more time."

2.

Until then, it had been a very good year to be a living legend. In March 1985 baseball's new commissioner, Peter Ueberroth, had reinstated Mantle and Mays, rescinding the fiat imposed by his predecessor and landing himself on the cover of *Sports Illustrated* between the parolees. Mantle's autobiography, *The Mick*, became a best seller. Larry Meli, the general manager of SportsChannel, the Yankees' cable outlet in New

York, made good on a promise to himself that someday he would repay The Mick for what he had meant to him as a child. Meli had recoiled at the sight of Mantle making a living by cutting ribbons at gas station openings—"*That's* what became of Mickey Mantle?"

After witnessing Mantle's delirious welcome home at the Stadium on opening day, his first appearance since August 1982, Meli hired him to join the SportsChannel broadcast team.

In November, Mantle and Ford hosted their first fantasy camp in Florida, dividing the profits equally among their seven children. The worrisome lump on Mantle's neck that reminded him of the onset of Billy Mantle's disease had proved to be no more than a buildup of calcium gravel. The *Daily News* hailed the diagnosis: "Mickey Mantle no longer fears he has cancer."

But Billy's prognosis was poor. After several years in remission, his cancer had returned in 1981, spreading from his lymph nodes to his liver and his bone marrow. "That's when he was angry at the world," Meli said. "It was supposed to be him."

Billy had entered an experimental program at M. D. Anderson, an agonizing and toxic regimen that attacked the cancer cells but also caused him to lose all the hair on his body. Merlyn moved to Houston to be with him. Two years later, Mantle was offered the job at the Claridge Hotel. "We needed the money," Merlyn told me. "We were paying for two homes. I think Mick was embarrassed he had to go to that to make a living."

In 1983, memorabilia madness was not yet certifiably mad. An old baseball was still just a baseball. When Glenn Lillie, the hotel's vice president of marketing, asked Mantle to bring some things from home to display at the Claridge, he showed up with two old grocery bags stuffed with home run balls—including one of the last thirty-six he hit—a glove, a pair of spikes, and a Babe Ruth Crown he won at the Top in Sports banquet in Baltimore. "He said, 'Is this the shit you wanted?'" Lillie recalled. "It wasn't memorabilia then. It was just shit."

Mantle's job at the hotel was to schmooze the high rollers and put a good face on the gaming industry. Sometimes he ate lunch in the cafeteria with employees whose checks he helped deliver—many of them didn't know who he was. Doorman Darrell Hammie saw how hard it was for Mantle to remake himself into a people person. "Mickey wasn't the most outgoing guy," he said. "Do you think if it was up to him to take this job, do you think that's in his character? It was like walking on eggshells every time he walked out the door."

He was a good employee, Lillie said, always on time

for visits to Grandma's House, a nonprofit shelter for young, pregnant, homeless women, or to the Pop Lloyd baseball stadium in Atlantic City, named for the great Negro League shortstop. He perfected his signature until it looked as if it had been carved. "By a surgeon," Lillie said.

There were perks that came with the job and Mantle took advantage of them. He had a succession of serious extramarital relationships after he retired from baseball. He had been involved with a woman from Florida when Roy True became his lawyer in 1969. She ended the affair after ten years, True said, because she wanted to get married and have children. In 1980, Mantle began seeing Linda Fetters, who later became a Hollywood stuntwoman and the wife of Ken Howard, the star of *The White Shadow.* He introduced her to Merlyn as his secretary and Howard says she was on the payroll. "Mickey always used the same line with these girls," True said. "'Would you like to be my business manager?'"

One weekend, Mantle brought both the women in his life to the Claridge, sharing his bed with his wife while his mistress/agent slept in the suite's second bedroom. Maris was also at the hotel that weekend. "Roger called up to the room, and Mickey was sitting on the couch between us," Howard recalled. "Roger said, 'So,

Mick, what you got goin' up there? A king and a rollaway?'"

Mantle was still seeing Howard in January 1985, when Greer Johnson arrived at the Claridge for a Super Bowl party on the arm of a VIP gambler who made the mistake of asking his genial celebrity host to entertain his girlfriend. For years, after she'd broken up with the gambler, after Mantle had called to ask permission to see her, after they had become an item, he'd say, "I'm still entertaining her."

In short order, the Georgia schoolteacher became Mantle's new agent and the woman with whom he would share most of the last decade of his life. She arrived at a propitious moment in the development of the nascent memorabilia trade. She would become his companion, his drinking buddy, his lover and employee, playing an indispensable role in the rebranding and marketing of The Mick.

Merlyn became convinced that the Claridge was her husband's downfall. She told me it was the beginning of "the bad drinking" and "the women thing." That surely was wishful thinking at best. But she clearly understood the corrosive synergy between his alcoholism and the heady reinvention of The Mick that began in Atlantic City. "I regretted him taking that job," she told me. "He'd just get drunk and talk to people."

"He'd get drunk so he can talk to people," said David Mantle, who recognized something valiant in his father's self-destructive work ethic. "I think he sacrificed himself for Billy. He did that for his son."

3.

It was after midnight and Mantle was still drunk when he got home from Maris's funeral. He woke Danny and told him he wanted to drive to the family's condominium in Joplin, Missouri. Danny didn't want any part of a 350-mile road trip in the shape he was in. "Screw you," his father said. "I'm going to load the damned car and go by myself."

Danny couldn't let him do that. He was twenty-five years old. He wasn't ready to just say no to his father. They drank all the way there. The rest of the family joined them later that week to celebrate Christmas and Mickey and Merlyn's thirty-fifth wedding anniversary. That was when she learned about Greer Johnson.

According to her account in *A Hero All His Life*, they began arguing about Mantle's one-night stand with the wife of a prominent country singer at a charity golf tournament in Nashville. He had left their hotel room before the banquet and hadn't returned until six the next morning. The fight escalated—Merlyn threw a wine bottle at his head, railing about Linda Howard. "Dammit, Merlyn," he said, "there's already somebody else, and you're still fighting with me about the last one."

Johnson said, "When she found out about us she called me and she told me 'I will never divorce Mickey. I like being Mrs. Mickey Mantle.'"

Howard said she had ended the relationship that August when he rejected her plea to get help for his drinking. "He told me, 'What would people think if Mickey Mantle went to rehab?' So I chose to leave. He called me one last time, begged me and pleaded with me to come to New York. I just said, 'If you're still drinking, I just can't do it.'"

(True recalled a different ending—a phone call from Mantle asking him to tell Howard it was over.)

After the fight with Merlyn in Joplin, Mantle disappeared for a couple of days. In his absence she resolved to travel with him more and drink with him more, which she did, resulting in scenes worthy of *Who's Afraid of Virginia Woolf?* Merlyn had a mild stroke— "stress-related." She was hospitalized for several days, and was prescribed blood thinners and blood pressure medication to treat a blocked artery in her neck. He flew home to be with her, but she was not in a forgiving mood.

There were separations and reconciliations. Once, after he'd been gone three months, he called in tears asking to come home. He stayed six weeks that time, some of which Merlyn spent in Arizona with Billy, who was in treatment for substance abuse. But the idyll didn't last. "Near the end he was really terrible to her, humiliating her," said a friend who spent time with them at the Claridge. "She'd say she wanted to go to the hairdresser and ask for money. Mick would peel off a $1 bill and hand it to her in front of people, making her grovel."

Many friends wondered why she stayed. Lucille McDougald, who knew her in their early days in New York, cited two reasons for her astonishing forbearance: "She really loved him, I mean, deeply, sincerely, truly, and accepted him with all his faults. The other one: she wanted the boys to have their dad. She wanted him in their lives, and she would do anything to keep that involvement, for better or for worse. It turned out to be worse. The boys ended up with the same addiction."

Eventually all four of their sons attended residential treatment programs for chemical dependencies, sometimes more than once. Merlyn drank, too—"not to the point the family did," she said—first trying to keep up with her husband and then just trying to keep him.

"Our holidays were all drinking," she told me. "Mick made eggnog. By the time I had dinner ready, they didn't want to eat."

During one stay in rehab, Billy told her he'd had his first drink at age nine. "And why wouldn't he?" Merlyn said. "We had the best-stocked bar in town. We made it so easy for them to do that."

As they told it in *A Hero All His Life*, Danny and David were chugging Mad Dog and Boone's Farm in the backyard by age thirteen. Billy, who was drinking and drugging heavily by age fifteen or sixteen, turned David on to cocaine; David almost killed himself one night when he snorted so much he couldn't breathe. Invariably, one thing led to another. "If I didn't drink, I wouldn't have done the cocaine," Danny told me. "You know you're screwed up when you drink a gallon of vodka and don't even feel it."

Everyone had an opinion about why Mantle drank. He drank because he was good at it; because he got paid to do it; because he had nothing else to do. "I think he'd drink to go to sleep," Clete Boyer said. "After he got out of baseball, he'd drink to wake up."

He drank, Mantle said, because "it was there all the time for me."

He drank because that's what alcoholics do.

Everyone saw what was happening. His friend Mike

Klepfer saw it at a dinner in New Jersey in the Eight-ies when "his face fell in a big bowl of pasta and we had to pull him out and clean him up." Roger Wagner, president of the Claridge Hotel, saw it when Mantle failed the Breathalyzer test the hotel planned for New Year's Eve guests. A photo op had been arranged so the local papers could promote the gimmick. "So Mickey comes down to do this photo shoot about ten o'clock," Wagner said. "He blows in the Breathalyzer, and it's about .2! At ten in the morning! Anyway, the Breatha-lyzer worked."

Mantle's doctor, Art DeLarios, a pal from Preston Trail, saw it in the blood chemistries that documented progressive liver disease. He warned him of the danger repeatedly—as had his previous physician. "You couldn't talk to him about getting sober," Merlyn told me. "Our doctor, Dr. Wade, tried to talk to Mick and he never went back."

When he overdid it and overloaded his system, he'd call DeLarios: " 'All right, double-M, come on down and I'll have a look at you.' I hospitalized him several times, just to get him IV fluids and get him off of the booze. He was tough. He was able to recover pretty quickly if we stopped the intake of alcohol and gave him some gastrointestinal medicine to relieve his indi-gestion."

DeLarios told Howard, "Linda, he's living on a rain check."

Everyone protected him, including the New York City police. "He drank 'til he fell out of his chair, then stumbled back to his suite at the St. Moritz," True said. "If he had trouble making it, the police would get us to the doorman. We could have passed out on the sidewalk, and someone would have picked him up."

Sometimes there was no rescuing him from himself. In the fall of 1986, the Claridge threw a birthday roast for Mantle and Ford. Twenty or so of their best pals showed up and played in the two-day golf tournament he hosted: Warren Spahn, Lew Burdette, Billy, Yogi, The Scooter, and Boyer. "By noon he's already drunk," Boyer said. "I mean, we all are. That night we had five or six hundred people there."

The banquet was larded with hotel bigwigs, local philanthropists, and high rollers. The drunken Mantle lurched toward the microphone. "We tried to tell him, 'Don't get up on that dais,'" Boyer said. "Sitting there, right in front of this podium, is Whitey and Joan and Merlyn. You can't hardly understand what he's saying. So now he gets serious and he says, 'I want to introduce my wife, Merlyn.'"

Whereupon Ford whispered something in his ear, Boyer said, and Mantle repeated it. "'Oh, Merlyn's not here? She's out in the bathroom? Oh, fuck her.'

" 'Now I want to introduce Whitey's wife, Joan. Oh, Joan went with her? Well, fuck her too.' "

Dinner was over. Riding down the escalator with Mickey and Merlyn, Burdette told the hotel's mortified PR man, "She's the one that should be in the Hall of Fame."

Wagner took responsibility for the fiasco. "We shouldn't have brought him up," he said. "We should have acknowledged him. Instead we invited him up to say a few words and he was not in the position to say the words real coherently."

Mantle's contract was not renewed. But by then he had little time for "the smallest but friendliest casino on the strand." He could make $50,000 on a weekend by signing his name.

4.

Baseball cards were the invention of American tobacco companies. The collection at the Library of Congress dates back to 1887. They didn't become the province of childhood collectors, card flippers, and bubble gum snappers until the mid-1930s. The protean pink goop that replaced tobacco in each pack of cards was invented by Walter Diemer, a twenty-three-year-old accountant for the Frank H. Fleer Company in 1928. He saw the future in translucent bubbles when he took five pounds of the pink glop to a Philadelphia grocery store on the day after Christmas and it sold out that afternoon. They called it Dubble Bubble.

The Topps Chewing Gum Company, makers of Bazooka, entered the trading card business in 1951, when Bowman was still the king of cards. That fall, after Bobby Thomson's Shot Heard 'Round the World, two Topps employees—Sy Berger, an assistant to the sales manager, and Woody Gelman, a commercial artist who had contributed to early Popeye animations—reinvented baseball cards. Sitting at Berger's kitchen table, they created a whole new look designed to appeal to baseball-crazed boys: their cards were larger and bolder in design, and saturated with color. On the front: the team emblem and a facsimile signature were added to each player's portrait. The year was deliberately omitted to create a sense of timelessness and a longer shelf life. On the back: a mini-biography with ht., wt., bats, throws, birthplace, birthday, and stats from the previous year, which allowed mathematically inclined Little Leaguers to compare themselves with their heroes. Mantle's card, the Topps #311 in a series of 407, was a study in intensity: the young slugger's hands tightly wrapped around a bat, his eyes shielded by the brim of an almost iridescent blue cap.

By the time it was released, football season had arrived. Orders were low. Sales were disappointing. The unsold cards were returned to the Topps warehouse in Brooklyn, where they remained until 1960. With storage space tight and even carnival brokers uninterested in purchasing the remainders, Berger needed to dispose of the inventory. He intended to burn it all, but when he got to the dump he saw scavengers going through the trash and said, "The hell with them." He called a friend whose father had a tugboat. One fine summer day, Berger climbed aboard a barge loaded with Topps' excess stock (some packaged, some loose cut cards) including a mother lode of Mantle #311s, and headed for open water, attached to the garbage scow by a safety wire. "All of a sudden, they pulled the thing, the floor opened up and they got drowned," Berger said.

He had no idea that the #311s he sent to their watery grave would become the most valuable baseball card of the postwar era; that the trading card frenzy would crystallize around the cardboard mug of Mickey Mantle. If he had known that, he would have kept one for himself.

In 1978, the cost of a Mickey Mantle autograph at a card show on Long Island was the price of admission—$3.00; $600 was a good price for a #311. Two years later, at the Philadelphia Card and Sports Memorabilia Show, in March 1980, three Mantle # 311s were put up for auction. According to Pete Williams's *Card Sharks: How Upper Deck Turned a Child's Hobby into a High-Stakes, Billion-Dollar Business*, the promoters, Bob Schmierer and Ted Taylor, prefaced the sale by recounting the tale of the drowned #311s. When the gavel dropped at the end of the day, they had sold the first card for $3,100; and the second and third each for $3,000. "It was incredible," said Robert Lifson, owner of Robert Edward Auctions, who was there. "It went from $600 to $3,000 practically overnight."

The proud buyers of the three #311s, Rob Barsky and Bob Cohen, were deluged with interview requests; the stories were picked up by the wire services. The #311 went viral.

A year later, the Major League Baseball Players Association went on strike for fifty-one days. Terry Cashman penned his hymnal to the good old days— "Talkin' Baseball," with its "Willie, Mickey, and The Duke" refrain. The song became a prospectus for the fledgling memorabilia industry and a sinecure for the onetime minor leaguer who has since written lyrics for all twenty-eight major league teams.

The confluence of nostalgia for a less mercenary time and Baby Boomer upward mobility was a boon for The Mick. The boys who had sent him their old scrapbooks hoping for an autograph in return had grown up and grown rich. And they were willing to spend hundreds, thousands, sometimes hundreds of thousands of dollars on talismans from a gilded era when baseball was played for the love of the game. Savvy card dealers quickly grasped the emotional calculus. As Mike Berkus, cofounder of the National Sports Collectors Convention, put it: "To get the card back was to get childhood back."

For a couple of years after the ground-zero sale in Philadelphia, the market for #311s waxed and waned. Then, in 1986, a card dealer named Alan Rosen found a trove of about 5,000 high-number Topps '52 cards, including about 75 pristine #311s. Most sold for about $2,000; the last one garnered twice that. Three years later, the same card sold for $40,000. Woody Gelman's original art fetched $110,000 in 1989.

The Hobby, as it was then known, quickly became big business, an investment opportunity that spawned a cadre of competing professional authenticators. "The Topps #311 both fed and was propelled by the astronomical growth" of the industry,"said T. S. O'Connell, editor of Sports Collectors Digest.

"Mantle led that revival," said Dominic Sandifer, who became Mantle's handler when he signed an exclusive deal with Upper Deck in 1992. "He *was* the demand."

He traveled the autograph circuit for a decade, sometimes with Willie and The Duke (they were at the Claridge in December 1983). More often he took his old pals, Moose and Hank and Blanch, along for a sorely needed payday. *You want me? You take them, too.*

Greer Johnson earned 20 percent of whatever bookings she arranged. He grew dependent on her to make the travel arrangements, to make sure he got where he had to go and got back in one piece. The restaurateur Bill Liederman, former owner of Mickey Mantle's, called her "the tour guide."

Johnson says Mantle didn't know how much money he had, what bank it was in, or how to write a personal check. All he cared about was having a fat wad in his pocket. Some friends, among them Jim Hays, a buddy from Joplin, thought she ran him into the ground. "She worked the hell out of him is what she did," Hays said. "I mean, she was makin' money. Little Mick called her 'Greed.'"

But Danny Mantle acknowledged the family's debt to her industriousness in more than one conversation with Sandifer. "He said, 'The truth is, Greer helped him make a lot of money, which helped us,'" Sandifer said. "Whether it was Greer Johnson or Roy True or Dominic Sandifer, somebody was going to help Mickey

Mantle capitalize on that. That's not a difficult job to do. That's as easy as picking up the phone. While she certainly arranged those things and made sure that he got there and did his job, that's what it was."

The money was crazy, and it loosed a kind of lunacy that Cashman first observed when Mays, Mantle, and Snider taped *The Warner Wolf Show* on October 2, 1981. "People were screaming, 'Sign my arm!' 'Sign my eyelash!'"

Mantle was often nonplussed: "The effect I seem to have on people makes my hair stand on end."

He never understood it, and he never got used to it. "Why can't they get over me?" he'd ask Glenn Lillie.

When one of his old baseball jerseys sold at auction for $71,000, he offered to take off his undershorts. When somebody called Bill Liederman at Mickey Mantle's Restaurant wanting to hire him to appear at a bar mitzvah, Mantle said, "If they pay me $50,000, I'll do it." They did.

When the Say Hey Kid stood up Paul Simon the day he was to begin shooting the music video for "Me and Julio Down by the Schoolyard," producer Dan Klores reached out to Liederman, who reached Johnson. Sure, Mantle would pinch-hit. "There were three conditions," Klores recalled. "'He needs $1,500 in cash in $100 bills in a brown bag.' That wasn't a problem. We paid everyone $1,500 in cash—the others didn't get it in a brown paper bag. 'He needs a stretch limo,' and he wouldn't sign any autographs.

"We shot it in a playground in Hell's Kitchen. At the shoot, you couldn't have gotten a better guy. He was great with Paul, great with the crew. The script was, Paul would pitch to him. Mickey would hit a home run. We were playing with a Spaldeen. He hit a pop-up. The camera made it look like a towering home run."

That's a wrap.

Mantle took his paper sack of Benjamins and went home. His parting words: "Suddenly, it's like I'm Mickey Mantle again."

But just beneath the giddy veneer of supply and demand lay incredulity and loathing. "He said several times he felt guilty," Danny said. "He said, 'I wish I had a job. This is the only thing I can do.'"

Mutt did hard business with the tangible; he toiled with a Sharpie. "He hated the whole card show deal," Merlyn told me. "He felt like a whore 'cause they hired him out."

He vented by abusing baseballs again, inscribing them with a potpourri of vile epigrams. "Have a Ball Cocksucker!" "Tough shit, asshole." "Fuck Yogi." A ball signed "I fucked Marilyn Monroe" sold for $6,700. Still, he was besieged.

Dave Ringer, his internist in Georgia, saw a posse of golf carts chase Mantle down on a fairway in Lake Oconee. *Oh, God, Mickey Mantle! Oh, man, can I get you to sign this?* "Nobody cared about Mickey Mantle," Ringer said. "They cared about his name. They didn't want to know who he was. He was so lonely. He even commented about that. He didn't have any friends. He basically had 'Mickey, I want.'"

5.

On Christmas night 1989, Billy Martin was killed when his pickup truck slid off an icy, isolated road near his farm in Fenton, New York. Neither he nor the driver, identified as William Reedy, was wearing a seat belt. Reedy was charged with driving while intoxicated. At his trial the following September, Reedy testified that he hadn't been behind the wheel that night. He said he had lied to the police and to a priest at the scene of the accident to protect Martin, not knowing he was already dead. The jury found him guilty of driving with a blood alcohol content above the legal limit of 0.1 percent.

Mantle had been the best man at Martin's fourth wedding and was an honorary pallbearer at his funeral. He sat in the front pew at Saint Patrick's Cathedral in

New York between George M. Steinbrenner and Richard M. Nixon. Afterward, he told the *Washington Times*, "I have no idea why I liked him so much. We never could figure it out."

Ford and Mantle hosted their last fantasy camp a few years later. Ford withdrew from the partnership— he and Mantle's sons cited interference by camp director Wanda Greer as the source of tension between the old friends. "She wanted me out so bad," Ford said. "I was getting sick of it anyhow. We never, never had a problem over the fantasy camp."

With Martin dead and Ford estranged, Mantle discovered his sons. They became his roadies and drinking buddies. Danny later said he was afraid of his father until they started drinking together—not physically; it was that *look* that unnerved him. Mantle's fear was different. He didn't know how to talk to his boys. "To go out and have drinks with his sons was communicating," said Roy Clark. "He had nothing to talk to them about. He thought, 'If I do nothing, I'm not hurting them. If I don't go see them, or call them, then it's okay. If I do call them, I have nothing to say.'"

He had always traveled with a minder. Roy True was the first, then Linda Howard, Greer Johnson, and finally Dominic Sandifer. And he needed minding more than ever. Mickey, Jr., David, and Danny took turns traveling with him. When his health allowed, Billy worked in the oil fields, driving a big rig. "He treated his boys like teammates," Sandifer said. "He treated me like a teammate. Everybody was on the Mickey Mantle team."

It was an opportunity Mantle's boys couldn't resist. It was the first time he had paid them any attention. And they had nothing better to do.

In his absence from Dallas, they had adopted his reckless example. Their lives devolved into a montage of drugs and alcohol, motorcycle crashes and car wrecks, near misses and shoot-outs. David described the Mantle speedball he and his brothers concocted to Bob Hersom of the *Oklahoman*, in 1995: "Budweiser, cocaine, and sake at four in the morning. We'd heat it up in a coffee cup in the microwave. We'd go for about three days straight."

Guns were a way of life in rural Oklahoma, and they became part of family life in Dallas, too. Mantle kept a gun in the glove compartment of his car; Merlyn slept with a .38 under her pillow. One night, Billy went around firing his gun at the ceiling of his apartment and almost shot a man in the nearby swimming pool.

Once, David wrote in *A Hero All His Life*, "when I lived in our house on Durango, I got a call from someone who threatened to come over and kill me. He said

I was dating his former girlfriend. I didn't know who it was, so I called Mickey Jr. and he came over with a couple of his friends. We had a couple of guys crouched behind the bushes and four or five of us inside the house, and guns were everywhere.

"The lights were down, the curtains closed, and we're peering out the windows. It must have looked like the Dillinger gang was hiding out. And stumbling up the front walk, here comes Dad. He takes a few steps, stops, studies the house, and yells out, 'Hey, y'all, it's me. Don't shoot.'"

Though David didn't think of himself as suicidal, he told Paul Solotaroff, author of "Growing Up Mantle," an award-winning 2003 story in *Men's Journal,* that he'd get high on cocaine, chug some beers, and play Russian roulette while watching *The Deer Hunter.* He also cut himself with razor blades and burned himself with cigarettes. He said it didn't hurt. The point was to feel anything at all.

Pain has been a constant for him since he and Danny wrecked their father's brand-new Cadillac one night en route to a concert. A friend was driving. Danny, who was sitting up front beside him, suffered a broken rib. David almost died. The whiplash he suffered when his head hit the backseat caused fluid to accumulate in his brain. Doctors drilled a hole in his skull to relieve the pressure. He was unconscious on and off for three days. Eventually, his spine was fused at his neck. Roy True, who said he was "always going out and picking the boys off the lawn," was dispatched to tell The Mick he needed a new car.

David left Baylor University after three years. According to his transcript, he is eighteen credits shy of graduating. "That's what it says, but I got kicked out because of drinking and drugging," he said. "I feel bad and embarrassed that I wasted Mom and Dad's money. Dad was so proud that I was first to go to college."

The football coaches encouraged him to go out for the team as a walk-on. "They said, 'We can put fifty pounds on you.'

"I said, 'I'm not doing steroids.'

"I look back now and think, 'Yeah, but we all do cocaine and drink.' "

After he left Baylor, he worked as an assistant manager at a McDonald's in Joplin, making $12,000 a year. He sold and serviced air conditioners and hired Danny to work with him. When the trading card industry exploded, they pooled their savings and opened a card shop in the Prestonwood Mall. When their dad was in town, he'd drop by sometimes and say, "C'mon, let's get a beer."

"We're working," they'd protest, before closing up for the rest of the day.

Danny attended one year of classes at Brookhaven

Junior College in Dallas. He worked on a pipeline in an oil field as a welder's helper, started a landscaping company with some friends, and dabbled in real estate development—none of them jobs he says he "ever would have wanted to do for a life."

Mickey, Jr., went to junior college, sold life insurance, worked for an oil company, got a real estate license, which, like his father, he never used. And like The Mick, he played a lot of golf, only better. He was the first of the boys to join the Mickey Mantle team and the first to start drinking with him. He returned calls, filled orders for Mickey Mantle collectibles, and filled out foursomes.

After Mickey, Jr., got married, Danny took his place on the road. It was a revelation. "To the women, my dad and those guys were rock stars and I was their roadie, catching the overflow," he told Paul Solotaroff. "I was fifteen, getting laid like a rug and pounding mixed drinks with Everclear or a topper of 151. The next morning, I'd sober up and realize, 'Hey, Dad's tagging a lot of trim.' As kids, we'd just assumed he was working when he was gone, doing endorsements and stuff. Now I saw that he wasn't the family man we'd always thought he was."

They covered for him and lied to their mother, badly, when she called their room asking if anyone else was in his, David wrote. "We all played our parts."

*   *   *

It couldn't last. The gold rush inspired by the Topps #311 peaked in the early 1990s, when trading card companies papered the country with untold billions of cards, so many no one could count them all. The mad money led to fraud, forgery, and two FBI investigations. Operation Bullpen and Operation Foul Ball exposed massive counterfeiting operations, card doctoring, fraudulent authentication, and fraudulent autographs.

It wasn't hard to do. One time at a card show in California, Snider and Mays met a man who claimed, "I can sign your name better 'n you." He could. And he could do Mantle, too. Willie and the Duke recruited him to sign The Mick's name to the photographs they had autographed for each other. "I gave it to a friend," Snider said. "He never knew it was a fraud. We paid the guy forty dollars apiece."

Mantle told Clete Boyer that when Senator Edward M. Kennedy asked him to come to Capitol Hill to talk about the trading card industry, he replied, "I'll testify about trading cards when you tell me about Chappaquiddick."

Pete Rose spent five months in jail after pleading guilty in 1990 to tax evasion on memorabilia earnings. Five years later, a Brooklyn judge sentenced his boyhood hero, Duke Snider, to two years probation and a $5,000 fine after he pleaded guilty to criminal conspiracy and failing to report $97,400 from card shows. Willie McCovey pleaded guilty the same day.

By 2008, what had been a $1.2 billion-a-year industry was doing only $200 million in sales, according to *Sports Collectors Digest*, the industry paper of record. "Baseball card shops have gone the way of the blacksmith," said Vin Russo, owner of Mickey's Place, a memorabilia shop across the street from the Hall of Fame in Cooperstown. "When Mickey died, he took the Hobby with him."

In the world of cardboard, O'Connell said, "Mantle has been and continues to be in a class by himself. Among postwar players, Mantle lives in different zip code." Which is surprising to Mike Berkus, because Boomers aren't exactly booming and, he says, "the one thing you can't do is pass on your heroes."

The Topps #311 remains the Holy Grail of postwar cardboard: according to its spring 2010 "Population Report," Professional Sports Authenticators, the industry leader, has graded only 970 #311s. Among them: three gem mint 10s, seven mint 9s (and two with qualifiers), and thirty near mint 8s. According to Vintage-CardPrices.com, the authoritative baseball card price guide, a PSA 9 sold for $282,000 in 2006 and a PSA 8 sold for $112,000 two years later.

The only nude photograph of The Mick—"Yes, you read correctly, a NUDE photo of Mickey Mantle!"— was sold on eBay for $25,000 in 2005. Astonishingly, the proud owner of the Arthur Rickerby candid preferred to remain anonymous.

Mickey, Jr., said in *A Hero All His Life* that he never intended to make his father his career. Marketing his memory—and weeding out counterfeit merchandise—became Danny and David Mantle's life work. In 1997, the FBI estimated that 70 percent of all sports autographs were forgeries. The Mantles told the *Financial Post* in 2009 that they believe 90 to 95 percent of all Mantle signatures are phony. Some people, friends and relatives, wonder about their career choice. "I know that Danny and David have probably never done nothin' in their lives," their uncle Larry Mantle said. "Mickey, Jr., never did. All he did was follow his dad around. And then Billy died so young, he didn't have a chance to do anything. I think they've sold just about everything they can sell. But, I mean, they gotta do what they gotta do."

Others, including Bill Liederman, the former owner of Mickey Mantle's Restaurant, have a different take. "The kids went into the family business," he said. "That's America."

6.

Late one November afternoon, Mantle was at the bar at the Pierre Hotel in New York, telling the same old stories to the same old guys. Only one problem: he had a load on, and a flight to catch at JFK, and if he didn't make it he'd catch hell from Merlyn. Billy Martin called on George Lois to get Mantle out of the bar and into a cab. Lois guided him to the coat check to claim his bag. "We go up to the check room woman, a good-looking woman in her forties," Lois said. "He peels off a hundred-dollar bill—a lot of money back then. He put it on her chest. I grabbed him—'Mickey, calm down, you're drunk.'"

Lois was already seeing 72-point headlines in the morning papers. But his agita proved unnecessary. All she wanted for her trouble was an autograph for her eight-year-old son, whom she had named Mickey. Naturally, Mantle was his hero. He scribbled his name, and Lois shoved him through the revolving door onto Fifth Avenue. By the time Lois got done making amends—she kept the $100—Mantle was nowhere to be found. He had disappeared into the rush hour dusk. Raw as it was, Lois was schvitzing: "Oh, my God, I've lost Mickey Mantle."

Frantic, he surveyed the pedestrian bustle. It didn't

occur to him to look down. Mantle was lying in a swell of slush by the curb. "Literally, his face was in the gutter," Lois said. "I said, 'Holy shit, Mickey, are you hurt?'

"He says, 'Fine place to be for America's hero.'

"I swear to God—it was beautiful. He was very aware that everybody was looking at him and he had made a fool of himself."

Everybody, including sometimes Mantle himself, could see what was happening. But nobody did anything.

Not his family. "We tried," David said. "But we were so into our disease."

Not Major League Baseball.

"They could have done something," Bouton said. "They *should* have done something. It's just stupid not to take care of its greatest heroes. Mickey Mantle still markets baseball. He's going to be marketing the game forever, and they didn't do anything."

Not his teammates.

Some had the same problem; others were as financially dependent on him as they had been for their World Series checks. Many people had a vested interest in his continued success. As his pal Mike Klepfer put it: "He was the big poppa bear."

One night, his friend Jimmy Orr, the football player,

was at the Greensboro, Georgia, condo Mantle shared with Greer Johnson when Whitey Ford called. Orr answered. "Whitey says, 'Take care of him, 'cause I ain't made enough money off his autographs.'"

Ford was kidding. But Orr wasn't: "He's pretty hard to take care of," he told Ford.

Mike Ferraro, who bunked with him one season at the St. Moritz and later ran Mantle's fantasy camp, said, "To me, Moose and Hank are the guys who should have said, 'You need to take better care of yourself.'"

But that would have been a violation of Baseball Code. "We don't interfere," Tony Kubek said. "We'll go out and drink with him. But no, we don't tell him the one important thing: 'You're drinking yourself to death.' I have personal regrets. We all have regrets. I think we were all enablers. Players, including myself, feel guilty, and they should feel guilty."

"We should have done more," Carmen Berra said. "What could we do?"

In the Eighties, Kubek recruited Sam McDowell to organize an intervention. "There had been three or four different people who had called me over the years about trying to step in and help Mickey," McDowell said. "Just by sitting down with an alcoholic, you're not gonna help him. No way. It has to be orchestrated. The individual has to be put into a corner in which his only answer is 'Yes, I want help,' then instantly taken to a rehab.

"When I got the call to help, I was the director of the sports psychology program in the employee assistance program for the Texas Rangers and the Toronto Blue Jays. We contacted many individuals that Mickey had known, two other past players that were close to him. They could come up with certain stories, which you have to have that are directly related to the alcoholism."

Kubek recalled, "Some guys wouldn't do it. Sam said, 'It might work, but chances are good it won't. And it if doesn't, chances are Mickey'll hate you.'"

In the end, Mantle was tipped off by friends McDowell declined to identify. "On one individual's part I would say it was strictly ignorance," he said. "On the other person's part it was protecting a relationship. The assumption was that if Mickey ever got sober, he would pull away from these people, which he probably would."

So the party continued. Johnson says she tried hard to get him to quit, or at least cut down, talking to bartenders behind his back to "make him think he was getting a strong drink when actually he wasn't, claiming I had a headache so he'd go back to the room and quit drinking."

But every white guy of a certain age wanted to be able to say, "I bought a round for The Mick." And then laugh or look the other way when he made a fool of himself. "I asked him once if he ever met Elvis," his filmmaker friend Tom Molito said. "He didn't understand why I asked. Mickey Mantle was not destroyed by alcohol. He was destroyed by celebrity."

One night at a restaurant in New York, McDowell ran into Mantle at the bar, where a big-shot sportscaster was buying him round after round. "I turned to him, and I said, 'Don't you understand that you're killing that man?'

"And he said, 'You don't understand, we're friends.'"

18

February 5, 1988

TOP OF THE HEAP

1.

On opening night at Mickey Mantle's Restaurant on Central Park South, the hero came in the back door. The owners, Bill Liederman and John Lowy, set him up on a raised platform at a six top in the back room, where he received his public in a natty Yankee blue tuxedo.

When Liederman and Lowy first showed him the space at 42 Central Park South, they thought it might be too uptown for his taste. They didn't know that it was the former site of Harry's Bar, Mantle's home away from home during so many baseball seasons when he lived at the St. Moritz next door. "I spent so much time here they used to call it Mickey's Place," he said at the ribbon cutting. "And now it is."

He wanted it to be a grown-up joint like Toots Shor's. He didn't have to put any money down. "When you hit 536 home runs, you don't have to invest any money," Roy True said.

The agreement called for him to receive $100,000 the first year and $80,000 each year thereafter, plus 7 percent of the equity. At the press conference announcing the deal, he promised, "Whenever I'm in town I'll be here, passed out at the bar."

Opening night was anarchy. Everybody showed— Yogi and Billy and Whitey, The Scooter and George Steinbrenner, Sly Stallone and Frank Gifford, Dan Rather and Walter Cronkite, Raquel Welch and Angie Dickinson, Bill Murray, Dan Aykroyd, Bruce Willis, Barbra Streisand, Bob Costas, and Howard Cosell, who let it be known, "Without me there would be no Mickey Mantle."

The line at the front door stretched down the block to the Plaza Hotel. The crowd grew cold and unruly and large—at least three or four times the maximum capacity of 215. Not long after the doors opened, the New York City fire marshals threatened to shut them. "We didn't hire security, which was stupid," Lowy said. "We took turns trying to keep people away from him. Everybody wanted a piece of him. There were so many writers. Channel 9 and 11 were doing live broadcasts."

The only no-shows were DiMaggio, who wouldn't come, former New York governor Hugh Carey, who couldn't get in the front door, and Merlyn, who wasn't invited. She had wanted to accompany him, but so had Greer Johnson.

Mantle and Johnson dressed for the party down the block at the Park Lane Hotel, where their friends Mike and Katy Klepfer had a suite. They had gotten to know each other when Klepfer hired Mantle to appear at the thirtieth-anniversary luncheon for his trucking company in Binghamton, New York, five months earlier. It was a $20,000 gig—Mantle gave $5,000 each to Martin and Ford. When Danny Mantle located Mantle and Johnson at the Klepfers' suite, he said he was going stag. Mantle was distressed about lying to his wife, who had also been left out of Martin's recent wedding. He told Katy, "My father would be upset that I didn't take Merlyn to these things."

She told him, "Your dad's been gone a long time, Mickey. You've got to make up your own mind."

Katy sensed Mantle wanted the kind of family life she and Mike had and wanted to tell him "to just go back home." Finally, one day she did. He said, "That's where I belong."

Mantle was nervous and sober when they left for the party, a result, Mike Klepfer said, of Johnson's edict: "You drink before an event, and I leave." He was also conflicted about Merlyn. "The whole night he felt miserable about it," Liederman said.

Klepfer was a former New York State trooper and trucker who, at six feet seven, had the presence and heft to help establish order in the chaos. "Myself and a New York City policeman, in civvies, we just held people back," Klepfer said. They funneled invited guests and crashers through the front door, all of whom wanted to be able to say they were at Mickey Mantle's the night he reclaimed New York.

Many of the notable invitees took one look at the mob scene and left. The crush was such, Lowy said, that "the hors d'oeuvres never made it past the kitchen door."

Mantle and his party did not last long. A snarky item in the *Daily News* reported, "Mickey Mantle apparently had to get away from it all during the wild preview party Friday for his new restaurant on Central Park South. At the height of the comings and goings by the curious, The Mick was nowhere to be found. Where was he? At another restaurant, our spies report—an eatery in the nearby Park Lane Hotel. With Mickey were Billy Martin and his bride, Jill, and some mutual friends. They chatted away while a couple of doors away Mickey's guests kept looking for him."

The next morning, they went back to survey the wreckage. Between breakage, booze, and cigarette burns in the protective plastic coating on the brand-new wooden floor, Liederman figured the evening cost $25,000. "We were all cleaning," Lowy said. "Myself and Bill and Jill were on our hands and knees. Billy and Mickey were sitting at a table, saying, 'Hey, you missed a spot.'"

2.

After nearly four decades and four sons, the "arranged" marriage of Mickey and Merlyn was irretrievably broken. Long after his career ended, Mantle continued to live a ballplayer's existence—home and away. For much of the last decade of his life, he divided his time between Dallas, where he built a new house that he would share with Danny and his wife, Kay, and Greensboro, Georgia, where he shared a condominium on a man-made lake with Greer Johnson. The name on the front door was hers. "It was the last real relationship outside his marriage," True said.

Mantle had told Merlyn he was leaving a month before the opening of the restaurant. Their marriage counselor had recommended a separation, telling Merlyn, "Mickey is totally controlled by fear."

In *A Hero All His Life*, Merlyn offered a remarkably candid portrait of their relationship and its dissolution. She wrote that as she helped him carry his belongings to his car, "He said, 'You will always be my wife. I know I'm giving up the best thing I've ever had.'

"I said, 'I don't even understand what you're talking about. What do you mean? I will always be your wife, but you don't want a life with me?'"

He left without another word, Merlyn wrote, returning later in the day—she wasn't sure why. While he was there, Johnson called. "She said, 'I really love Mickey.' I said, 'I really love him, too.'"

There had been so many women, so many blondes, including, he had bragged, Angie Dickinson and Doris Day. "I remember Mickey telling me one time that he woke up in bed and there was a blonde there," Pat Summerall said. "He didn't remember who she was. He picked up her hand, and it was Merlyn. She said, 'You sonofabitch, I married you.'"

David told his mother the day I visited: "People called you Mrs. God."

"No," she replied firmly, "they called me a saint."

* * *

Merlyn put up a good fight. He never hurt her, though, he wrote later, "she put a lump or two on my head." With good cause. One year at dinner during fantasy camp, with Mickey and their sons, he asked their waitress for her phone number. She took Merlyn aside and told her about it. Back at the hotel, Merlyn changed out of her party dress and put on a tracksuit—fighting clothes. She threw a bar stool at him, breaking the glass coffee table. He was sprawled out on the bed, too drunk to notice. "I jumped on the bed and straddled him. I started slapping his face, from one side to the next, like a windshield wiper. All the boys were in the room, and they were unsure how to go about pulling me off him."

"Hey, Merlyn, that hurts."

The guests in the next room called the police, who sent them to separate quarters. He called a few minutes later, pleading, "Honey, come back and sleep with me." She ran down the hallway to be with him.

One thing Greer Johnson and Merlyn Mantle agreed on: he had no respect for women. Johnson traced that back to Mutt. "He always indicated to me that his dad really didn't have any respect for women," Johnson said.

Mantle also led her to believe that Mutt had an eye

for the ladies. Merlyn said he told her about one paternal indiscretion. "I think it was a brother's uncle's wife," she told me. "I don't think it was where he thought about leaving the family. It could have been just a one-night stand."

Mantle had plenty of those.

"Mantle loved women," Roy True said, "then treated them like crap."

He astonished his old friend Marjorie Bolding by calling from hotel rooms asking her to say hello to whoever happened to be sharing his bed. "Those sexual encounters were absolutely nothing to him," Bolding said. "They were like an exercise for him."

His language reflected that. George Macris, a fantasy camp umpire, recalled the indignation of an irate camper after Mantle had insulted his wife. She had been hanging around too much, getting in the way. "Her husband goes to Whitey and complains. 'I'm paying $5,000. I shouldn't have to put up with this.' So at the banquet Mickey gets up and says, 'I want to apologize. I wasn't a gentleman. I called Mrs. X something I shouldn't have.' Then he leans into the microphone and says, 'For the rest of you who don't know, I called her a fucking cunt.'"

When Johnson chastised him for his language at a Bible Belt autograph show, he turned and looked at her.

"Almost like he's hurt," she said. "He said, 'Nobody has ever told me that before.'"

He could be verbally and publicly abusive to her, too. Tom Molito attended a planning session at the restaurant for the 500 Home Run Club video he produced in 1989. Johnson was the only woman at the table. One of the men asked, "What's your role?"

"Mickey steps in," Molito said. "He says, 'She sucks my dick.'"

"She said, 'Yeah, Mick, but not tonight.'"

Because he was The Mick, Molito didn't say anything when Mantle made a pass at his wife one night at dinner.

Johnson says she and Mantle talked about getting married. With or without a piece of paper, she had come to regard herself as the de facto Mrs. Mickey Mantle. The real Mrs. Mantle received this odd piece of news from her husband the day he left her: "I don't want a divorce but you can have one if you want it."

Quickly, the separation turned nasty. She hired an attorney. He tapped her phone. Mickey, Jr., was so hurt by the separation, Merlyn wrote, that "he just sort of faded into the scenery" for a while. At one point, David didn't see his father for almost a year. She turned increasingly to alcohol for solace. Though she could no longer look to True for legal advice, he told her in confidence, "Merlyn, don't get a divorce. Mick doesn't want one."

Mantle offered different explanations to different people for his odd notion of marital fidelity. "He really loved Merlyn," True said, echoing the sentiments of many friends and relatives. "He said, 'I love her. She's like my sister, though.'

"He told Greer, 'I'm never going to divorce my wife.' Greer would end up angry. They had great battles over that."

He told Liederman he couldn't divorce Merlyn because "the only thing she ever wanted to be was Mrs. Mickey Mantle."

He told his Georgia friend Ron Wolf over a drink at the Harbor Club, "I never have any intentions of ever divorcing Merlyn. You can understand why, I guess, when I provide her with two condominiums in downtown Dallas, an American Express card, a nice Mercedes. Why should she ever want to leave?"

He told his Dallas attorney friend Troy Phillips, "I didn't want to put her through the embarrassment of divorce."

"I said, 'My God, Mickey, how can you embarrass her more than you have?'" Phillips replied.

He told Pat Summerall that he did not have "the courage to get divorced or separated from the boys.

But he was telling Greer at the same time that he was very much in love with her and wanted to marry her."

He told Carmen Berra, "I don't want you to worry about Merlyn. I will never divorce her. She will get everything. Greer is an employee. She is a ten percent woman."

3.

Mantle's first panic attack occurred in the air. In April 1987, he had to be carried off a plane in Dallas—he thought he was having a heart attack. His heart was breaking, but not physiologically. The center could no longer hold.

One night, sitting at his booth in the restaurant with Liederman and Lowy, a woman approached his table. Autographs were on the house for diners, and Mantle kept a stack of postcards with him for that purpose. "He's in a great mood," Lowy said. "He's the perfect Mickey. He starts to write the autograph. The pen runs out of ink. He takes it and hurls it into the wall. He threw it so hard and he was so furious, she just stood there and she started to cry. She finally had to leave.

"He could be one person one moment and a totally different person the next and go back and forth—though usually once it flipped into the bad Mickey, it didn't flip back. When he was the good Mickey, he was funny, friendly, generous, kind, gracious even. But he could turn on a dime."

The two Micks were as puzzling as they were unnerving. Lowy thought, "There was something from a very early age that happened to him."

Something Mantle had never confided to his wife. One night, long after they had separated, they spent an evening in Dallas watching a TV movie about a child who had been sexually molested. "That happened to me," he said.

He told her that often when Mutt and Lovell went out to a Friday-night barn dance, her teenage daughter, Anna Bea, babysat for her half siblings. He was four or five years old when she began molesting him, pulling down his pants and fondling him while her friends, "teenagers and older," giggled and smirked. "They started playing with him," Merlyn told me. "And, of course, he got an erection. They laughed at him. He remembered how embarrassed he was." That was the only time they ever spoke about it. "It could have been why he turned out the way he did," she told me.

"He had kept a secret to himself for nearly his entire life," she wrote in *A Hero All His Life*. "That night I thought I understood more clearly than I ever had why his ego was so fragile. He was a loner who loved a crowd, when they cheered from a distance. He never respected women. He demonstrated it in the ladies he chose for his one-night stands, in the crude way he talked and acted in front of women when he drank. And in the way he treated me, with too much credit for raising our sons and too little for being an adoring and faithful wife."

In fact, he had not kept the secret entirely to himself; nor was Anna Bea his only abuser. He had confided pieces of the story in Linda Howard, Greer Johnson, Larry Meli, and Mike Klepfer. One evening at Mickey Mantle's, Meli was watching a football game with him at the bar. Meli was troubled by Mantle's decision to take Johnson to the house in Dallas when the boys were there and have her sleep in his wife's bed. "I told him how wrong it was," Meli said.

Mantle told him about Anna Bea. "It was almost like 'Screw the sons, look what I had to go through!'" Meli said. "He was sober and melancholy. Mickey was so embarrassed that it was a half sister that he said it was an aunt!"

Years earlier Mantle had raised the subject with Klepfer over a hand of gin rummy. Theirs was an intense two-year friendship late in Mantle's life, so close that the Klepfers outfitted an apartment in their basement for Mantle and Johnson. Mantle fished in their pond, tucked into Katy's home cooking, and learned how to build a fire in the fireplace. On trips to New York, he and Mike would play cards while Katy and Greer went shopping. The two men had a natural kinship. Like Mutt, Klepfer's father, Ellis, was a miner—he went to work in the coal mines of Kentucky at age nine. Like Mutt, whose mother died when he was eight years old, Ellis got little maternal nurturing; his mother, Mike said, was "a coal-town whore."

He told Mantle about his father's wretched childhood. "That's when he broke out talking about his early childhood. He started telling me, 'If you wanna know about abuse as a child . . .'

"And then he more or less shut up, went quiet, and said, 'I can tell you stories about that. I just know what it's like.'

"He asked, had anybody ever fiddled with me?"

Klepfer had never told anyone what had happened to him on the porch on Doubleday Street in Binghamton, New York. But he told Mantle the whole story. "They were playing with my johnson. I didn't know what was going on because I was so damned scared."

"Yeah, I know about that," Mantle said.

He listened closely as Klepfer described his inability to fend off the assault. Like Little Mickey, Klepfer was a small boy who grew into a hulk of a man. Mantle had osteomyelitis; Klepfer was asthmatic. "I couldn't fight anybody because I couldn't breathe. And Mickey laughed and said, 'Everybody talks about my arms and how strong I was. I was a piss-ass sissy, too.'"

Mantle told him that an older boy in the neighborhood had pulled down his pants and fondled him and that it had happened more than once. "He told me, 'Well, that's how I learned how to run like lightning,'" Klepfer recalled. "'If I got wind that something like that was going to happen, I got the hell outta there. And I could run.'

"What happened to him as a kid drove him nuts. He lived with his situation where he was being abused for a while, long enough for it to be indelible. It was something that he had never forgotten."

Mantle alluded to the abuse by Anna Bea but never told Klepfer the details. Anytime the subject came up after that, Klepfer said, "He would just get drunk. *Massively* drunk. The drunkest ever. There'd be no talking about it the next day."

There is no way to know how often or for how long he was abused by the neighborhood boys or by his half sister—he told Merlyn that it continued until Anna Bea moved out of the house. She worked in the bars around Commerce, married, and died young. By 1956, when New York scribes descended on Commerce to document his all-American childhood, Anna Bea had been written out of the family script.

In high school, he was seduced by one of his teachers, Merlyn told me. "She just laid over him," she said. He took her to Independence, Missouri, to meet his roommates his first year in the minors. "She was a hot date," Jack Hasten recalled.

Mantle laughed when he told Greer Johnson about her decades later: "That's how I got through high school was screwing the teachers. That's the only way I was able to graduate."

No doubt the "Mrs. Robinson" attentions of an older woman made him the envy of the Independence Yankees. But the seduction was no joking matter. It may have assuaged lingering, unarticulated hurts and insecurities but it was also a violation of innocence and trust, an exploitation of a hormonally charged teenager who wet his bed until he left home to go away that season.

Richard Gartner, a New York psychologist and the author of the definitive work on the subject, *Betrayed as Boys*, says the incidence of abuse does not necessarily determine its impact, nor does the age at which it occurs. "I've treated people who have had one relatively mild incident and yet were deeply affected all their lives, sometimes more than people who were chronically abused," he said.

Abuse by an older sibling is a violation of the gravest taboo—incest. Abuse of a heterosexual boy by other boys undermines an emerging sense of manhood. Abuse by an older woman in a position of authority is an abuse of power, even if, Gartner says, it "made him feel like a man."

Every boundary had been crossed—familial, gender, professional—which could account for why Mantle crossed so many lines of behavior and decorum. If it was okay for others to violate his boundaries, it was okay for him to violate those of others.

To experts in the field, Mantle's story is consistent with a cluster of symptoms often seen in survivors of childhood abuse: sexual compulsivity or extreme promiscuity; alcoholism or substance abuse; difficulty regulating emotions and self-soothing; bed wetting; a distorted sense of self; self-loathing, shame, and guilt; a schism between a public image and the private self; feelings of isolation and mistrust; and difficulty getting close to others.

Those deeply held feelings of isolation and shame abide. Mantle nodded tearfully when Bob Costas told him in a 1994 interview that he had always sensed a deep sadness in him. "I don't get close to people," Mantle replied, dabbing his eyes with a handkerchief. "I'm weird or something, I guess."

One night, over a candlelight dinner at the Klepfers'

house, after grace had been said, Mantle looked up at his friends and asked: "Why do you people have anything to do with me?"

Today, many victims of childhood sexual abuse are diagnosed with complex post-traumatic stress disorder—the term used to distinguish long-term trauma. "A sense of a foreshortened future" is one of the clinical criteria for making a diagnosis of complex PTSD.

Like many abused boys, Mantle may well have downplayed these early traumatic experiences. He would have had good cause to do so. As Gartner points out, American culture leaves no room for men to see themselves as victims; if they are victims they are not men. Nowhere would it have been more essential to hide those feelings than in a major league locker room.

4.

On Labor Day, 1988, Mantle returned to Cooperstown—as a paying customer—with Greer Johnson and Mike and Katy Klepfer. He told them that he hadn't been back to the Hall of Fame since his induction. In fact, he had filmed *A Comedy Salute to Baseball* there with Billy Crystal in March 1985. As they got into the car with a thermos of Bloody Marys—Katy always poured light—Johnson remembered the five-by-seven-inch autographed cards she always brought along when they went out in public. She went back into the house and got a thick stack of them, anticipating a swarm in Cooperstown.

Mantle was wearing a white Oklahoma Sooners windbreaker, a white Sooners cap, and a pink golf shirt. Though he had seen his plaque when it was presented to him in 1974, he had never taken the time to visit the Hall of Fame gallery on the first floor, where the earliest inductees are honored. Klepfer hung back as Mantle read every plaque, squinting through the dollar cheaters that Klepfer had purchased at the drugstore. Then he put his hands up to his face and cried. " 'Y'know,' he says, 'until today I thought I was a pretty good ballplayer.' "

Klepfer thought, "He was humbled by greatness."

Or perhaps they were "what if" tears. Mantle articulated his regret in a private conversation with Costas: "He said, without a thimble-full of bravado, but wistfully and with affection and respect for the other players involved, 'I know I had as much ability as Willie. And I had probably more all-around ability than Stan or Ted. The difference is none of them have to look back and wonder how good they could have been.' "

No one disturbed him. No one asked for an autograph. No one recognized Mickey Mantle—not even

582 · JANE LEAVY

the staff at the Hall of Fame Museum. "Toward the end, he took his hat off and said, 'Well, maybe if I don't have my ball cap on they'll recognize me,'" Klepfer said.

They posed outside on the front steps, trying to attract attention. *Hey, Mickey, a little over to the side, Mickey. Hi, Mickey! How are you, Mick? You're here in Cooperstown!* No one noticed. They had lunch at the Otesaga, the old inn on the lake where inductees annually gather for the Hall of Fame Weekend. Klepfer tried one more time. "'Do you have a reservation for Mantle?' The girl says, 'No, I'm sorry, we don't. You don't need a reservation. You can go right in.'"

19

February 4, 1994

GETAWAY DAY

1.

As he waited for Mickey Mantle's flight to arrive at the airport in Palm Springs, Mark Greenberg reminded himself, "He's just another drunk." Greenberg was an administrator at the Betty Ford Center, and Mickey Mantle was his childhood hero. When the plane taxied to the gate, he steeled himself, "Don't get sucked up in who this guy was. Get sucked up in who he is."

Five days earlier, Greenberg had found a note on his desk saying, "Call Mickey Mantle." Since then, he had been on the phone negotiating Mantle's admission, finding a bed, and making concessions he didn't want to make to celebrity. But nothing in his professional training prepared him for the moment Mantle stepped off the plane. "All he was was a broken-down alcoholic," Greenberg recalled. "For Mickey to have to come face-to-face with the fact that he's being admitted for his disease that's torn himself and his family apart—just really breaking down at that airport—I stood there. I didn't know what to do."

Mantle had hit bottom. An MRI showed what his friend Jim Hays already knew: "His liver looked like a doorstop." He would need a transplant eventually.

Ten years of warnings and hospitalizations, abnormal blood tests and panic attacks, binges and blackouts—*I did what?*—had coagulated into a forgotten decade, which was just as well, given his most egregious behavior. Even his dearest, most forgiving friends were sometimes appalled. Barbara Wolf, who lived at Lake Oconee, cooked for him the way his mother did— "none of that plate art stuff"—chicken, and biscuits, and beans, and a tropical fruit trifle made with angel food cake, sugar-free, fat-free vanilla pudding, fat-free milk, Cool Whip, and tropical fruit cocktail. "All out of cans," she said. "But he loved it."

One night, after dinner, he blew his nose in one of her good linen napkins. Dismayed, she reprimanded him. "When I called him on it, he handed it over to me and said, 'Do you want me to sign it for you?'"

He was invited back, but thereafter she used paper napkins.

The wife of his Georgia physician, Dave Ringer, would not allow him to be around their children. You never knew what he might say—like the message he scrawled on a ball autographed for one young boy: "You're lucky. Your mom has nice tits. Mickey Mantle."

He raised $220,000 for the Make-A-Wish Foundation of Oklahoma through the annual golf tournament he hosted at the Shangri-La Resort in Joplin, Missouri, according to former president and treasurer Mike Bass. But, one year he was so offensive that Jimmy Dean, the country singer, summoned Jim "Mudcat" Grant, the pitcher turned poet, to the microphone for some soothing verse. "Well, Mud, I think you better read your poem at this time to calm the people down," Grant recalled him saying.

Moodiness devolved into nastiness and paranoia. He started to get "belligerent and mean in 1992," David wrote, calling his sons by "hyphenated names." He squeezed Bill Liederman's testicles every time he saw him at the restaurant. "Just enough so it hurt a little bit," Liederman said.

The restaurant staff saw what he was doing to himself, starting at ten in the morning with his "Breakfast of Champions"—brandy, Kahlua, and cream—but they were too intimidated to say anything. "God forbid," said onetime chef Randy Pietro. "This is Mickey Mantle."

Some nights when he was in Dallas, he would call his sons to come get him but couldn't say where he was. True said sometimes the Dallas police drove him home to get him off the road. He lost a rental car he didn't remember parking. He bought a '49 Plymouth convertible at his own charity auction that he didn't remember buying. Sometimes he locked himself up in the house at night, not to prevent intruders from getting in but to keep himself from leaving. Because when he went out, Hays said, he might pour two shots of vodka into a glass of wine. Or decide he liked the looks of the concoction the folks at the next table were drinking at the Harbor Club—lime shooters, he'd never had one of those—and tell the bartender to bring twenty of them.

There were moments when the 100-proof scrim lifted. Like the night he called Tony Kubek from a Dallas restaurant where he was having dinner with Merlyn and the boys—a $700 to $800 bar bill wasn't uncommon. "He looked at the check, which he never did, and saw what it was for—round after round of drinks, bottle after bottle after bottle of wine," Kubek said. "He said that's when he knew there was a problem, and it was all of them."

But when his filmmaker pal, Tom Molito, screened footage for him that he had shot at a banquet of Mantle at his drunken worst in an effort to make him confront his alcoholism, it made no impression.

He told True and Liederman that he was having trouble performing sexually. Sometimes he spoke of suicide. "I'm not sure he cared if he died," Merlyn told me. "Mick just felt guilty. He wanted to lay down on the railroad tracks."

"The misery would be over," David said.

One day he asked True to meet him for a drink. "He told me he was thinking of killing himself," True said. "Yeah, it scared me. I stayed with him and talked to him and talked to him."

The moment of crisis passed. "We got off on other subjects because he was drunk," True said.

The breaking point came in December 1993 at a fund-raiser for the Harbor Club's Christmas Fund that Mantle and Johnson had organized to raise money for gifts for needy children in Greene County, Georgia. Mantle was holding court at a table surrounded by good ol' boys and enablers, some of whom worshiped at the church of the preacher Wayne Monroe, who administered the fund. Monroe stopped by the table to thank Mantle for his efforts. As he walked away, "there was a lot of noise—all of a sudden a lot of laughter," Monroe

588 · JANE LEAVY

said. "I never heard anything that was said. But Greer come running up to me afterward and said, as I was leaving, 'Oh, please, forgive Mickey for what he said. He didn't mean it. He's drunk.'"

What he said was: "Here comes the fuckin' preacher."

What upset Monroe most was the complicity of the other revelers. "Because they were sittin' with Mickey Mantle, they just felt great and wonderful," he said. "And they were all laughin' at what he said. 'Yeah, Mickey! Tell that guy!'"

The next morning, Mantle had no memory of the incident. When Johnson told him what happened he was distraught because he thought Monroe had heard him. "He said, 'I need professional help,'" Johnson recalled. "And I said, 'Yes, you do. I can't help you, and you can't help yourself.'"

It was a sobering realization and perhaps, Dominic Sandifer said, the first adult decision of Mantle's life. In making it, he was following his youngest son's example.

One night that fall, while on a trip to California for Upper Deck, Danny had disappeared from the hotel where he and his father were staying. He went out for a drink and blacked out for three days. When he came to, he put himself on a plane and checked into the

Betty Ford Center in Rancho Mirage. "Dad never saw me again for a month," Danny recounted in *A Hero All His Life*. "I ducked out on him."

It was the only way to save his life and his marriage. "It ruined our lives," he said of his addiction. "You wake up, you're thirty-five. It's such a waste of time and life. When I went to Betty Ford, I was thirty-three. I already had a little bit of liver damage. I had been drinking pretty hard for ten years."

His wife, Kay, would join him at the Betty Ford Center soon after. Furious and threatened, Mantle refused to attend Family Week, a five-day educational and therapeutic program, during week three of the monthlong stay. Merlyn went alone, fortifying herself on the plane with vodka to steady her nerves in the air. She didn't drink during Family Week but ordered a stiff one at the airport as she got ready to board the flight home. A high school friend who had come to see her off, a recovering alcoholic, asked whether she thought it was wise to drink in front of Danny and Kay. In the air, she began to chastise herself. "Here you are so happy that your child got sober and here you are still drinking," she wrote in *A Hero All His Life*.

That was her sober date: November 2, 1993. She joined Al-Anon, the support group for relatives and friends of alcoholics, and got herself a sponsor. "Al-

Anon saved my life," she told me. "I was so angry when I got there. It helped get rid of the anger."

Danny's example was deeply affecting. Roy True and Pat Summerall, who received a new liver after his own stay at Betty Ford, had urged Mantle to check in. At lunch one day with Danny and Sandifer, Summerall told Mantle, "Mickey, you need to get help." That led to a series of conversations. After a round of golf one morning at Preston Trail, Mantle asked Bill Hooten to join him and Summerall at their favorite watering hole. "He asked Pat some very serious questions," Hooten recalled. " 'What's it like after? Was it tough getting through? Do you miss it?' "

He also asked, "Did you have any fun?"

"I said, 'Mick, I don't think that's what you go to do,' " Summerall recalled. "He said, 'Do they get into any religious stuff?' "

"I said, 'Yeah, that's part of recovery.' "

"Finally one day he asked me if I could get him in. I called a friend and said I needed a bed. He said there were no beds."

Summerall told him who the patient was. A bed became available on Friday, January 7, 1994.

Hooten was surprised to learn that Mantle had registered under his own name and asked, "You think that's wise?"

Mantle told him about seeing Ryne Duren on TV talking about his recovery. "That guy, when he was playing ball, was a wreck and he whipped it. He goes around talking, and he does a lot of good. If I can go out there and come back and the fact that I've whipped the drinking can help somebody else, then, sure, I want that known."

When he met Mark Greenberg at the airport in Palm Springs, Mantle reiterated the pledge.

2.

At the Betty Ford Center, he was just the guy in Room 202, and he didn't much like it. He called the other patients inmates. He referred to treatment as doing time. He was contemptuous of gays in the program, and when he was assigned a roommate after having had a room to himself, Greenberg said, "He took all the coverings off the bed and put it in the bathroom and slept the night in the bathroom."

The life of a patient at the Betty Ford Center is as structured as that of a major league baseball player. Counselors tell you where to be and when to be there: 6:30 A.M. wake-up, breakfast, morning walk, therapeutic chores (making beds, doing laundry, setting and clearing tables), group therapy, individual therapy, spiritual counseling, and Alcoholics Anonymous meetings. There are rules against fraternizing with other patients and rules about contact with the outside world.

For the first five days patients are not allowed to make telephone calls. Mantle broke the rule almost immediately, Greenberg said, walking to a building on the campus of the Eisenhower Medical Center, "requesting to get the hell out of here."

He called Summerall three or four times. "I said, 'You're not thinking about leaving, are you?'" Summerall recalled. "He said, 'If you ever see me taking another drink, I want you to promise you'll kill me.'"

The Betty Ford Center was founded by the former first lady after she was successfully treated for alcoholism at the U.S. Naval Hospital in Long Beach, California. By 2009, the center had treated more than 90,000 patients—among them Liza Minnelli, Elizabeth Taylor, and Chevy Chase. Greenberg laughed when Mantle asked for a pass to do an autograph show for Upper Deck. "He was saying that he was going to lose a lot of money," Greenberg said. "I said, 'There is absolutely no way that Mickey is getting a pass to leave treatment to go to work and sign autographs.'"

In Greenberg's view, Mantle was a classic example of what Robert Millman, a New York psychiatrist and former medical adviser to Major League Baseball, calls "acquired situational narcissism"—a syndrome peculiar to the inhabitants of a celebrity caste, who are attended to by flacks, bodyguards, agents, lawyers, personal assistants, and groupies in whose rapt eyes they see the reflection of their own importance. At Betty Ford, Mantle was looking through the mirror crack'd.

Seeing himself in the despairing gaze of the other broken human beings there was devastating. Far easier to hide in the bathroom or behind polished empty sound bites than to introduce himself at an AA meeting with the words everyone in recovery must utter: "Hi, I'm Mick, and I'm an alcoholic."

"Take the pain"—that was the fatherly advice he always gave his sons. That's the way he had led his life. But tracing the path that had led him to Betty Ford was excruciating. The therapists wanted him to talk about his father. He had steadfastly refused to go beyond pat bromides in conversation with friends and teammates, with Herb Gluck, the ghostwriter of *The Mick*, and with Angelo Pizzo, the screenwriter his family had approved to write the film version of Mantle's autobiography. The most Gluck could elicit was an apocryphal tale about his teammate Bobby Brown, whose father also invested his thwarted ambitions in his son. " . . . when you get totally wrapped up in dreams about somebody else's future," Mantle wrote, " . . . they become your own. That's the kind of background Bobby had, same as mine, insofar as father-son relationships go."

But finally, the guy in Room 202 had to come face-to-face with his father. If Mutt was the central character in the Mantle narrative, "the appropriateness of his grief" for his father emerged as a central theme in his treatment, Greenberg said. He had never said goodbye to Mutt. He had not called or written during the two months Mutt lay dying at the Spears Chiropractic Hospital in Denver. "Up until coming in to the Center, Mutt was still very much alive in this guy's life and still controlling him in his actions," Greenberg said.

At the Betty Ford Center patients are required to write "grief letters," addressing past relationships and patterns that control or contribute to their disease. Some patients write to their addiction, others write to their drug of choice. Mantle addressed one letter: "To the drunk who shares my body" but didn't complete it. He finished his letter to Mutt.

Grief group is sacred ground at Betty Ford, a place where all is bared and everything is confidential. Spiritual counselors pushed Mantle hard to account for the feelings of self-loathing he jotted down in the margins of the journal he had to keep. *Embarrassed, angry with myself, angry, humiliated, foolish, ashamed, stupid-ity, inadequate, exasperated.* The answers are almost always in the distant past, counselors say, rather than in recent experience. They demand rewrites if they think the letters aren't honest enough. Mantle didn't need a do-over.

He read his letter to Mutt aloud to the group and then burned it, David Mantle said, as patients are urged to do in an attempt to leave the past behind.

In the confessional account he gave to *Sports Illustrated* upon his release from treatment, Mantle described the letter as an apology for not living up to his father's expectations. He said he was sorry for the things he did and for the things he hadn't done, the things Mutt might have done had he been afforded the chance he had given his son. *Mutt wouldn't have turned his sons into drinking buddies. Mutt would have done the exercises prescribed for his knee. Mutt would have been Lou Gehrig if he'd been a Yankee. Maybe even the next Babe Ruth.*

He told Mutt how much he missed him, how he wished he had lived to see him play better than he had his rookie year. He told him that he, too, had four sons. And he told him, finally, that he loved him.

Some years later, with the permission of Mantle's family, Pizzo spent a week at Betty Ford researching

Mantle's experience for a new script for *The Mick*. One of Mantle's counselors, Louis Schectel, described Mantle's progress in rehab. "Louie said every time he started talking about his dad he started sobbing," Pizzo said. "He didn't think Mickey had it in him. In the first couple of weeks he was too shut down. He said, 'His defenses were so high, I thought maybe we wouldn't get him. That letter proved me wrong.' "

It took ten minutes and a lifetime to write. When he was done, he called Mike Klepfer, his friend in upstate New York, with whom he hadn't spoken in years. "He was crying like a baby," Klepfer said. "He said, 'They made me write Mutt.' He didn't even say 'my dad.' And he said, 'They made me tell him what I wanted to tell him.' "

They had spent many hours and many long nights talking about their fathers—"two strong men who dominated us in our early life," Klepfer says. And what he concluded from Mantle's call was this: "He wanted to tell Mutt to stop running his life."

The grief letter afforded him that opportunity. "He went back to the famous story about the hotel room in Kansas City where Mutt grabbed his stuff and said, 'Okay, come on, you coward, we'll go home and you'll go in the mine.' And he said, 'I shoulda grabbed my dad right then and told him, "Hey, I'll make my own decisions. Get outta here." But I was too young then.'

"I got off the phone, and I remember thinking, 'Wow, this letter wasn't saying he was sorry but was telling Mutt that he shoulda gotten outta his life and let him fall on his butt.' In other words, if he had taken control earlier and not lived his whole life based on what Mutt wanted him to do, he would have been better off. Somehow he found that in Betty Ford."

Mantle described the contents of the letter to Linda Howard, with whom he had remained friends, and to Greer Johnson, in much the same way.

In saying goodbye to Mutt's ghost, Mantle had to face the here and now—and what was left of his family. "He was terribly guilt-ridden about his behavior towards his family," Greenberg said. "And appropriately so."

Perhaps that's why Mantle invited Johnson to Family Week instead of his wife or sons.

She thought it was because he didn't feel close to them. "It was safer than having to face my family," Mantle explained in the chapter he wrote for *A Hero All His Life*.

The collateral damage was enormous. Danny, who had some liver disease by the time he went to rehab, was born with a congenital abnormality in his pancreas—two openings instead of one. When it becomes inflamed, the pain that visits him in the wee hours of the morning is so intense, he told me, "I'm just rolling up on the floor, trying to eat crackers."

The condition was diagnosed at age thirty-eight, when he had his gallbladder removed. He's had surgery to remove blockages from his bile duct and precancerous polyps removed from his colon. Doctors follow him closely. "I guess we just weren't put together too good," he once said.

For years after the crack-up in their daddy's Cadillac, David required pain injections up and down his spine; he has migraines resulting from a motorcycle accident and getting hit in the head once too often, as well as advanced hypertension that caused a stroke-like episode in 2002. Spread across the width of his back is a tattoo comprised of Japanese characters, which, he told me, he calls his spiritual shield.

No one was more damaged than Billy Mantle, who lived with non-Hodgkin's lymphoma for seventeen years. His life spiraled into a nightmarish succession of debilitating chemotherapy treatments and drug treatment programs. Three times his cancer went into remission; four times he went into residential rehabs. In 1993, the cancer staged its third assault, and treatment was excruciating. He became addicted to the painkiller

Dilaudid. He had one prescription in Houston and another in Dallas. "He didn't have to do without," David said.

Billy used the catheter port doctors had put in his chest for chemotherapy to facilitate his drug use. One day, his father walked into his trophy room in Dallas and found Billy grinding pills into a powder, which he mixed with water to make an injectable solution. "He was putting it in the syringe and shooting up," David said. "He got completely cured from cancer and it left him an addict."

As his drug use increased and diversified—speed, crack, and alcohol—his parents grew desperate and scared. Mantle summoned Billy Martin, then managing the Texas Rangers, to have a talk with his namesake. Martin arrived so drunk he was too tongue-tied to speak. When their house was burglarized, Merlyn suspected his drug friends of the crime. She sold the house she had picked out in 1958 and moved into a condo.

One overdose left Billy in the hospital, on a ventilator. Mantle forced his brothers to go see him. He hated drugs. On at least one occasion that David recalled he tested them for drugs. He always reminded them that drinking wasn't against the law.

Billy's attempts to get sober received little familial

support. "Every time he went and would come back, we'd go to dinner and we're all shit-faced and he knows we're getting ready to do cocaine," David said. "He would last about three months and then go off. But who can blame him? We all said, 'Billy you can't do this' while we're doing it."

"I had a lot of guilt," Danny said. "You're not supposed to put a person back in the same element where it got started."

In 1994, David became the third Mantle to enroll at the Betty Ford Center. "Mom was already in Al-Anon," he said. "Danny and Kay had gone. Dad had gone. I had familial reinforcement."

Billy was still living at home with his mother when she had a heart attack in 1993. "What took her so long?" was Yogi Berra's response to the news. She was still in the hospital recovering from double bypass surgery when Billy went missing. His heart was damaged as well. He had been admitted to the same hospital for a double valve replacement and a double bypass. He had a stroke on the operating table that left him partially paralyzed; he would drag his left leg for the rest of his life.

He overdosed soon after he went home. They found him in bed bleeding from his nose and mouth. After the ambulance took him to the hospital, a nurse found his stash hidden beneath the mattress.

In an interview with Roy Firestone on ESPN after leaving the Betty Ford Center, Mantle tried to express the guilt he felt about the boys' addictions. "When I think of what I did to my sons . . ." His voice trailed off. He couldn't complete the thought—because it was unthinkable to him.

A decade later, David was very clear about the lines of responsibility. "That was one of Dad's regrets later on—that we became his drinking buddies," he told me. "That's not his fault. It's not Mom's fault. It was my fault. I made the choice to drink. I chose to do drugs. I don't blame anything on my childhood. It's me."

3.

As he was leaving the Betty Ford Center, counselors warned him: "Your first test will be in the airplane sitting in first class and they offer you a drink," Johnson recalled.

Mantle just laughed. They flew back to Atlanta. A limousine sent by the governor of Georgia was waiting at the airport.

Three weeks later, Billy Mantle died in jail. He was in custody because "he had gotten too many DWIs," Merlyn told me. "He had been arrested for driving while intoxicated."

He was incarcerated at the Dallas County Judicial Treatment Center in Wilmer, Texas, a state facility described on a government website as "a residential, correctional treatment center for the diversion of drug-involved felony offenders from long-term incarceration."

On Friday, March 11, 1994, he complained of chest pains. In the morning, guards took him to Parkland General Hospital, where he was examined, and released back into the prison population. "He was walking to lunch when he clutched his chest and fell to the ground," Merlyn told me.

Billy was dead of a heart attack. He was thirty-six years old—his father's age when he played his last game for the Yankees.

Troy Phillips was in the clubhouse at Preston Trail when he got a call asking him to keep Mantle there until one of the boys could bring the awful news. He had just gotten out of the shower when Danny arrived. He had a glass of water in his hand. Danny only had to say, "Dad, Billy."

"He knew," Danny told me. "He just dropped the glass of water and started walking in circles."

He went to Merlyn's condominium to tell her that their son was dead. That was the last night they spent together.

They had been expecting the worst for so long—

every time Billy left the house, in fact—that death came almost as a reprieve. "All those nights you sit there wondering, 'Is he going to get killed tonight or kill somebody else,' " Danny said. "I hate to say it, but it was a relief. We know where he is."

The next morning, when Mantle was late arriving at the funeral home, they feared the worst again. In a nationally televised interview three weeks later with Bob Costas on *NBC Now*, Mantle tearfully acknowledged the temptation he felt then. Chest heaving, tears seeping through a veil of heavy pancake makeup, he said: "I'll tell you, it was pressure, a lotta pressure."

His old teammate Tom Sturdivant had remained close to Merlyn and the boys, offering her solace and advice during her estrangement from her husband. His son Tommy, David's best friend, died of a heart attack in David's arms. Mantle was not happy to see Sturdivant at Billy's funeral. "What the hell are you doing here?" he said.

Sturdivant replied, "Mick, I saw Billy more in the last five or six years than you did."

Somehow the tension evaporated and they began to reminisce. Sturdivant reminded him of the time Billy had left a condom filled with buttermilk hanging from the door to Sturdivant's room at Mantle's golf tournament at Shangri La. Finally, they laughed.

Danny and Kay were married three weeks after

Billy's death. That presented another challenge for the father of the groom. Could he take happiness straight up? He could.

Danny and David attributed his sobriety during the last eighteen months of his life to stubbornness. "Made his mind up, it's done," David said.

Johnson thought it had to do with the weight of expectation—he was used to accomodating those. She says he drank because people expected Mickey Mantle to drink. "After Betty Ford, he said it was that people expected him *not* to drink. He felt like people were watching him all the time to see if he was gonna take a drink. So it was the public expectations that made him drink or not drink."

That made Mark Greenberg wonder how he would have fared long term: "I think that if it wasn't for all the pressure, I wouldn't have given him much hope in staying sober. He wasn't going to the meetings. He wasn't following up with any counselors. He was basically doing this on his own—'white-knuckling it.'"

David Mantle went to AA meetings after leaving Betty Ford. "You like going?" his father would ask. "Dad, I need to go," he'd reply.

Anonymity is the bedrock of AA. "Mickey M" wasn't going to fool anyone. His recovery became America's recovery. "What meeting could Mickey Mantle go to?" asked Dr. Kenneth Thompson, medical director of the Caron Foundation and Alcohol Treatment Centers.

"Any meeting," came the tough-love reply from former counselors at Betty Ford.

Mantle was as stunned by the public outpouring of affection and support as he was by the absence of his thirst. He had his moments; he told Jimmy Orr about working out at Preston Trail one day when the course was closed and no one was around. "He says, 'I think I'll go just have one glass of wine' over at some bar he always went to," Orr said. "Says he got in the parking lot. Said he must have sat there fifteen minutes, debating whether to go in. And finally he says, 'I know if I go in I'm going to have more than one.' So he cranked up and left."

Reliable statistics on rates of recovery are hard to come by. People who relapse aren't in a hurry to report it. Thompson cited composite figures gathered from many different programs to predict the likelihood of a relapse: 25 to 50 percent will relapse within the first three months; 50 to 75 percent will relapse within the first year. "We know that the longer you stay sober, the more likely you are to stay sober," Thompson said. "Somebody at eighteen months would not necessarily be considered long-term recovery, but it is something special."

With or without continued twelve-step support,

Mantle had learned enough at Betty Ford to offer words of counsel to his friend Rhubarb Jones, the host of his favorite country-western radio show on Y106.7 in Atlanta, at the celebration of Zell Miller's second inauguration. There was wine on the table. Jones, who had not yet begun to grapple with his own drinking problem, asked Mantle, "Mind if I have a glass of wine?"

"He said, 'Go ahead, I'm done with that.'

"I said, 'Maybe I should be doing that.'"

Mantle recited the mantra of recovery: "If you're sick and tired of being sick and tired, don't drink it."

It took Jones another three years to reach that point, but he credits Mantle with planting the seed that saved his life.

The sober Mantle discovered the joys of domesticity—he took delight in cleaning up the kitchen. He bragged to Danny and Kay about the jobs he had been required to do at Betty Ford. "Like it was a thrill," Danny said.

The sober Mantle was helpful, autographing a ball for a woman whose boyfriend had commitment issues: "Marry this woman," he wrote.

The sober Mantle was forgiving. He made peace with Jim Bouton, who had sent a condolence note after Billy's death. "I hope you're feeling okay about *Ball Four*," he wrote.

Mantle had never read the book—excerpts and sec-
ondhand reports were sufficient reasons to shun its
heretic author. One day, in the spring of 1994, Bouton
came home to find Mantle's familiar twang on his an-
swering machine: "Hey, *Bud.*"

Mantle thanked him for the letter, said yes, he was
okay about *Ball Four*, and assured him that he was not
responsible for Bouton's banishment from Old Timers'
Day. "I always felt somehow, I dunno, guilty having
done a bad thing and that Mickey would never, ever
forgive me for that and that I would never be able to
make it right with him," Bouton said. "To have him
effectively forgive me, tell me, 'It's okay'—it's just a
wonderful thing for him to do. On some level he didn't
want me to carry that."

The sober Mantle was contrite. He welcomed newcomers to his last fantasy camp by saying, "For you
first-time guys, it's the first time for me too." And he
apologized to the returning veterans who had fallen
into a black hole in his memory. Encountering Bob
Sheppard at the ABC studios in New York when he
arrived for an appearance on *Good Morning America*
on June 8, 1994, Mantle blanched. "He said, 'Bob, did
they think I wouldn't make it?'" Sheppard recalled.
"I said, 'Oh, Mickey, no, no, they were sure that you'd
show up.'"

Sheppard explained that he had been asked to introduce him, and he did so with the familiar intonation: "In center field, number seven, Mickey Mantle, number seven."

Mantle told the show's host, "Every time I heard Bob Sheppard say that, shivers went up and down my back."

Sheppard replied, "Mickey, every time I said it, shivers went up and down my back."

Sam McDowell hadn't seen Mantle since the abortive intervention he had organized years earlier at Tony Kubek's behest. He wasn't sure what to expect when he ran into him that winter at Sardi's, the theater district restaurant in New York. Given Mantle's blackouts, McDowell wasn't sure whether Mantle would remember him at all. "He was sitting there with a whole bunch of different friends, which I thought was interesting to begin with," McDowell said. "I went over and shook his hand and said, 'Hi, Mick. How are you?' And I told him who I was.

"He looked at me and there was total silence. Then he stood up and he came over. There was tears in his eyes. And he hugged me. It was like two lost brothers who'd never seen each other before. Apparently somebody had told him about the intended intervention. And all he said was 'I thank you, brother.'"

They talked that night, McDowell said, the way only two alcoholics can. "He was very, very proud of his sobriety. Bottom line is that he was one of the rarities. He was able to have internal peace. He was able to have honest happiness before he died."

It was baseball's worst off-season. The players' strike that forced the cancellation of the 1994 World Series depressed the memorabilia industry—and everyone else. Upper Deck was hurting and demanded that Mantle give up more than $2 million from his three-year contract. Roy True filed suit for $5.5 million, claiming breach of contract. Upper Deck countersued, claiming that Mantle's treatment for alcoholism had diminished his marketability.

In January 1995, tickets to the annual dinner held by the New York chapter of the Baseball Writers Association of America were a tough sell. They needed—baseball needed—Willie, Mickey, and The Duke again. An award was created in their name to lure them to the intimate black-tie dinner for a thousand disaffected baseball fans.

It was a long trip around the bases from Lou Walters' Latin Quarter, where they had first shared a stage as members of *Look* magazine's 1954 All America Team, to the dais at the New York Sheraton. Mantle had grown remorseful; Mays had grown cranky;

Snider was growing old in California with the same girl he had stolen from Pete Rozelle in high school, which is a whole other metric for deciding "Who's better?"

Terry Cashman sang their song, and Mays got up to speak. "He talked about what an honor it was to be remembered with Duke and Mickey," Cashman recalled. "He said, 'When I die, I want a plane to fly over New York and have my ashes sprinkled there.' The whole place was in tears, including Mickey."

Mantle followed Mays to the podium.

"If I knew we were going to be dying, I'd have prepared different remarks," he said. He acknowledged his long absence from the annual event and the untoward remarks that had scandalized a previous gathering—he had called George Weiss a "cocksucker." "I don't remember anything from the last time I was here. All I know is what I heard from other people, and I know I never got invited back until now. I guess after I got out of Betty Ford they figured it was okay to invite me back."

Then, he turned to Mays. "I'm often asked who was the best of us . . ."

He had answered the question many times—when they posed together for the cover of *Esquire* magazine in 1968 and when the Yankees retired his number a year later—but never so publicly or soberly. He and

Snider had become good friends—"there was a close-ness there," Snider said. Mickey Mantle's had hosted The Duke's book party and Mantle autographed a photo of the two of them that night: *Duke, you were the best—until I came along.* Now the two old friends, New York's two other center fielders, exchanged a look and Mantle said, "We don't mind being second, do we, Duke?"

PART FIVE

Riding with The Mick

ATLANTIC CITY, APRIL 1983

In the backseat of a limousine, I made Mickey Mantle cry.

We were headed to the airport in Philadelphia for his next flight in a life of endless flights. "Still looks like winter up here," he said, peering at the world through another smoked glass window. "I always thought Atlantic City was right by New York. Someone told me it's right on the Mason-Dixon Line."

Disorientation was the inevitable consequence of a life long divided into road trips and home stands. And that hadn't changed, except that now the destinations weren't always big league. He figured he was making *ten to twelve appearances a year, gigs that paid a mini-mum of $5,000; he wouldn't take less. Still, he worried that he was pricing himself out of the market. They only pay so much to lead an apple harvest parade. The Claridge was planning a couple of card shows with Whitey, Billy, and Rog, his three favorites, maybe Willie and The Duke, too, and there was talk of an au-tobiography. "To get even with Bouton," he said.*

He got to thinking about how much things had changed, which led to a discussion of women in the locker room, an indignity he was spared by retiring before people like me barged in. He grinned. "I told you—they called me Pee Wee."

Turned out I had remembered all too well his part-ing words by the elevator when I said I wasn't going upstairs with him. He was good at cutting himself down to size.

He shrugged. "You can't be worse than what's-her-name."

What's-her-name—Diane Shah—had dared to quote him accurately in a 1980 profile in New York magazine. He had been a hero to her as well, and she, too, was nervous when they met, blurting the first question that came into her head. "Do you still hunt and fish?"

"Yes," he told her.

"What?" she asked.

"Puss," he replied.

The rules of engagement with the fourth estate had changed, but he hadn't. And, he grumbled, she didn't even quote him right. "I thought I said 'cunt.' Puss don't even rhyme with hunt."

I tried to explain the new journalism. "She was trying to give a flavor of the way you are."

"But she didn't say I was kidding," he protested. "I didn't care except that it hurt Merlyn's feelings, and whenever it hurts Merlyn's feelings it hurts me."

He was pensive for a moment, then asked, "Was it in there about the Mickey Mantle look-alike contest?"

It was—there were two contests at the Longview Mall in Longview, Texas, one for kids, one for grown-ups. Only one adult entered. He looked like a middle-aged Howdy Doody, Shah reported. A dozen or so blond, freckle-faced boys vied for a $100 mall coupon. "I picked the kid," Mickey said, a boy named Stanley. For this he earned $2,500.

Nobody looked more like Mickey Mantle than his oldest son and namesake. I had spoken to Mickey, Jr., as well as Yogi Berra's son, Dale; Carl Yastrzemski's son, Mike; Don Shula's son, Dave; Hank Aaron's son, Larry; Joe Frazier's son, Marvin; Vince Lombardi's son, Vince; and George Allen's son, George, for a story about fathers and sons in sports. "With this name you aren't sneaking by anyone," Mickey, Jr., had told me.

"Did you really talk to Little Mickey? Did he tell you about when I took him to spring training?"

"No."

"Well, then, what did he say?"

In 1977, Little Mick was in his mid-twenties and, like his three younger brothers, trying to find a place for himself in a world that had room for only one Mantle. He spent a few weeks at a rookie camp where Mike Ferraro, one of several Yankees once projected as the "next Mickey Mantle," was working as a coach. Ferraro had to tell Mantle that Mickey, Jr., did not have what it took. "He showed skills," Ferraro told me. "Mostly, he showed he didn't play a lot."

They assigned him number 76—his father's two numbers—but neither brought him any luck. He soon washed out of the Yankee system and spent the rest of the year with the Texas Rangers Class A team. He was traded to the Alexandria Dukes of the Class A Carolina League the following season. His father told him to take whatever he could get. They offered $500 a month.

For the Dukes it was a loss leader. For Mickey, Jr., it was a shot in the dark. One night on the road, after striking out four straight times, he heard a voice in the crowd: "Go home and tell your daddy how you're playing." Little Mick went into the stands after the guy.

"Well, what did *he* tell you?" Mickey asked.

"He told me, 'I thought I'd try it, more or less for his'—your—'sake but I never really loved it, which you have to do. I wasn't really disappointed.'"

I summarized Mickey, Jr.'s scouting report on Mickey, Jr. "He said he obviously couldn't hit a curveball, and he knew it and everybody knew it and that was it. He said he played because he knew how much you loved it and how much you wanted one of your kids to play. How he did it for you."

Mickey considered Little Mick's act of filial devotion. "If he would have had my dad for a dad he would have made a major league ballplayer. I know he would have. Hell, he never played high school or Little League or anything. All of a sudden he's twenty-five years old, says, 'Dad, I'd like to play ball.'

"I said, 'Fuck, I'll take you to spring training with me.'

"I'll tell you how good a coach I was. I didn't even know how to work the batting machine. I put a ball in it and I had it turned up too high and the sumbitch was going a hundred twenty miles an hour and hit him right on the knee, the first pitch. Damn near broke his leg."

Mickey, Jr., batted .070 in seventeen games for the Dukes (4 for 57, 4 1B, 0 BB) and hung 'em up. By the time we spoke he was working in public relations for an oil company in Texas. "I play a lot of golf," he told me. "I go out to dinner a lot."

He had followed in his father's footsteps after all.

"I'll tell you who's pretty good, Dale, Yogi's son," Mickey said. "He didn't tell you no Yogi stories, did he? 'Cause they won't hardly tell them. Carmen, if she's had a couple of drinks, she can tell you some funny stories. Like one night they were watching Steve Mc-Queen in a late-night movie and Yogi said, 'He musta made that before he died.' "

Mickey always cracked her up. She loved to tell about the time he was hurt and hanging around the hotel pool with the wives. This was after George Steinbrenner bought the team. Mrs. Boss was late for lunch with the girls. She couldn't find a place to park and didn't want to tell anybody she was Mrs. Steinbrenner. "I don't blame you," Mickey said when she arrived. "I wouldn't, either."

He seemed to relax some, what with the Yogisms and the vodka, and I figured the time was as good as it was ever going to be. "Listen, I've got to ask you about something."

Of course, I didn't have to ask him about his dying son. I wanted to ask him. I wanted a scoop. I wanted what the Post newsroom called a "Holy shit story."

Mickey looked at me expectantly.

"Given what happened to your Dad, what's happening with your kid must be really tough."

For the next two minutes, the only sound was the rasp of rubber meeting the road on the Atlantic City Expressway. My tape recorder timed his anguish. "What?" he finally stammered.

"What's happening with Billy must be really tough."

It was almost Shakespearean in its diabolical symmetry. Billy had received the death sentence Mickey had spent so much of his life anticipating. After chemo and remission, and more chemo, and another remission, cancer was all over his body. Merlyn said the doctors gave Billy a 25 percent chance to live. "Yeah. I don't want to talk about that, okay? We're hopin' he gets okay and I think he will. I certainly hope he will."

A tear formed and fell. The heretic gene had given The Mick a pass but it hadn't spared him. "It's worse for me than with my dad" was all he could manage by way of reply.

I told him I was sorry, that I sympathized, that everyone would, especially when they understood he had taken the job at the Claridge to pay for Billy's medical bills. "Everybody is. I don't like to talk about it because I don't like to act like a pussy. It really . . ."

"I'm sorry."

"It's a bitch and a half. Billy is sick. We're fightin' it."

"I didn't mean to upset you."

"It's just really hard for me to talk about."

"*You obviously love him a lot. I wish I could say something to make you feel better.*"

"*Sometimes I tell my wife it's better to die in an automobile accident.*"

A tower of ice cubes collapsed in his tumbler. He put down the glass and flexed his hands in an effort to control unwelcome emotion, his fingers scaling the shaft of an imaginary bat, the one sure thing in his life. "Did you read the sports pages this morning? Where are the Yankees? Do you know? They got beat by somebody. KC. They're in KC."

The discreet chauffeur glanced in the rearview mirror and gently changed the subject. "What's Whitey up to?"

"*'Bout like me," Mickey said. "Waitin' to die.*"

Mickey poured himself another drink.

I asked where he was headed. "Louisiana," he said. "For an eightieth birthday party."

It wasn't the first time he'd been paid to do that. The party was for Lorraine Green of Ferriday, Louisiana, whose daddy taught her: "If you believed in God, you also believed in the Yankees and the Democrats."

She loved Mickey dearly, enough to make the effort to see his twin brothers play minor league ball back in 1954. "They asked her what she wanted for her eightieth birthday and she said, 'I want to have lunch with

Mickey Mantle.' So I guess her daughter or grand-daughter called, found out how to get a hold of me. So I'm going to go down and have lunch and give her a painting."

It was a framed 14" x 16" oil painting of his own self, the work of a former University of Oklahoma football player named Tommy McDonald, autographed and personalized by The Mick. Lorraine's grandson made the arrangements through Roy True. Before leaving for the party, Mickey called ahead to ask what to wear.

"You're a funny guy," I told him. "You do all these nice things and yet you don't want people to know."

"Well, I can't go to everybody's eightieth birthday. If it was known I would do that I would get a lot of requests."

"Or maybe you'd rather not be known like that."

"Naw, I'd rather people think I'm a nice guy than an asshole. I'll tell you what I like. I like that everybody calls me Mick. It makes me feel they like me better if they say Mick."

"Maybe you don't want people to see the other side of you."

Briefly Mickey considered the proposition, then pondered the fluttering heart of the birthday girl. "Maybe I can fuck her, who knows?"

The punch line drove me into the plush velvet

corner of the town car, as far away from him as I could get. The driver glanced in the rearview mirror. That's what he wanted: to make me recoil, to drive me away, anything to make me quit asking all these questions. So he went for the knockout. "She might be a good blow job, who knows?"

Probably he didn't figure I'd laugh. He eyed me with those surveillance eyes. I wasn't about to let him off the ropes. "You're weird. You say all this crazy shit, vile things. And on the other hand you go take a painting to an eighty-year-old lady who wants to have lunch with Mickey Mantle."

He shrugged.

"There's two sides to you."

"See that duplex there? I'm gonna buy it for Merlyn."

"Am I wrong? Am I wrong to think you hide that other side?"

"Well, I don't go braggin' about it."

"The question is, why do you do it?"

Mickey took a gulp of his vodka. Somewhere, sometime, he figured out that the best defense isn't a good offense—it's being as offensive as humanly possible. He deflected scrutiny like an unhittable pitch hacking away, until he got something he could handle. Most people never saw through it. The rest quit trying.

Mickey poured me a drink. "You're slowin' down."

I pointed out that I was drinking Perrier.

"I've got cirrhosis of the liver," he said. "How do you like that?"

Now I was at a loss for words. "I don't like it. Do you like it?"

"Fuck no, but it ain't gonna worry me none. I'll die of somethin' else before cirrhosis. One time I was talking to Billy"—Martin, of course—"I said, 'How's that spot on your liver?'

"He said, 'Now I got a little liver on my spot.'"

"Does it hurt? Are you in pain?"

Stupid question to ask The Mick. "Fuck no. When you lead the league in the clap six straight years . . ."

"You want me to put it in the paper?"

"I'm sorry they didn't put it on my card."

One more time, he trotted out the tired line about the statistic left off his baseball card. "And my wife was second in the league four times."

"If I write it . . ."

"I'd say, 'Another goddamn Diane Shah.' I talked to Merlyn a while ago. I told her you were doing this story. She said, 'This ain't gonna be another one of these Diane stories.'

"Merlyn said the same thing you said a while ago—'You're too truthful, Mick.'"

"You say these things but you don't say, 'Don't write this' and then you get upset."

"I don't get upset. It pisses me off. But I would never call you and say, 'You rotten little cunt.' If you wrote it you wouldn't care what I thought anyway, right?"

"I would care. I care too much whether people like what I write."

"Will you send me a copy?"

"Yeah, I will. And I'll be scared about whether you'll like it."

"I'll like it," he said. "I can tell. I'll like it."

The limo pulled up to the curb outside the terminal and the driver got out to open the door. Mickey put on his cowboy hat and ducked his head the way he always did rounding the bases. He said goodbye and smiled that smile, the one part of him that never got old.

I don't remember whether I kept my promise to send him the story. I do know I thought long and hard about whether to do so. I didn't have to think about whether to write what he'd told me—he said it was off the record. Eleven years later, when I heard that he had entered the Betty Ford Center for treatment of alcoholism, I wondered if I'd done the right thing.

20

August 13, 1995

THE LAST BOY

1.

On March 19, 1995, Mickey Mantle went home to
Commerce to bury his mother. Lovell's surviving children—Mickey, Ray, Roy, Larry, Ted, and Barbara—
gathered at the cemetery, where an empty space on
Mutt's headstone had been waiting for her for forty-
three years. Her children didn't see much of each other,
maybe two or three times a year, except, of course, the
twins, who worked back-to-back shifts in the casinos
in Vegas for years and lived in houses with adjoining
yards. In college football season, they consulted Larry,
the football coach, on the point spread.

Barbara hadn't seen her famous big brother since he had gotten out of rehab. She was glad to see him looking so well and made a point of saying so.

He had made it clear that he didn't want a lot of people at the funeral just as he hadn't wanted them at Billy's, which puzzled her. Maybe, she thought, he didn't want people to see him cry. Maybe he was just tired of burying people. "I hope I'm next," he told Danny's wife, Kay, "because I'm not going through this shit again."

He had said the same thing when Roger Maris and Billy Martin were buried.

For a long time, he had kept his distance from Lovell, which wasn't hard to understand, given his schedule and her cantankerous ways. Years passed without a visit, and those did not always go well. Once she took a swat at her grandson David, which did not sit well with his daddy at all. She had never remarried, and lived for many years with Barbara's family in Oklahoma City. In the car on the way back to Barbara's after a visit in Dallas, she asked her famous son to arrange a tryout with the Yankees for Barbara's son. He demurred, saying he had never seen the boy play and didn't know how good he was. "She said, 'Well, I'll tell you one thing, he's better 'n you were!'" said his friend Joe Warren, who went along for the ride. "We got his mother to his sister's house. We took her suitcase in.

His mom walked in the door, and Mickey said, 'Let's go.' We walked out the front door, and first thing he said, 'Did you hear what she said?' I think that really stung him."

Lovell moved in with Faye Davis, Ted's widow, for six months before Alzheimer's disease forced them to move her into assisted living. She was a big woman, and it took two people to get her out of bed. "I don't think he had a great lot of motherly love for her," Davis said, and she could understand why. "The mother-in-law from Hell," she called her. "After she got Alzheimer's she was as nice as she could be."

Mantle paid for her care and paid a lot more attention after he got sober, arranging for flowers to be delivered to her room every day. After she was buried beside Mutt, he and his siblings posed for a family photograph just as they had at the funeral for Barbara's first husband. The shutter clicked and Mickey said, "Well, guys, I guess I'll see you at the next funeral," Larry recalled.

Mantle knew time was short when he left the Betty Ford Center. He peppered Danny and Kay with questions before they went out for the evening and sometimes begged them not to go. *How do you call an ambulance? What do I do if I can't breathe?*

He told his minor league teammate Jack Hastens,

"I'm gonna die."

"I said, 'No, you're not.'

"He said, 'Yes, I am. I've led a terrible life. I've done too many bad things.'"

He set about righting the things he could in the time he had left. He left everything to Merlyn—including his name. According to his will—executed after they separated in 1989—Mantle provided that his entire estate would go into trust for Merlyn's benefit during her lifetime and thereafter to his sons and their heirs. At the time of his death, he and Merlyn jointly held assets of $6.9 million in cash and property, half of which by law automatically went to her; the other half constituted his estate. The will, on file in public documents in Texas, provided that his sons would be able to license his name and his memory in perpetuity.

He also gave them a glimpse of the good heart they so much needed to see and wanted the world to know. "He had a lonely heart and it was a good heart and it was open and it was big, and the mantle he put around him was not his name," his friend Margie Bolding said. "It was a facade, and he did not like it to be penetrated and it wore him out finally."

Now, the facade was crumbling. Now he let his boys in. On Father's Day 1994, the first David could remember sharing with him, Mantle apologized for not being the daddy he should have been. For the first time, he told his boys he loved them.

Danny and David accompanied him to his last autograph show in Atlantic City in early May 1995. The event at Donald Trump's Taj Mahal was billed as a convocation of Yankee legends—Mantle, DiMaggio, Berra, Rizzuto, and Mr. October. Mantle and DiMaggio shared a private signing room off the main hall. Mantle's boys had noticed how much more patient he was with the fans who assembled before him, waiting for a purchased signature and hoping for eye contact. He was too accommodating for Joe D., who sent a representative over to Mantle's table with a request. "He said, 'Joe D. wants you to quit shaking hands and being nice,'" David recalled. "Dad said, 'Tell Joe D. to get another room.'"

One thing hadn't changed: his good ol' boy sense of humor. At fantasy camp in November 1994, he sported a baseball cap with two pins fixed to the brow: a cross for Jesus and a decal that said BE HEALTHY EAT YOUR HONEY.

Steve Donohue, the Yankee trainer, ministered to aching boomer hamstrings. That week, as he wrapped Mantle's ravaged knee, the patient grumbled: "All that's holding this damned thing together is skin." Doctors had warned him that his liver was too damaged to withstand the stress of knee replacement surgery. He told Donohue he was scared to have another operation. Donohue thought he was scared he wouldn't wake up.

Mantle entered the Betty Ford Center six weeks later.

One night the winter after Mantle got out of rehab, Donohue brought his wife to Mantle's restaurant on Central Park South. After he got sober, he sipped diet soda or iced tea at the bar, when he was there at all, maybe an occasional O'Doul's non-alcoholic brew, which, he said, tasted "a little bit like beer." Donohue was glad for the opportunity to introduce his wife to The Mick, flaunting their relationship by making small talk. Mantle interrupted the patter. "He told me he loved me," Donohue said.

Donohue later counted himself among the enablers who were more interested in having a story to tell than in actually listening to him, who wanted to "protect him, not change him. You saw his pain and you wanted to help but you wanted a piece of him more."

So he plowed ahead with his unfinished thought. "Mickey, did you see Paul?"

"He said, 'No, Steve, I *love* you, *Steve.*'"

2.

Mantle was on the golf course at Preston Trail with Mark Zibilich when he complained of pain for the first time in his life. "It feels like something wants to crawl out of me," he told Zibilich, a radiologist on staff at Baylor Hospital.

Indigestion had long been a constant companion. He gobbled antacids like Cracker Jack but had little appetite for anything more. Barbara Wolf noticed that he no longer ate the comfort foods she made especially for him. His ankles were swollen. "Big enough that he tried to hide it with his pants," said Ryne Duren, who had noticed the telltale sign of liver failure at a golf tournament in 1994, when Mantle sat down and his pants leg hiked up.

At Darrell Royal's annual tournament that spring, he was scheduled to team up with Mickey, Jr., but left his room only once to fulfill a charitable commitment: he raised $13,000 for families who had lost loved ones in the Oklahoma City bombing by daring all comers to better his tee shot on a par-3 hole. "He was in pain most of his last year," Mickey, Jr., wrote later.

Liver failure can make just getting out of bed feel like a forty-hour workweek. His liver had been ravaged by alcohol and by the hepatitis C virus. Both cause cirrhosis or scarring; either can kill. In combination they are a lethal one-two punch. For a patient with hepatitis C, drinking alcohol is like pouring an accelerant onto a smoldering fire.

Mantle had been diagnosed with alcoholic cirrhosis in the early 1980s. When and how he had contracted the virus was unclear; the diagnostic test was not developed until the early 1990s.

His doctors assumed it was the result of a blood transfusion given during one of his many orthopedic surgeries. David Mantle wondered if it could be traced back to a dirty needle in Dr. Feelgood's office in 1961.

Mantle had long ignored the warnings of his internists, Dave Ringer and Art DeLarios. Ringer couldn't begin to guess how many drinks ago—how many doctors ago—Mantle had first been told, "The next drink you take may be your last."

"He'd come in the examining room and just talk and talk and talk," Ringer said. "It was like, 'You're gonna die. This is stupid. Why are you doing this?'"

Then: "all hell broke loose," Ringer said. "All of a sudden one day he started bleeding and bruising. His platelet count was going crazy."

Platelets are small, colorless cells essential to blood clotting. A normal count is between 150,000 and 350,000 per microliter. A platelet level below 20,000 is considered life-threatening. Mantle's count, Ringer said, was about 10,000 per microliter. His bone marrow, damaged by years of alcohol abuse, was unable to produce sufficient platelets to prevent bruising and bleeding. Reviewing the numbers in the hall outside the examining room, Ringer told him, "Okay, this is it, man."

He needed to be evaluated immediately. Ringer suggested either Emory University Hospital in Atlanta or, if he preferred, there was a smaller facility in nearby Athens, Georgia. "No, I want to go back with my family," Mantle said.

He entered Baylor University Medical Center in Dallas on May 28, 1995. Roy True was with him when the test results came back. The cancer he had been waiting for all his life had struck where he was most vulnerable: his liver. Blood tests showed an extremely elevated level of alpha-fetoprotein, a marker for liver cancer. That spot on his liver he and Billy Martin liked to joke about appeared as a cancerous mass on a CAT scan. The tumor had blocked the duct that drains bile from the liver, leaving pockets of bacterial and fungal infection. A transplant was his only hope. "He sat up, his legs were hanging over the bed, barefoot," True said. "He said, 'You know what, partner? I think I really fucked up.'"

His condition deteriorated rapidly. His liver function was almost nonexistent, and his kidney function was worsening. His stomach became so distended that visitors couldn't see his face when they stood at the foot of his bed. Jaundice had turned his skin yellow. "School bus yellow," Danny said.

Like most transplant centers, Baylor required six months of sobriety for an alcoholic patient to qualify for a new organ and required patients to sign a contract promising to return to treatment in the event of a relapse. Dr. David Mulligan, the senior fellow on the surgical team, now the chairman of the Division of Transplant Surgery at the Mayo Clinic in Arizona, said doctors spent four days tracking down and discounting rumors that Mantle had been seen drinking beer on a golf course before they put him on the transplant list.

He was listed as "priority status 2," the designation then used for patients who had been hospitalized in acute care for five days or longer. Only those deemed incapable of surviving a week without a transplant were listed as status 1. By June 7, he had gone to the top of the waiting list. He was the sickest patient of his blood type and weight in an area a third the size of Texas. Doctors told reporters that the wait for a new liver could be as long as a month—in part because Mantle didn't want sportswriters all over the country taking bets on when he would die. In fact, the average waiting time for a patient with his status and blood type in that region was 3.3 days.

Göran Klintmalm, the Swedish-born head of Baylor's Regional Transplant Institute, did not know who Mickey Mantle was when he was placed on the waiting list, an inadvertent deterrent against favoritism. "He was treated exactly the same because I just didn't know any better," Klintmalm said.

By the time a match was found two days later, Klintmalm had figured it out. "Oh, shit," he said.

"He knew there would be a tidal wave of outrage and everyone would claim Mantle got favorite treatment," True said. "Klintmalm knew it would actually harm the program because someone famous got it."

In fact, Klintmalm said, "When the call came about the donor, I was actually tempted to turn it down. He had just been listed so recently there would immediately be suspicion that we had rigged it. But then it was also obvious that I couldn't—for ethical reasons—turn this donor down, because he needed the transplant."

In short, it would have been as unethical to penalize Mantle for his celebrity as it would have been to reward him for it. Anticipating a furor, Klintmalm asked officials at the Southwest Organ Bank (now the Southwest Transplant Alliance) to rerun the listing and asked the director of the United Network for Organ Sharing (UNOS) to conduct an independent review of the listing and allocation. "When they reconfirmed that he was number one for that particular donor in this region, I then accepted it," Klintmalm said.

Mark Zibilich was on call the night before Mantle's surgery. Mantle's room was on the VIP floor, 16 Roberts. A sign on the door discouraged visitors: " 'Private, contagious, don't come in w/o a nurse's permission,' " Zibilich said. "It's around midnight. I figured family would be there. But I knocked on the door, and there he is in bed. We talked for about half an hour, about golf and Oklahoma—and then he said, 'Doc, tell them I will do anything 'cause I want to get rid of this thing and be healthy again.' "

Zibilich hadn't expected Mickey Mantle to be all alone.

The surgery began at 4:30 A.M. and took nearly seven hours. Robert Goldstein, the senior attending surgeon, was assisted by Mulligan and a general surgeon. Liver transplants are among the most difficult for patients and their surgeons, Mulligan says. In Mantle's case, the task was complicated by scar tissue left by previous surgeries, pockets of pus, and, most ominously, the unforeseen extent and type of his cancer. At first they saw no evidence that it had spread beyond his liver. Lymph nodes tested during surgery were clear. The backup patient, who had been prepped for surgery, was sent back to his room. (That patient later received a transplant.)

The surgeons proceeded, removing Mantle's liver, which was swollen, lumpy, and hard. As the new liver was being sewn in, Mulligan said, the pathologist arrived with dire news: his cancer had spread down the central bile duct into the pancreas. And it was an exceptionally rare and voracious form of the hepatocellular carcinoma commonly associated with cirrhosis—anaplastic carcinoma.

"Oh, shit," Mulligan said.

Number 7's luck had run out. Fifteen years later, Klintmalm said, "We have never had anything like that before or after."

There were only two choices: to continue with the operation, removing as much of the bile duct and pancreas as possible to eliminate metastatic tissue, or to let Mickey Mantle bleed to death on the operating table. "When the liver's removed, that's past the point of no return," Mulligan said. "The new liver is going in while the pathologists are studying the old liver. So actually, we don't know how bad things are until we're well into the new liver getting perfused and him getting a new chance at life. At that point, they come back, and it was a surprise. It's, like, 'Holy cow, do you guys realize this was there?'

"I was pretty worried about Mickey. I knew this was bad, that his likeliness of recurrence was high, and that he wasn't going to do really well long term."

When Klintmalm briefed the press the next morning, questions focused on the success of the transplant and the identity of the donor, not on the extent or rapaciousness of Mantle's cancer. "Well, is the donor alive?" a reporter asked. "Can you tell us that?"

The donor was a twenty-eight-year-old Caucasian male, a small-town bank teller, who had donated seven vital organs to six dying patients. "About a week before his death, the young man was at a lake near his hometown when a man trying to swim to an island started to go under," said Pam Silvestri, the public affairs director for the Southwest Transplant Alliance. "This young man, the donor, was a water skier and former lifeguard. He swam out and pulled the man a long distance to shore. About a week later, he woke up with a headache. Before leaving home for work, he collapsed. A blood vessel had ruptured in his brain. He died of the aneurysm that afternoon."

After a year on the transplant list and decades of heart failure, Dillard Worthy, a chicken farmer from Pittsburgh, Texas, received the young man's heart.

Asked thirteen years later how his new ticker felt, Worthy said he couldn't tell that much difference between it and the old one when it was working right. There is no better argument for organ donation.

The day of the transplant was Silvestri's first day on the job. She had worked in sports administration and understood the potential of a story that would appear in 18,000 newspapers across America—with countless spins on fact and taste. Case in point: New York's *Village Voice* published a bogus advertisement for "the ultimate in baseball memorabilia—MICKEY CHARLES MANTLE HEPATIC ORGAN."

There was so much competition for news about his condition that hospital switchboard operators were offered twenty-dollar bribes to lose messages from rival TV stations. Reporters scoured obituaries and lobbied local funeral homes for information about newly arrived bodies.

In early August, Silvestri arranged an anonymous telephone interview with the donor's mother to put an end to the ghoulish goose chase. "He'd never been sick before," she told the *Dallas Morning News*. "He'd never been in the hospital before."

She did not know that her son had designated himself as an organ donor until the day he died.

Despite Klintmalm's preemptive measures, despite testimony from UNOS officials and from Dillard Worthy and other transplant recipients, the public outcry was loud, immediate, and lasting.

The fix was in. It had to be. How else could he have gotten a liver so fast? And even if it wasn't fixed, he shouldn't have gotten it anyway because he destroyed his own liver.

"The truth is more mundane than what people think happened," UNOS spokesman Joel Newman said.

Several factors worked in Mantle's favor, Newman says—none of them his fame. He had the most common blood type, O, shared by 45 percent of the population, increasing his chances for a match. The population density in the Dallas metropolitan area made the Southwestern Transplant Alliance one of the highest-volume organ-procurement organizations in the country. And, Newman said, the prominence of the transplant program at Baylor, then one of six centers in the U.S. participating in an experimental protocol to treat liver cancer patients with transplantation and chemotherapy, "translated into higher liver donation rates than in other parts of the country."

Fifteen years later, Mantle's liver remains fodder for conspiracy theorists, medical ethicists, sports radio loudmouths, and sitcom wise guys, among them Larry David, who worked riffs on the transplant into a two-episode story arc on *Curb Your Enthusiasm* with the comic Richard Lewis entitled "Lewis Needs a Kidney." With each new "celebrity" transplant the blogosphere erupts with renewed howls of protest. The *New York Times* attributed the resurgence of "dark theories" occasioned by Steve Jobs's 2009 liver transplant to a "holdover from the case of Mickey Mantle."

Arthur Caplan, the director of the Center for Bioethics at the University of Pennsylvania, says the uproar over presumptive favoritism obscured other equally important questions about organ allocation. "I don't believe he jumped the queue, although I think arguably he could have been too sick," Caplan said. "The combination of alcoholic cirrhosis and hepatitis C might have led some transplant teams to say, 'Yeah, he's sick, but he's way too sick.' "

Klintmalm said every possible diagnostic test then available was done prior to surgery to detect whether Mantle's cancer had spread. "We kind of smelled that there could be a tumor there, and we did virtually everything in the world to find that tumor, but we couldn't find it," he said.

The bile duct, an organ no bigger than a pinky finger, is difficult to biopsy, especially when a patient lacks enough platelets to clot blood. "It's something you never get close to with a needle," Klintmalm said, "and if you do, the patient usually bleeds like a pig and dies."

With today's enhanced imaging technology, he said, "We probably would have had a much better chance to make a positive diagnosis. If we had found the tumor before the transplant, we would not have done it. If we could have figured that out [during surgery], we would have interrupted, closed him up, and brought back a backup patient."

In 2002, UNOS instituted a new system of organ allocation that ranks every patient in the country waiting for a liver on an objective numerical scale that determines the risk of dying within ninety days without a transplant. Patients with cancer in danger of progressing get additional points and move up the list. But if a tumor is found to be beyond a certain size or there is evidence of metastatic spread, that score drops. The goal is to give organs not just to the sickest patients but to "the patients who have cancer with the best possible prognosis," Mulligan said.

Added Caplan, "They're really more oriented toward outcome than they were then. It doesn't make sense just to say, 'Who's sickest?' If they're almost dead, then you're not going to rescue them."

In Caplan's view, Steve Jobs, a California resident who received a transplant in Tennessee, illustrates the advantage of celebrity: money. "It's not that they get a liver," Caplan said. "It's that they get admitted to the waiting list in the first place. You don't even get considered if you're some poor homeless guy whose liver's blown out because you're drinking wine in the park. You're going to be dead. That's where Mantle or David Crosby or Evel Knievel, the drug-damaged livers of the Fifties and Sixties, had an advantage."

3.

When Mantle awoke after surgery, doctors told him what they had already told his family: the operation was a success, but the prognosis was uncertain at best. It was a bittersweet result—the gift of a new life might prove to be no opportunity at all. In the hospital, he was inundated by an outpouring of public affection—20,000 cards and letters were delivered to him c/o Baylor—and a torrent of outrage. Robert Goldstein told reporters that Mantle wouldn't have gotten out of the hospital alive without the surgery. Mantle told the New York *Daily News* that doctors had told him he had just one day to live.

Though Merlyn complimented her husband on his newly trim shape—she hadn't seen his stomach that small since he was twenty—Mulligan saw little of the joy he usually sees in transplant patients, an indication, perhaps, that Mantle sensed how little time he had. But when Klintmalm asked him to go public with his support for organ donation, he readily agreed. "Mickey was absolutely committed," True said. "I told him, 'If you promote it, you'll save so many lives it will eclipse baseball.' He said, 'I know.'"

True began making plans for "Mickey's Team." Mantle coined the slogan for a campaign he wanted to announce at Arlington Stadium at the end of August: "Be a Hero, Be a Donor."

He left the hospital on June 28, exactly one month after he checked in. He went back to the house he shared with Danny and Kay and to the accustomed routine, organizing his days around the clubhouse at Preston Trail. At first he was strong enough to ride a stationary bike a bit. But two weeks later, the stomach pains returned and there were new ones in his chest. It was hard to eat; he subsisted on protein drinks and lost forty pounds. One Monday morning, he called Pat Summerall from the clubhouse and said, "Get your ass over here, I need to talk to you."

It was an hour's drive. When Summerall arrived, he was dumbfounded by Mantle's request. "He said, 'I want you to have a look at my ass.' He pulled his pants down. It was all bruised and black and blue. He said, 'I used to have a good ass. I wanted you to have a look because you remember what a good ass I used to have.'"

Chemotherapy left him anemic and in need of transfusions. He returned to the hospital for treatments, and surgeons implanted a catheter in his chest, like the one Billy Mantle had used and abused.

At the end of the first week of July, as Mulligan was packing up his apartment to move to Cleveland for his new position at Case Western Reserve University, a reporter for the *Dallas Morning News* came to interview him for a profile of his newly famous mentor, Göran Klintmalm. Mulligan had been up all night doing his last transplant at Baylor. When the reporter asked how Mantle was doing, Mulligan showed him the data on survival rates of patients with his cancer. An unexpected and unwelcome headline appeared in the morning paper: "Mantle's Outlook Uncertain." Mulligan was quoted as saying, "There was tumor left behind that was unremovable. There was no way we could remove all of it."

Goldstein tried to quell the furor, dismissing Mulligan's comments as those of a "junior" surgeon who hadn't seen enough cases to justify such pessimism. Klintmalm's protégé had performed 104 transplants during his two years at Baylor.

On July 11, Mantle held his last press conference. It

was a standing-room-only crowd. He greeted reporters he knew, including Jerry Holtzman from Chicago, and acknowledged Barry Halper, a limited partner in the Yankees, a memorabilia maven. "Hey, Barry, did you get my other liver?" Mantle asked.

His comic timing was still acute, but the robust physique, the Popeye biceps, and the untroubled face of American plenty were gone. His tracksuit hung on his desiccated frame. His face looked like a dry riverbed. The band on his blue-and-white 1995 All-Star Game cap couldn't be made tight enough to fit his skull. He looked like death. In fact, he looked a lot like Mutt in the family photograph taken just before he went off to Denver to die.

When he pointed his thumb at himself, it seemed as if his chest might collapse. "God gave me a great body and an ability to play baseball," he said. "God gave me everything, and I just . . . *pffttt!*"

What would be remembered most was the anguished plea directed at children: "I'd like to say to kids out there, if you're looking for a role model, this is a role model. Don't be like me."

Some scorned the tortured mea culpa as the moral equivalent of a death row confession. "The most decent thing he ever did may be the only decent thing he did that ever mattered," the sports columnist Jerry Izenberg said. But even the most hard-boiled scribes agreed it was his finest inning. David Mantle said the family received hundreds of letters, saying, "We used to hate you, but you doing that on national TV, apologizing and making amends, you have got new fans."

A reporter asked Mantle if he had signed a donor card. "Everything I've got is worn out," he said. "Although I've heard people say they'd like to have my heart . . . it's never been used."

Two days later a CAT scan showed cancer in his right lung.

One day, not long after, Halper pulled up in front of Mickey Mantle's Restaurant and asked Bill Liederman to come out to the curb. Halper said he had "some shit in the car" he wanted Liederman to see. "He actually had a bag of shit," Liederman said. "Mickey sent him a bag of shit." And a pair of signed shit-stained examination gloves. "He was going home to put it in the freezer."

Liederman thought it was Mantle's final comment on the memorabilia industry that had made him a rich man, a very Mantle way of saying, "You're all full of it."

Ten thousand donor cards were distributed at the All-Star Week FanFest at the Ballpark at Arlington. Interest in organ donation surged. "Before the transplant and the news conference most organ donation agencies were getting a call a week, if that," Silvestri said. "Post–Mickey Mantle, calls were in the thirty-to-forty-a-week range. People were saying 'Where can I get that Mickey Mantle donor card?' 'What do I have to do to be a donor?' Or 'I'm ripping up my donor card because of what happened to Mickey Mantle.' Just getting people to talk was our goal because it wasn't a conversation people were having."

That was enough to persuade her that The Mick had done his job. "At a time when he didn't feel good and his family didn't feel good, they opened their lives to anyone who wanted to talk about organ donation because they felt like they could do some good," she said. "And the bottom line is, they did some good."

Arthur Caplan has a different view: "Nothing dries up altruism, the willingness to donate organs, faster than the perception that the distribution of the organs is unfair. We wanted to believe something good came of this. There's this hagiography that The Mick helped boost donor rates. I don't think there's any evidence that that's true."

Donations to the Southwest Transplant Alliance actually tailed off in 1995, according to UNOS figures, from twenty-two in February to thirteen in August. In the fifteen years since, donations from living and deceased donors have increased dramatically throughout the United States—they are not quite double what they were in 1995—but the numbers are still achingly small.

In 2008, nearly 28,000 people in the United States received transplants of vascular organs (kidney, liver, heart, lung, pancreas, and intestine), 7,990 of which came from deceased donors; nearly 50,000 transplant candidates were added to the queue; 6,600 patients died waiting for organs. "The tragic thing is, Mickey died of cancer before he got the opportunity to do the saving grace," True said. "That would have been the happy ending."

4.

By the time Greer Johnson arrived in Dallas in mid-July, Mantle had begun to say his goodbyes. She didn't realize that was the purpose of her visit. Her disconnect from the gloomy reality of Mantle's illness was so profound that when she joined him at Pat Summerall's house, which he had offered for a rendezvous, she thought he would be returning with her to Georgia to continue chemotherapy. They would take a planned trip to Hawaii. They would build a house and a life.

Their business relationship changed after he had signed an exclusive marketing deal with Upper Deck in 1992, and Dominic Sandifer assumed many of her daily duties. (Roy True was the executor of the contract.) Ed Nelson, her pastor in Greensboro, thought he had noticed an increasing distance between Mantle and Johnson. He wasn't alone in that observation. Kathleen Hampton, Roy True's assistant and office manager, thought he didn't want to be married to anyone. Ron and Barbara Wolf had concluded that he was quite content with his two separate lives. When Nelson tried gently to allude to the possibility "that Mickey was on his way out. Ohhh, she don't wanna hear that."

Nelson spotted the couple sitting in the Amen Corner of his church on Mother's Day 1995, when he delivered a sermon about what a father owes the mother of his children. Glancing at Mantle, he thought, "He's going back to Mama."

He had noticed other changes in Mantle, who had begun attending services at a non-denominational church that Danny and David went to in Dallas. One day on the eighteenth hole of a round of "best ball" golf, Mantle had jokingly wagered his soul on the deciding putt. "Mickey said, 'If that's that preacher's ball up there on the green, he can baptize my ass out there in the pond,'" Nelson said.

Nelson made the putt and won the bet. "Mickey threw his golf club straight up in the air," Nelson said. "He said, 'You can't beat God!'"

The preacher declined the opportunity to save Mantle's soul among the water moccasins in the fetid hazard.

Increasingly, the hereafter was on Mantle's mind. In June, Mantle had called Bobby Richardson, the Yankee second baseman whose baseball afterlife was as a Christian pastor, to tell him about the transplant. He asked Richardson and his wife, Betsy, to pray with him. She took the opportunity to remind him "there was someone else who had died so that he might live."

In July, Mantle asked Summerall if he could arrange a baptism. When he called Mike Klepfer, his old friend barely recognized Mantle's voice. "He sounded like Billy Graham, not Slick," Klepfer said.

By the time Tom Molito reached him in late July, he just sounded resigned. Molito asked if he was going to attend Old Timers' Day at the Stadium. Mantle said, "The doctors won't let me travel."

"I said, 'Why don't you tape a message?'

"He said, 'Tell the Yankees it's your idea 'cause they think I'm a dumb fucker.'"

On July 22, the Yankees celebrated Babe Ruth's hundredth birthday and played Mantle's taped farewell on the JumboTron. "I feel like Phil Rizzuto in Babe Ruth's uniform," he said, a spectral figure in disembodied pinstripes looming over center field.

5.

Mantle checked back into the hospital on July 28. Nothing was said to the press. Doctors showed Danny and Mickey, Jr., the results of a new scan and offered a grim prognosis: "Ten days, two weeks tops."

Mantle did not want to live the last weeks of his life as a public figure. He told doctors not to release any information about his condition. He watched the O. J. Simpson trial (he wasn't crazy about Marcia Clark and thought Simpson would get off) but didn't want to hear updates about his condition on TV. Doctors finally persuaded him to videotape a statement in order to forestall inevitable leaks. It aired on ABC's *Good Morning America* on August 1. "Hi, this is Mick," he said, identifying himself in the way he liked to be greeted. "About two weeks ago, the doctors found a couple of spots of cancer in my lungs. Now I'm taking chemotherapy to get rid of the new cancer. I'm hoping to get back to feeling as good as I did when I first left here about six weeks ago."

Rumors swirled. Grief junkies and profiteers gathered. Security guards stationed outside his suite nabbed hospital personnel in stolen scrubs trying to steal his blankets; one tried to sell his MRIs on eBay. They were returned after legal action was threatened. The Reverend Jesse Jackson, who was in Dallas to attend a political summit organized by Ross Perot, asked to pray with The Mick and was turned away. "He just come to see what was going on, I guess," Merlyn told me. "Just wanted to get on camera."

Mantle turned other visitors away as well. "Let me get my legs underneath me," he wrote in a note to Billy Crystal and Bob Costas. "Maybe next week," he told Dominic Sandifer. But he couldn't forestall the bad news Mark Zibilich brought after a new biopsy: there was cancer in his new liver and both lungs, his pancreas, and even the lining of his heart. "He said, 'I would still like to see all of my buddies,'" Zibilich recalled. "That's all he said."

On Monday, August 7, when doctors told him there was nothing else they could do, he said he didn't want to know how long. He had signed a living will instructing them not to perform any heroic measures. He told his sons, "Somebody call Roy."

The singer made arrangements to fly to Dallas.

Mantle's doctors promised to make him comfortable, but it wasn't easy. His legs were so swollen the skin cracked. Flesh was hanging from his arms. Fluid had to be aspirated from his stomach.

Visitors were urged to come quickly. Although she was still in Texas, at Pat Summerall's house, Greer

Johnson did not receive an invitation. That was Mantle's decision. "He did not want a showdown," she told me. "He didn't want a scene, and that's what it would have been."

They spoke on the phone, Johnson said, "right up until maybe a week before, when he started going in and out of consciousness."

But no one had told her the gravity of the situation. "Of course, I'm not wanting to think that he's gonna pass away," she said, recalling his leave-taking from Summerall's house for what she thought was a routine chemotherapy appointment. "And I see Danny and Mickey drive off in the car, not having a clue that's the last time I'm ever gonna see him again. I never got to say goodbye."

Merlyn was staying in an adjoining room in Mantle's hospital suite, asserting her wifely prerogative over what little remained of her marriage. By then, she told me, "We were probably more like good friends." But she had never given up hope that they would live together again under the same roof.

On Wednesday, his condition was downgraded from stable to serious. He was sitting up in a wheelchair when Barbara, Ted, and the twins arrived. The tube feeding him morphine got tangled up in the wheels, and Barbara rushed to his aid. When she touched him,

she knew the end was near. "I said, 'Boy, you're cold,'"
she remembered. "He was so cold."

Larry came alone. He was proud to be the youngest brother of the man he considered "the best baseball
player that's ever played"—though it might have been
nice if Mickey had called him something other than
Butchie. Larry would regret that Mickey never saw
him as an adult.

After he left for college in 1960, their lives diverged,
so much so that four decades later he was hard pressed
to say whether he missed him. Still, it was hard taking
his leave from the hospital. "Before I left and pretty
much said my last goodbye, he started cryin'," Larry
said. "He said, 'Hey, Butch, I may be cryin', but everything's all right. We got all this shit under control.'"

As always, Mantle rallied for his teammates Hank
and Blanch and Moose, whom he called every day after
his double-bypass surgery that spring. They talked
about golf and the home run he hit off Ray Herbert in
1964, when he threw down his bat in disgust and broke
his gamer, thinking he hadn't gotten it all.

They lifted him out of his chair and helped him to
the bathroom. When Bauer saw the color of his urine,
he knew time was short. "We bring him back and he
lays down in bed and he closes his eyes and I said to
Moose, 'Let's get the hell out of here,'" Bauer said.

656 • JANE LEAVY

"And he woke up and said, 'You guys aren't leaving already, are you?'

"I said, 'Oh, God!'"

They stayed a little longer. Blanchard averted his eyes. When it was time to go, they "hugged up," Skowron said, and told him they loved him.

Whitey Ford arrived with a baseball autographed by the 1995 New York Yankees. They had spent a lot of nights together, none like this. Whitey fudged the digital numbers on the pain dispenser, trying to make the time between the doses of morphine tenable. "He'd say, 'How much time, Slick?'" Ford said. "And he'd ask me a few seconds later, and I'd say, 'Three seconds' when it really was about five or six. I was just trying to make him feel like it was going to happen any second. Then, when it would happen, he was able to talk. He just looked up and said, 'If I knew this was going to happen two years ago, I wouldn't have quit drinking.'"

Back in Georgia, Ed Nelson was fretting about Mantle's soul. He called the chaplain's office at the hospital repeatedly to ask permission to make a pastoral call but got no reply. So he was relieved to learn later that Bobby and Betsy Richardson had been by his side. By the time they arrived, Betsy said, there wasn't much time for talk. "He was already on morphine," she said.

She knelt at his side, offered her testimony, and

when he volunteered his "Good News," she asked how he knew he would spend eternity with God in Heaven.

"We're talking about God?" Mantle said.

"Yes, God," she replied.

She reminded him of God's unconditional love and he recited the passage from the Gospel of John that attested to his faith. That, she thought later, was the moment of "realization that he doesn't have to perform to be loved."

Some friends were skeptical about his newfound religion. But True thought, "When Bobby Richardson came to Mickey and said, 'You're dying, here is a way to make peace,' he embraced it. He wanted to be forgiven."

After the Richardsons left, his friend Joe Warren recalled, he asked Merlyn, "Why doesn't He go ahead and take me?"

The chemotherapy session scheduled for Friday, August 11, was canceled. For the next two days he drifted in and out of consciousness in a morphine haze. Merlyn and the boys took turns keeping him company. She had written him a letter, telling him how much she loved him and how good he had been to her, but by the time she brought it to his attention he was too sick and too tired to listen. "Oh, honey, can it wait?"

He would go to his grave with her letter in his suit

pocket and whatever words of apology he might have expressed unsaid. "Apologize?" Merlyn told me. "No, he didn't."

David Mantle's depression was profound. He was having trouble coming to grips with his father's mortality. "I said, 'Wait a minute, he's supposed to be a god. This is not supposed to happen to *him*.'"

But it was, and the reality made him desperate. One night, Danny found him at home playing Russian roulette. He had pulled the trigger twice by the time Danny wrestled the gun away. The gun fired and a bullet lodged in the wall.

David had written a letter, too. After watching his father gazing out the window of his hospital room at what lay beyond the green grass and the Dallas sun, David had a vision of him as a boy, running through the fields with his dog, young and whole and fast. He wrote it all down and intended to read it aloud to his father. "But then he died, so I stuck it in his jacket," David said.

David and Merlyn were with him at the end. "He started running his feet under the covers like he was trying to fight it," David said. "Mom was over on this side of the bed. It was like chaos and everything. He was just lying there looking at us. He raised up his hands like this. We grabbed them. He smiled. It was like all the air went out of him."

Mickey Mantle died at 1:10 A.M. Central Time on August 13, 1995. "Just a little past midnight," David wrote later. "An hour he knew so well."

6.

Mickey Mantle's Restaurant was packed the afternoon of the funeral, four deep at the bar, the windows draped with black and purple bunting, televisions tuned to the live feed broadcast on ESPN2 and on WCBS Channel 2. "It was like being in a funeral parlor," the chef Randy Pietro said. "People were coming in off the street, crying and hugging each other like they were all related to him. Some people came in just to sit and run a hand on the bar, a way of saying goodbye to him."

Lovers Lane United Methodist Church in Dallas was overflowing with 900 mourners in the main sanctuary and 700 more in an adjacent fellowship hall watching on a video screen, with another 100 or more standing in the hall and still others milling around outside. Joe DiMaggio did not attend, issuing a statement instead that made mention of Mantle's rookie year visit to the minors. Willie and The Duke were also absent. Stan Musial, recently diagnosed with prostate cancer, came alone. Ralph Terry drove in from Kansas. Crossing the bridge over the Spring River in Baxter Springs and the park where Mantle had been a Whiz Kid when the whole world was still ahead of him, Terry noticed a black wreath lying in the grass where home plate had been.

Whitey Ford, Yogi Berra, Moose Skowron, Hank Bauer, Johnny Blanchard, and Bobby Murcer were the honorary pallbearers. "I seen more Yankees than Andrew Jackson," Faye Davis told her brother later.

Kathleen Hampton arranged for the service to be held at her church and for a dry wake later at True's home. Mantle's friend Jim Hays said Mickey, Jr., greeted mourners, saying, "Welcome to the house my dad built." Hampton stationed friends at every door of the church in an effort to intercept Greer Johnson. She arrived with Pat Summerall and Georgia governor Zell Miller, and, she says, with True's knowledge and approval. Hampton orchestrated the family's departure from the service during the Lord's Prayer so that Johnson would not be able to follow them to the private burial.

When an usher offered Summerall a seat up front near the family, he declined, staying with Johnson in a section near the back of the church reserved for players and their wives. "I couldn't leave her," Summerall said. "She couldn't sit up. She couldn't look at the service. I held her in my arms the whole time."

Years later, Johnson told me, "My best friend, my lover, my employer, my everything was gone. It was like I was gone."

Bobby Richardson officiated, keeping the promise he had made to Mantle at Maris's funeral a decade earlier. Roy Clark sang "Yesterday, When I Was Young," telling the congregation he hadn't expected to have to keep his promise so soon. When Richardson whispered Merlyn's unexpected request for a chorus of "Amazing Grace," Clark blanched. Arthritis in his fingers made guitar picking hard, and he wasn't sure he remembered all the words. "I'm going to try this," he said. "As Mickey would say, 'When you're on the golf course with a two-foot putt, just get it close.'"

Bob Costas gave the eulogy, speaking for the child he once was, the children we all were before Mickey Mantle forced us to grow up and see the world as it is, not as we wished it to be. Costas remembered him as "a fragile hero to whom we had an emotional attachment so strong and lasting that it defied logic."

Mickey Mantle did not go home to Commerce. He was laid to rest in the Sparkman Hillcrest Memorial Park, a posh Dallas cemetery that is also the final resting place of the cosmetics queen Mary Kay Ash, Cowboys coach Tom Landry, Alice Lon, Lawrence Welk's champagne lady, and Judge Sarah Tilghman Hughes, who administered the oath of office to President Lyndon Baines Johnson aboard Air Force One on November 22, 1963.

Directions to his plot (NE-N-D14–15) in the Saint

Matthew Mausoleum are posted online at www.ehow.com/how_4392292_visit-mickey-mantles-grave.html and its coordinates (N 32° 52.064 W 096° 46.857) at www.waymarking.com. Findagrave.com invites visitors to send e-flowers and e-mails to The Mick. On the fourteenth anniversary of his death, he received fifty-four digital notes and bouquets.

In this august, austere, and climate-controlled precinct of death (crafted by fine Italian artisans from rare marble and granite, the Web site says), the footsteps of the living resound like the trumpets of angels, loud enough almost to wake the dead. Mantle is interred in a crypt illuminated by flickering sconces and graced with plaster angels whose wings shelter cards and letters left by his fans. Fixed to the wall is a plaque, not unlike the one that marks his place on another wall in Monument Park at Yankee Stadium:

MICKEY CHARLES MANTLE
October 20, 1931 · August 13, 1995
A magnificent New York Yankee,
true teammate and Hall of Fame centerfielder
with legendary courage.
The most popular player of his era.
A loving husband, father and friend for life.

He was buried in pinstripes.

EPILOGUE

F ar from the chilled silence of Mickey Mantle's tomb, the living continue to trade on his memory. On the day he would have turned seventy-seven years old, a "tiny hair strand speck from Mickey Mantle" was offered to the highest bidder on eBay. In 1997, Greer Johnson auctioned off more than two hundred lots of personal effects, including his birth certificate, a neck brace, four prescription bottles, and a clump of barber-bagged hair.

She says she tried to contact his family and received no reply other than notice of a lawsuit, which successfully blocked the sale of thirty-three of the most personal items. After the final hammer came down, she went to Mickey Mantle's to lift a glass to The Mick, and was escorted out of the restaurant by the maître d'. Johnson says she placed the money, $541,000 minus her executor's commission, in a charitable remainder trust, all of which was invested and all of which will go to the American Cancer Society and Baseball Assistance Team upon her death.

The clump of Mantle's hair sold for $6,900.

Long after his death, he remained his family's sole breadwinner. In 1996, an arbitration panel awarded his estate $4.9 million after finding that Upper Deck had unfairly breached his contract. In December 2003 Merlyn emptied the attic—and some of the prized trophy room—selling off bits and pieces of the past to provide for the future education of her four grandchildren—Mallory, Marilyn, Chloe, and Danny's son, Will. Some things were not for sale. Will, who has his grandpa's lips and maybe some of his talent, will inherit his Hall of Fame ring.

The auction was held in Madison Square Garden, inside a vault of undifferentiated utility space where the elephant cages park when the circus comes to town. At the preview, strangers ogled the relics of an original American life while Will bounced a pink Spaldeen under the watchful eye of a former NYC cop happy to do his bit for The Mick. Locked behind temporary glass display cabinets were expired passports and credit cards, canceled checks, Yankee pay stubs, coasters from Mickey Mantle's Country Kitchen, a 1961 commendation from the Texas Board of Corrections for efforts on behalf of the Texas Penitentiary Rodeo, and twelve of those childhood scrapbooks in which he failed to recognize himself.

There were also big-ticket items, including his 1957 and 1962 Most Valuable Player awards, which sold for $275,000 and $250,000, respectively; his 1962 World Series ring, which brought $140,000; and a motorcycle he never rode, an homage Harley tricked out with faux Louisville Slugger wood grain, painted pinstripes, and a leather seat made from one of his baseball gloves. The bike, inscribed with a poem written by David, was purchased for $55,000 by Randall Swearingen, who made it the centerpiece of his Mantle shrine in Houston. The auction netted $3.25 million.

Five days after Mantle's death, his family announced the establishment of Mickey's Team. The campaign he envisioned on behalf of organ donation would be funded by the foundation he had created in Billy Mantle's name. Carl Lewis, the Olympic sprinter, whose best friend had died while waiting for a kidney transplant, presented a check for $25,000. Six million baseball card–style donation forms were distributed. But momentum flagged. Mickey's Team needed The Mick. The donation card began showing up on eBay. "We found out people were selling them for ten dollars apiece," David told me at Madison Square Garden.

"We talk about cancer now. My dad had that, too."

By the end of 2009, the Mantle Family Fund for the American Cancer Society had raised $125,673, which was donated to the Hope Lodge, a home away from home in Manhattan for cancer patients and their families. Danny and David dedicated the Mickey Mantle Suite on September 29, 2009, with the help of Rudy Giuliani.

The family continues to be haunted by the disease.

Mickey Mantle, Jr., entered the Betty Ford Center in October 1995. A patient who had been in treatment with The Mick told him that his father had expressed the hope that Little Mick would come and get sober. Mickey Elven Mantle died of cancer in December 2000 at age forty-seven, telling his brothers, "Can you believe I sobered up for this?"

Danny relapsed after Little Mick's death, then reclaimed his hard-won sobriety. Merlyn Mantle won custody of Mickey Jr.'s daughter, Mallory, rising at six every morning in the golden years of her life to care for another teenager. She was the first of Mantle's grandchildren to go off to college.

Roy Mantle died of non-Hodgkin's lymphoma at age sixty-five in September 2001.

Merlyn's hometown was wiped off the face of Google Earth at the close of business on August 31, 2009, when the municipality of Picher, Oklahoma, ceased operations and city hall closed its doors. Two businesses defied the decree: the drugstore and Paul Thomas, the undertaker who buried Mutt Mantle. Thomas stayed open for six months before reluctantly accepting a government buyout. On March 27, 2010, he auctioned off the mining relics he kept in his garage, including a portrait of Mutt's crew at the Blue Goose Mine. That month the last derrick in Picher, which stood over a void as vast as the Astrodome, was torn down.

Merlyn Louise Mantle died of Alzheimer's disease in July 2009. In her last days, she sometimes confused Danny and David with their father. She was laid to rest above her husband. Below him rest two of their sons.

ACKNOWLEDGMENTS

At first blush, Mickey Charles Mantle and Charlotte von Mahlsdorf, the hero/heroine of Doug Wright's Pulitzer Prize–winning play, *I Am My Own Wife*, are unlikely teammates. Charlotte was a man living as a woman in Nazi and, later, Stasi Germany. Mickey solicited advice from pals about whether he could get AIDS by sharing the hot tub at the Betty Ford Center.

But when my literary aspirations foundered on the shoals of hero worship, my non–baseball loving friend Carole Horn insisted: "You have to see this play. This is your book."

So I went.

In the course of his research, Wright discovered that his hero/heroine wasn't what he/she seemed. In fact, she was a collaborator. Disillusioned and disabused of his faith in the icon of gay culture, he found he could no longer write the play he had envisioned. In fact, he couldn't write at all.

The iceberg of writer's block wouldn't dissolve until six years later, when a friend said, "Whatever you do, don't write a play about history. Write a play about your love affair with Charlotte von Mahlsdorf."

"For the first time, the play's structure dawned on me," Wright said in his preface. "It wouldn't be a straightforward biographical drama; it would chart my own relationship with my heroine. I would even appear as a character, a kind of detective searching for Charlotte's true self."

When the curtain fell, I knew what to do and whom to thank for it. Carole had given me permission to write about my guy my way, my love affair with my hero, Mickey Mantle. I would even use myself as a minor character, a kind of every-fan detective, negotiating the swells of adulation while searching for Mickey's true self.

Mistress of the arts of healing and semantics, Carole also kept my head on straight and the moving parts moving, bandaging psychic wounds and ministering to sentence structure throughout the long march.

Leo Tolstoy wrote *War and Peace* in five years. This took longer. I'm not proud of that. "Well," Mantle's cousin Max said kindly, "Mick was kind of like war and peace."

This campaign was a slog for my family and friends.

During the most difficult time in my life, I was supported by a cadre of stalwarts without whom I couldn't have put one foot in front of another, much less one sentence after another. First among the cohort: my children, Nick and Emma. Nick never lost interest and is the only person, finally, from whom I could tolerate the hated inquiry, "When are you going to be done with the book?" Emma never lost patience. She knows me better than anyone and tolerates me more. She abides in loveliness and grace.

Hal and Marilyn Weiner (Karen, Andrew, et al.) are my family of choice. Hal is a Yankee fan from Brooklyn, the older brother I never had. He was the first to insist that I make Atlantic City a part of the narrative; Marilyn calls every day. Nothing more or better can be said of a friend. Sid and Diana Tabak are my anchors. Sid (God bless you) makes me look better than I should; Diana, you ground me. Dick and Caren Lobo are my ports in the storm. Andy Levey and Roberta Falke are family (even if he spells his name funny). Linda Greenhouse and Gene Fidell sustain me with constant friendship and wise counsel. Michael Nussbaum and Gloria Weissberg make me brave. Arlie Schardt and Bonnie Nelson Schwartz give me hope for the future. Steven Phillips has always been there. Kim Sammis and Jim Ulak are always there—they live next door—but they would be, anyway. Ada Vaughn and Alba Hernandez were there when I needed them most. Leslie Harris and Peter Basch, Mary Brittingham and David Plocher, Christine Tan and Al Acres make me glad I live where I do. Rhonda Schwartz and Steve Wermeil took me in, and took me away from it all when I needed taking care of most.

Singing backup in sweet harmony, my All Star girl group: Maria Applewhite, Sharon Alperovitz, Sherry Berz, Jordana Carmel, Claudia Cormier, Toni Cortellessa, Aviva Crown, Julie Crown, Donna Grell, Caroline Herron, Kate Hill, Caroline Janov, Annette Leavy, Cuyler McLaren, Jane Shore, Carey Tompkins, and the redoubtable Amy Katz, Jane Moffett, and Barbara Weinschel.

The boys in the band keep me from losing a beat: Philip Bahnson, Lenny Croteau, Michael Holt, Harry Jaffe, Scott Kahan, Sandy Koufax, Robert Pinsky, Dennis Roche, Jon Rupp, Norman Steinberg, my main man George Vecsey, and J. P. *the* Man—you know who you are and what you mean.

By all accounts, Mickey Mantle was a good, loyal, and generous friend. Thanks to him, I have gained friends for life: Brad Garrett, gumshoe extraordinaire, and his intrepid partner, Elisa Poteat; Alan M. Nathan, professor emeritus of physics at the University of Illinois–Champaign-Urbana and the chairman of the Sabr Committee on Science and Baseball, who tutored me in the physics of bats and balls, and whose patience with my mathematical ineptitude was endless; Ed Keheley, whose efforts on behalf of the people of the Tri-State Mining District and willingness to educate me about their plight is nothing short of heroic; Larry Meli, who graciously made introductions to the Mantle family; Preston Peavy, the irrepressible and indefatigable hitting coach–kinetic whiz who contributed hundreds of hours of his time; and Marty Appel, the man who escorted me to Commerce, Oklahoma, and stayed with me throughout this journey.

Among the hundreds of Mantle fans, friends, and family members who met my intrusions and inquiries with tolerance and insight, the following deserve a special shout-out: Yogi and Carmen Berra, Marjorie Bolding, Jim Bouton, Terry Cashman, Billy Crystal, Ryne Duren, Carl Erskine, Nick Ferguson, Frank Howard, Greer Johnson, Mike and Katy Klepfer, Bill Liederman, Glenn Lillie, Phil Linz, Jerry Lumpe, Bob Mallon, Gil and Lucille McDougald, Tony Morante, Ed Nelson, and especially Claude Osteen, Bob Wiesler, and Jerry VonMoss.

Understanding what made Mickey Mantle Mickey Mantle required a remedial education in physics, biomechanics, and neuroscience. Home run sleuth Bruce Orser and Greg "The Hit Tracker" Rybarczyk always made themselves (and their smarts) available. Psychologists Richard Gartner and David Pelcovitz educated me about the scourge of childhood sexual abuse. Nobel Prize winner Dr. Eric R. Kandel expanded my understanding of how a body remembers. My very own, hands-on PT goddess, Amy Engelsman, rehabbed every cranky part of this aging jock's body and schooled me in the nuances and complexities of sports physiology.

Dr. Stephen Haas and Dr. Benjamin Shaffer, Washington, D.C., sports orthopedists extraordinaire, made sure I didn't make a fool of myself in the effort to explain the breakdown in Mantle's body. Dr. Kenneth Thompson, medical director at the Caron Foundation in Wernersville, Pennsylvania, expanded my grasp of the plague of addiction. "Sudden" Sam McDowell eloquently elaborated on the long, sad history of alcohol and baseball. Dr. David Mulligan of the Mayo Clinic in Arizona and Dr. Göran Klintmalm of the Baylor Regional Transplant Institute helped me understand the last months of Mantle's life. Pam Silvestri of the Southwest Transplant Alliance and Joel Newman of the United Network for Organ Sharing (UNOS) taught me about the system of organ allocation in the United

States and the vast and urgent need for organ donors. It's not too late to join Mickey's Team.

My far-flung colleagues have been generous beyond description with their time and recollections, chief among them: Allen Barra, Bill Brubaker, Pete Cava, Bob Costas, Bob Creamer, John Hall, Bill Handleman, Dick Heller, Jerry Izenberg, Dave Kaplan, Mark "The Rabbi" Langill, Bob Lipyste, Marty Noble, Phil Pepe, Diane Shah, Matt Shudel, John Thorn, Bob Wolff, and Phil Wood.

"Honest" and "generous" are the words Mantle's friends most often used to describe him. I experienced those same traits in the members of his family who shared their time with me. Danny Mantle offered help to a member of my family in need. Merlyn Mantle expressed concern for my safety when I traveled to her imperiled hometown, Picher, Oklahoma. Larry Mantle, Mickey's youngest brother, spent hours in a hot, dusty trailer in Lawton, Oklahoma, recounting his family history. His sister, Barbara DeLise, who has little in the way of big brother memorabilia, gave me a Styrofoam cup embossed with a photograph of the young, smudged Mick standing by an Eagle-Picher pickup truck. I treasure it. Max Mantle drove one hundred miles to the gravesite of Grandpa Charlie to verify his age at the time of his death.

I cannot predict how his loved ones will receive this effort. I hope they feel I kept my promise to return his humanity to him in full.

Documenting this life was a team effort. Victoria Torchia, Bill Leggett, Jon Gaither, and Claire Ulak assisted me with humor and tenacity. The intrepid and uncomplaining Paul Janov at the Library of Congress located every historical document no matter how obscure. Tim Wiles, Ted Spencer, and the retired rock 'n' roller Russell Wolinsky at the Hall of Fame were tireless in their efforts on my behalf. Larry O'Neal at the Baxter Springs Heritage Center and Museum in Kansas provided a guided tour through the archives and the landscape of the Tri-State Mining district.

Louis Plummer and Mike Owens at PhotoAssist helped me find the images that make words superfluous. Bill Richmond in Dallas helped secure legal documents. Marti Hagan, proprietress of Word Wizards in Atlanta, made the inaudible audible and translated it into the printed page with kind dispatch. Stu Hancock and Rick Prescott saved me from permanent computer angst.

Major props to the copyediting staff at Harper-Collins—Lynn Anderson, John Jusino, Tessa Roush, and especially mensch David Koral; Beth Silfin, legal bodyguard; Kate Blum, P.R. maven; art director

Archie Ferguson and senior designer William Ruoto; George Quraishi and especially Barry Harbaugh, who always had my back.

The indefatigable and irreplaceable Antonella Iannarino at the David Black Literary Agency always sounded glad to hear from me, even when she had no reason to be. Emily Parliman, office intern, is my candidate for Rookie of the Year. Ann Gerawa Hess held down the fort at home.

Knute Rockne had Four Horsemen. I have Four Davids. Dave Smith has been an invaluable reader and trusted friend since our serendipitous meeting, communing over Sandy Koufax's perfect game. He read and reread, vetted and revetted every statistic in this book and managed to sound interested and make improvements with each reading.

David Black, adviser/agent/consigliere/friend, is my go-to guy. He cares about books and words and the people who write them with a passion that is nothing short of heroic in a world that has little time or patience for either.

Dave Kindred, ink-stained wretch, wordsmith extraordinaire, first reader, is my best pard.

My editor and old friend David Hirshey, executive editor at HarperCollins, is a tireless advocate, chivalrous and loyal, a Jewish knight errant in a fraught literary world. This damsel in distress will be always grateful for his protection, patronage, parsing, and patience. To press box cognoscenti, David will be forever known as Dr. Deadline. To me, he is the good and loyal doctor of letters.

Twenty years ago or so, I made an abortive attempt to write a memoir about life as a female sportswriter. I discovered the discarded first draft when I exhumed my 1983 Mantle files. The unfortunate lede went like this: "We both wore Number Seven, Mickey Mantle and me."

In a moment of weakness, I sent the premature effort to Gerri Hirshey, who had the distinction of editing the first newspaper story I ever wrote—a sixteen-inch piece about the Queens College women's basketball team. I didn't get a byline, but if I had, she should have gotten one, too. Came in the mail a letter explaining firmly and deftly why what I had written should never see the light of day. But, she added: "Trust your voice. You have a right to have a voice."

At a uniquely tender moment in my life, Gerri stepped up to the plate, wielding a sharp editorial eye and clout earned in over three decades in the business of words and friendship. She told me when to swing and when to lay off. She corrected flaws when my mechanics went awry. One day, near the end of the pro-

cess, I received an e-mail, "your self-righteous scythe."
She was anything but—gentle, bemused, invested,
and persuasive. I trust her more than I trust myself.
It's hard to say whether she's a better writer, editor, or
friend. Thank God, I don't have to decide. This is the
only part of the book she didn't get her hands on. Un-
doubtedly, she'll tell me it's too long.

JL, June 2010

APPENDIX 1: INTERVIEW LIST

NEW YORK YANKEES

Larry Allen
  (brother of Mel Allen)

Marty Appel
  (public relations)

Stan Bahnsen

Hank Bauer*

Howard Berk
  (front office)

Carmen Berra

Yogi Berra

John Blanchard*

Ron Blomberg

Jim Bouton

Clete Boyer*

Bobby Brown

Tommy Byrne*

Andy Carey

Bob Cerv

Tex Clevenger

Jim Coates

Jerry Coleman

Bobby Cox

Johnny Damon

Keith Darcy (batboy)

Bobby Del Greco

Joe DeMaestri

Art Ditmar

Al Downing

Ryne Duren

Mike Ferraro

Whitey Ford

Joe Gallagher (broadcast)

Jason Giambi

Jake Gibbs

Eli Grba

Angie Greenwade

(daughter of Tom
Greenwade)
Bunch Greenwade
(son of Tom Greenwade)
Bill Guilfoile (public
relations)
Whitey Herzog
Ralph Houk*
Arlene Howard
(wife of Elston Howard)
Janet Huie(daughter of
Red Patterson)
Reggie Jackson
Johnny James
Esther Kaufman
(sister of Mel Allen)
Steve Kraly
Andy Kosco
Herman Krattenmaker
(front office)
Tony Kubek
Tony Kubek III
(son of Tony Kubek)
Johnny Kucks
Bob Kusava
Don Larsen

Phil Linz
Hector Lopez
Jerry Lumpe
Lee MacPhail
(front office)
Elliot Maddox
Gil McDougald
Lucille McDougald
Gene Michael
Malcolm "Bunny" Mick*
(minor league coach,
instructor)
Tony Morante
(Yankee Stadium tours)
Ross Mosschito
Thad Mumford (ball boy)
Bobby Murcer*
Ray Negron
(special adviser to
George M. Steinbrenner)
Irv Noren
Bruce Patterson
(son of Red Patterson)
Gregg Patterson
(nephew of Red
Patterson)

Joe Pepitone
Frank Prudenti (batboy)
Pedro Ramos
Willie Randolph
Jack Reed
Mickey Rendine
  (clubhouse)
Bill Renna
Betsy Richardson (wife of
  Bobby Richardson)
Arthur Richman*
  (public relations)
Curt Rim (Yankee
  Stadium architecture)
Mickey Rivers
Pat Rizzuto (daughter of
  Phil Rizzuto)
Eddie Robinson
Art Schallock
Donna Schallock
Kal Segrist
Bobby Shantz
Rollie Sheldon
Bob Sheppard* (public
  address announcer)
Norm Siebern

Charlie Silvera
Bill "Moose" Skowron
Mel Stottlemyre
Elaine Sturdivant
Tom Sturdivant*
Frank Tepedino
Ralph Terry
Joe Torre
Tom Tresh*
Virgil Trucks
Bob Turley
Roy White
Steve Whitaker
Bob Wiesler
Bernie Williams
Stan Williams
Sybil Wilson
  (wife of Archie Wilson)
Hank Workman

MAJOR LEAGUE BASEBALL

Bernie Allen
George Alusik
Joe Amalfitano

Sparky Anderson

Ed Bailey*

Gary Bell

Clete Boyer*

Jim Brosnan

Ed Charles

Bill Clark (scout)

Joe Coleman

Alvin Dark

Brandy Davis* (scout)

Tommy Davis

Dom DiMaggio*

Johnny Edwards

Mike Epstein

Carl Erskine

Sam Esposito

Bob Feller

Bill Fischer

Paul Foytack

Herman Franks*

Jim "Mudcat" Grant

Clark Griffth (Washington
  Senators, Minnesota
  Twins ownership)

Dick Groat

Jim Hannan

Ron Hansen

Ray Herbert

Billy Hoeft*

Frank Howard

Mike Hudson
  (Washington batboy)

Bobby Humphreys

Monte Irvin

Larry Jansen*

Nobe Kawano
  (Dodger clubhouse)

Bowie Kuhn*
  (commissioner)

Jim Kaat

Al Kaline

Jim Landis

Tony LaRussa

Frank Lary

Whitey Lockman*

Mickey Lolich

Jim Lonborg

Jim Maloney

Juan Marichal

Jim McAnany

Tim McCarver

Mike McCormick

Willie McCovey

Sam McDowell

Denny McLain

Sam Mele

Marvin Miller
(Major League Baseball
Players Association)

Don Mueller*

Don Newcombe

Claude Osteen

Jim O'Toole

Camilo Pascual

Billy Pierce

Al Pilarcik

Boog Powell

Jim Price

Brooks Robinson

Frank Robinson

Ed Roebuck

John Roseboro*

Al Rosen

Red Schoendienst

Lou Sleater

Duke Snider

Alfonso Soriano

Bob Speake

Ted Spencer
(Hall of Fame)

Rusty Staub

George Steinbrenner*

Charlie Stobbs
(son of Chuck Stobbs)

Evan Stobbs (grandson of
Chuck Stobbs)

Joyce Stobbs (wife of
Chuck Stobbs)

Ron Swoboda

Wayne Terwilliger

Jake Thies

Bobby Thomson

Jeff Torborg

Dick Tracewski

Fred Valentine

Ed Vargo* (umpire)

Mickey Vernon*

Bill Virdon

Bob Willis

Eddie Yost

MINOR LEAGUE TEAMMATES AND OPPONENTS

Charles Ane*†

Tod Anton†

Walt Babcock

Bill Bagwell
Al Billingsly
Stan Charnofsky†
Dave Cesca†
Phil Costa
Rod Dedeaux*†
Justin Dedeaux† (batboy)
Lee Dodson
Dan Dollison
Dick Getter
Joan Getter
Tommy Gott
Jack Hasten
Herb Heiserer
Bob Hertel†
Ed Hookstratten†
Howie Hunt
Don Keeter
Carl Lombardi*
Tom Lovrich
Bob Mallon
Nick Najjar
Cal Neeman
Bob Newbill
Dave Newkirk*
Dave Rankin†
Tom Riach*†

Dean Rothrock
Dick Sanders
Lilburn Smith
Cromer Smotherman
Keith Speck
Joe Stanka
Charlie Weber*
Len Wiesner
Mike Witwicki*

OKLAHOMA

Joe Barker
LeRoy Bennett
Brent Brassfield
Charles Brinkley*
Paul Churchill
Ben Craig
Nick Ferguson
James Haynes
Jim Hays
Don Hicks
Donna Hicks
Lee Jeffrey
Billy Johnson*
Irene Keheley
Ben Lee

Charlene Lingo
  (wife of John Lingo)
Delbert Lovelace
Jim McCorkell
Mike Meier
Bill Mosely*
Lee Mosely
Howard Moss*
Larry O'Neal
Kim Pace
Wylie Pitts
Ivan Shouse
Sue Sigle
Corrine Smith
Marshall Smith
Paul Thomas
Colleen VonMoss
Jerry VonMoss
Brian Waybright
Bill Whipkey

FRIENDS/ASSOCIATES

Jim Abercrombie
Bart Alexander
Carmen Basilio
Ed Beshara, Jr.

Dick Biley
Marjorie Bolding
Tommy Bolt*
Roy Clark
John Crouse
Billy Crystal
Bill Dougall
Frank Gifford
Bill Grainger
Jack Hamlin
Darrell Hammie
Kathleen Hampton
Jickey Harwell
Bill Hooten
Linda Fetters Howard
Julie Isaacson*
Jack Jackson
Greer Johnson
Warren "Rhubarb" Jones
Katy Klepfer
Mike Klepfer
Bill Liederman
Glenn Lillie
George Lois
John Lowy
George Macris
John Matney

George Matson
Dave McLaurin
Larry Meli
Tom Molito
Wayne Monroe
Ed Nelson
Jimmy Orr
Don Perkins
Frank Petrillo, Jr.
Troy Phillips
Randy Pietro*
Dan Reeves
Lon Rosen
Darrell Royal
Dominic Sandifer
Martha Stewart
Pat Summerall
Roy True
Bobby Van
Lanny Wadkins
Roger Wagner
Joe Warren
Dale Wittenberger
Barbara Wolf
Ron Wolf

FANS/D.C. RESIDENTS

Bill Abernathy
Henry Akers
Don Arken
Johnny Barnes
Jim Barrett
Paul Berkman
Greg Bischoff
Gail Blackwell
Jack Bottash
Alfonso Brooks
Johnnie Brown
Steve Bryant
Alan Budno
Rosa Burroughs
Glenn Cafaro
Terry Cashman
Pete Cava
Sarah Chase
Will Corbitt
Nelson Diaz
Marv Diemer
Donald Dunaway*
Kenneth Dunlap
George Enterline

Sandra Epps
Alan Feinberg
Kimberly Fox
Emilio Furiati
Mike Green
Roberta Green
Kevin Hannon
Bobby Harper
Jim Hartley
Fred Heller
Jerry Holt
Lauretta Jackson
Jim Jay
Jerry Joseph
Walter King
Bob Kleinknect
Bobby Lane
Rob Liebner
Frank Martin
Cathy McCammon
Larry McCosky
Maxine McCollough
Walter McCollough
Steve Meeds
Len Melio
Joe Montanino

Paul Nuzzelese
John Nicolossi
Bill O'Connor
Bernadette O'Donnell
Bruce Orser
Jean Piper
Alicia Pratt
Rahmin Rabenou
Barney Rapp
Ed Rudofsky
Joe Saccoman
Bob Schiewe
Warren Sherman
Mary Ambush Smith
Cecil Stouts
Paul E. Susman
Randall Swearingen
Al Taxerman
Todd Ulitto
Cornelius "Nini" Wooten
Robert Wuhl
Larry Zaback

MEDIA/MEMORABILIA/ ARTS

Maury Allen
Richard Andersen
Dave Anderson
Dave Baldwin
Allen Barra
Jim Belshaw
Alex Belth
Sy Berger
Mike Berkus
Steve Borelli
Talmadge Boston
Sandy Brokaw
Charlie Brotman
Bill Brubaker
Judy Burr
Lonnie Busch
Tom Callahan
Fredrich Cantor
Tony Castro
Bob Costas
Richard Ben Cramer
Robert Creamer
Clay Davenport

Frank Deford
Bill DeOre
Anthony Dohanos
David Falkner
John Fox
Samuel Freedman
Warner Fusselle
Herb Gluck
Tom Goldstein
Hans Gumbrecht
John Hall
Arnold Hano
Bill Handleman*
Ernie Harwell*
Dick Heller
Clay Henry
Ed Hinton
James Hirsch
Phil Hochberg
Jerry Holtzman*
Tom Horton
Stan Isaacs
Jerry Izenberg
Steve Jacobson
Dave Jamieson
Bill Jenkinson

Richard Johnson

Phil Jordan

Christina Karhl

Peter Keating

Aviva Kempner

Erik Kesten

Dave Kindred

Dan Klores

Jack Lang*

Robert Lifson

Robert Lipsyte

Jeffrey Lyons

Bill Madden

Murphy Martin*

Jeffrey Marx

Gail Mazur

Bill McCaffrey

Terry McCaffrey

Peter Mehlman

Robert Moss

Richard Mueller

T. S. O'Connell

Tim Peeler

Phil Pepe

Robert Pinsky

Angelo Pizzo

Diane Prang

Wendell Redden

Richard Reeves

Vin Russo

Richard Sandomir

Ralph Schoenstein*

Matt Schudel

Diane Shah

Glenn Sheeley

Eddie Simon

Curt Smith

Dave Smith

Brad Snyder

Jill Lieber Steeg

Norman Steinberg

Charley Steiner

Glenn Stout

Andy Strasberg

Bert Sugar

Ozzie Sweet

John Thorn

Calvin Trillin

George Vecsey

David Vincent

Anvil Welch

Howard Williams

Warner Wolf
Bob Wolff
Phil Wood
Vic Ziegel*
Daniel Zwerdling

**MEDICAL/SPORTS
MEDICINE/PHYSICS/
BIOMECHANICS/
ENVIRONMENTAL**

Robert Adair
Mike Anderson
Dave Bary
Beth Bryant
  (daughter of Dan Yancey)
Arthur Caplan
Jennifer Coleman
Christine Courtois
Arthur DeLarios, MD
Steve Donohue
George Ehrlich, MD
Amy Engelsman
Robert Fine, MD
Richard Gartner
Andrew Gaynor (son of
  Sidney Gaynor)

Deborah Gaynor
  (daughter of Sidney
  Gaynor)
Rob Gray
Mark Greenberg
Stephen Haas, MD
Carole Horn, MD
Thomas Jacobson, MD
Scott Kahan, MD
Eric Kandel, MD
Ed Keheley
Göran Klintmalm, MD
Stan Krukowski
Kenneth V. Luza
David Mulligan, MD
Alan Nathan
John Neuberger
Joel Newman
J. Mark Osborn, MD
Preston Peavy
David Pelcovitz
Marilyn Pink
Don Porter
  (brother-in-law of
  Sidney Gaynor)
Cecil Priebe, MD

Dave Ringer, MD
Greg Rybarczyk
Don Seeger
Benjamin Shaffer, MD
Pam Silvestri
Merrie Spaeth
Frank Sundstrom, MD
Kenneth Thompson, MD
Barbara Weinschel, MD
David Whitney
Frank Wood
Dillard Worthy
Alice Yancey
  (daughter of Dan
  Yancey)
Mark Zibilich, MD

FAMILY

Faye Davis
Barbara DeLise
Pauline Klineline
Danny Mantle
David Mantle
Larry Mantle
Max Mantle
Merlyn Mantle*
Jimmy Richardson

*Deceased since interview, as of July 2010
†University of Southern California 1951 player/coach

APPENDIX 2: THE KINETIC MICK

Ted Williams was wrong. The hardest thing in sports isn't hitting a baseball—the hardest thing in sports is hitting a baseball equally well from both sides of the plate. That doesn't stop parents from begging hitting coaches like former major leaguer Mike Epstein, *Make my boy a switch-hitter.* "They all say the same thing," Epstein said. " 'Well, you know, Mickey Mantle . . .' "

Epstein tries to break the news gently: there was only one Mickey Mantle.

Joe DiMaggio was wrong, too, when he told the *San Francisco Chronicle* in 1951: "He's the only switch-hitter I ever saw that has the same stance and mannerisms from both sides of the plate—hits the ball the same way, strides the same, hits equally hard from left or right."

There may have been only one Mantle, but he was two very different hitters, right-handed and left-handed. In an effort to illustrate the differences—to show how he generated his power, as well as to highlight the strengths that distinguished him and some of the tendencies that contributed to his physical deterioration—I asked hitting coach Preston Peavy to produce a set of mini-films called kinematics for Mantle the way he does for students at Peavy Baseball in Atlanta, Georgia. Peavy, who works with youth, high school, and college players, uses his motion-analysis system to convert high-speed videotape into computer-generated stick figures that move through space like animated cartoon characters. Similar technology is used throughout the country at golf schools and by major league baseball teams. Kinematics allow coaches to see a hitter's form in its purest state, to offer guidance when "mechanics" go awry and to illustrate when things are working right.

Creating a kinetic Mick wasn't easy, which may be why it hasn't—to the best of my knowledge—been attempted before. Although Mantle was the most telegenic and most televised ballplayer of his time, the quality of much of the available footage wasn't good enough for analysis. Peavy screened hours of film and video that I collected in order to cull ten to fifteen of Mantle's best swings. To create the kinetics, he picked the most representative of them: a right-handed swing from the 1959 Home Run Derby with Willie Mays, and a left-handed swing from George Roy's 2005 HBO film *Mantle*.

Today, Peavy would have hundreds of swings from which to discern a pattern. Obviously, no analysis based on one swing can be definitive. But Claude Osteen, who spent eighteen years in the major leagues as a pitcher and fifteen more as a pitching coach, was impressed when he reviewed the kinematics at my request. He knew intuitively and from experience (Mantle batted .533 against Osteen with a double, a home run, and a slugging percentage of .800) that Mantle was a very different hitter from the left and right side. "You have a perception in your mind of the kind of hitter he was both ways, but the silhouettes prove it," he said. "The difference was extreme, but you couldn't see it with the naked eye. I didn't think the differences would be as extreme as they were."

According to the *Oxford English Dictionary*, the first use of the word *swing* as a noun, meaning "to stroke with a weapon" dates from 1375. The act of swinging—instinctive in children and habitual to manual laborers—must conform to the laws of physics in order to generate power and leverage. "Cavemen who needed to be able to throw rocks to protect themselves discovered that if they put a lever under a rock, it went farther," said Marilyn Pink, Ph.D., P.T., who has studied batting and throwing biomechanics in conjunction with team doctors. "The principle is the same in all ground reaction sports—pitching, batting, and golfing. All the power is generated from the lower leg. Good batters have huge butts because they have huge muscles back there. There has to be a stable tube, a rigid pipe, for the power to go through. The trunk muscles have to be firm so that the energy isn't dissipated through wiggly muscles."

The replication of the biomechanical ideal, what man-made machines can do absent annoying human deviation, is complicated by physical idiosyncrasy. "Style and technique are vastly different," said Mike Epstein, who runs a hitting school in Denver, Colorado. "Style is the individual and the technique is universal. All the great, productive hitters conform to the same technique. Their bodies fit into what I call that envelope. Before a hitter can get into that envelope, he goes through all the different gyrations that make him feel comfortable: hands high, hands low, hands out, hands in, bat straight up, flat bat, closed stance, open stance, wide stance, narrow stance, stride, no stride. If a good technique doesn't conform to the laws of physics, then you're essentially rolling a boulder uphill."

Viewing Mantle in kinetic form strips away nuance and reveals what was universal in his swing and also what differentiated him. To facilitate an understanding of Peavy's analysis, he created kinetic snapshots to insert into the text below. To view the full set of Mantle kinematics and the film clips from which they were drawn, go to www.peavynet.com or www.janeleavy.com.

From a coach's perspective, was Mantle's swing perfect? "Absolutely not," Peavy said, "but it was a thing of raw beauty, completely uninhibited," and "strikingly modern" in the way he shifted his weight to recruit all his available power, and positioned his elbow and moved his hands. "Add a top-hand release," he says, and you have A-Rod or Ken Griffey.

"I think he just stumbled on a real modern swing. Nobody was looking for that then. Nobody *knew* to look for it. This was pure, blue-collar, farm-boy aggressiveness. I can't think of anybody as naturally aggressive. He was allowed to be naturally aggressive. There are no similar guys. Especially from the left side, I can't think of anybody as violent as he was. Unbridled aggression is what made Mantle Mantle."

That zealousness was aided and abetted by sheer athleticism and a physique tailored for baseball. "His proportions are classical, ideal," Peavy said. "The length of his arms and legs enabled him to implement a modern swing. He's not particularly square; not particularly long in the torso; he wasn't a fireplug. He was a truly gymnastically proportioned athlete. If you wanted to build a baseball player from scratch, Mickey Mantle was it. He was built for this."

START POSITION

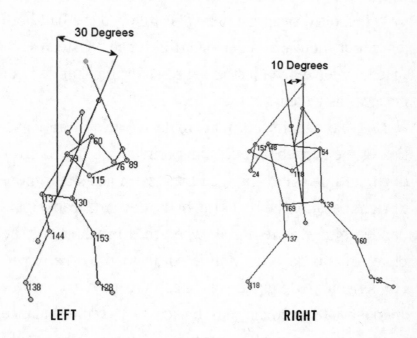

LEFT RIGHT

Mantle's right-handed swing was natural, easier on his body, and easier for him to perfect. He batted twice as often left-handed but hit the ball harder and more effortlessly right-handed. He was more energy-efficient as a right-handed hitter but his posture, batting left-handed, was more classical. Right-handed, he stood upright; left-handed, he assumed a feral crouch. Right-handed, his swing was compact; left-handed his swing was like a storm, a vicious prairie updraft.

"Brutish," Peavy called it. It had to be to generate the same power that came naturally right-handed. "He had to apply more force with his legs," Peavy said, and he had to work harder to achieve the same results.

"He looked beautiful when he attacked the ball both ways, but there was a decided difference," Osteen said. "He may have been just as aggressive right-handed but not near as violent."

Left-handed hitters tend to lean back in anticipation of the curveballs and sliders righty pitchers deliver down and in. Right-handed hitters tend to stand more upright because the breaking balls they face from right-handed pitchers are higher when they break across the plate. In this regard, Mantle conformed to the norm. His left-handed stance was sharply canted backward, an axis of 18 to 20 degrees, a necessary counterbalance to the vehemence of his swing. "Otherwise he would have fallen over," Peavy said.

Batting right-handed, his axis was only 10 degrees off vertical, and it became more so as his body moved forward in space. His shoulders and hips stayed level, enabling all the momentum he generated to rotate around a solid axis. "Like a barbershop pole," Osteen said.

When everything is working as it should, a batter's

hips rotate at 760 degrees per second. "Think of a spinning top," Peavy said. "When the axis is steady, the top spins like crazy. When the axis starts to wobble, the top rapidly loses speed and efficiency and finally stops spinning."

BAT AND LOWER HAND PATH

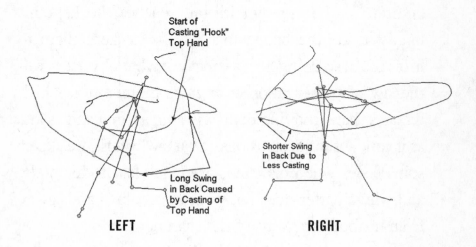

An average major league fastball drops three feet on its 60-foot, 6-inch journey to home plate, descending at a trajectory of 10 to 15 degrees. By swinging upward at approximately the same angle, a hitter increases the window of opportunity for making contact with the ball. That's why coaches preach swinging level to the pitch, not to the ground. The long, flat trajectory visible in the kinetic of Mantle's right-handed bat path meant, in coach speak, that he was "long in the usable plane of the pitch." Peavy estimates his hitting zone was 2 to 3 feet on the right side and 10 to 20 percent less when batting left-handed.

No hooks or loops detoured this right-handed swing. Mantle's hands moved with speed and an economy of motion, traveling through the hitting zone, Peavy estimates, in .11 to .14 seconds. "Tremendous bat speed," Osteen said.

And, in the argot of hitting coaches, he kept his hands inside the ball, which means he kept them to himself. In so doing he was obeying what physicists call the law of conservation of angular momentum, which dictates that a rotating body will spin as long and as fast as it can, absent protrusions. That's why a figure skater spins faster when his hands are held close to his body, and why a batter does not want to cast his hands away from his body the way a fisherman does.

Even the slightest hook diminishes bat speed. In the kinetic of Mantle's left-handed bat path, a slight hook is visible just before the bat dipped down toward the catcher. The looping, lasso-like uppercut meant that his bat was not nearly as long in the plane of the ball. Peavy estimates the extra length added perhaps .2 of a second to the time it took his hands to move through the hitting zone. "That's why he had to start his swing earlier," Peavy said.

And that made him more vulnerable to an off-speed

pitch. "It's like, 'Oh, shit, now I gotta wait for that big-ass curve,'" Peavy said. "And he can't do it because he's already moving his hands for the fastball. It's why pitchers can sometimes make him look stupid."

Of course, they knew that. "That's why we throw slow curves to almost all of the low-ball, back-legging fastball hitters," Osteen said.

STRIDE

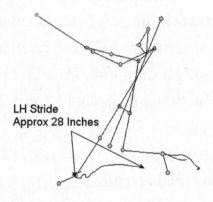

LH Stride
Approx 28 Inches

Mantle's unbridled aggression was best expressed and most visible in the length of his left-handed stride, which measured as much as 28 inches, a stunningly long distance, Peavy says, for his era and for his height. Ken Griffey's stride was approximately half that length. In most mortals, a 2-foot stride would cause a precipitous and calamitous drop in the center of gravity. Mantle minimized that drop through sheer athleticism and balance, moving forward, Peavy says, "like a big cat."

Maintaining a level center of gravity is essential if you want to keep your bat on the plane of the ball. "If your center of gravity dips, your eyes and head dip, making it much harder to put your hands in the right place and to keep your eye on the ball," Peavy said. "What's remarkable is that even with that huge stride, his head stayed relatively level."

Striding forward, Mantle threw all the accumulated force against his locked front leg, which is typical of left-handed power hitters. Those forces are considerable. By way of comparison, Peavy collected force plate data on a college player approximately Mantle's height and weight. His swing generated a force 2½ to 3 times his body weight.

As he shifted his weight from back to front, his hips and shoulders shifted up and down "like a teeter-totter," Osteen says. "I keep centering on the hip bar. Swinging from down to up creates a 45-degree angle in the hips from back to the front. On the left side, his back hip is much, much lower than on the right—I'd say 1 to 2 inches. That shoulder bar is even lower."

That posture made him a formidable low-ball hitter. "You don't want to throw the ball down in that power zone," Osteen said. "Usually you try to keep the fast-ball at belt-buckle height. If you go down, you prefer it to be near the dirt. If it is a breaking ball, it *has* to be in the dirt."

As he moved forward, he lifted his back foot was as much as 3 to 6 inches off the ground, prima facie evidence of a complete transfer of weight onto his front side. To the trained eye of a batting coach, the height of his toe isn't as significant as how far his back foot traveled. "My God, when you're striding 24 inches or more, you gotta bring that back leg with you," Peavy said, and Mantle did, sometimes as much as 1½ feet. In some pictures, you can see a small cloud of dust kicked up by his cleat as his front hip pulled his backside forward.

His right-handed stride was more compact, perhaps 12 to 15 inches, Peavy says, typical of a modern power hitter. It was also kinder to his knee; he didn't extend it fully, or lock it in place. "His back foot was off the ground 1 to 4 inches," Peavy said, evidence of a less vociferous stride. "And it didn't move forward much, if at all."

His shoulders remained on an even keel throught this swing, an indication of where the pitch was—waist high. "He probably doesn't swing at a lot of pitches down," Osteen said. "He would prefer to wait you out and make a mistake around the waist or higher."

Because his posture was upright, his head remained virtually stationary until after he hit the ball. "That's one of the secrets of hitting," Osteen said. "He had a quiet body and a quiet head."

The willful, learned aggression that made him such a fierce left-handed presence also contributed to his undoing. Watching film taken during Mantle's brief tenure with the Kansas City Blues in the summer of 1951, Peavy noted that Mantle's right front knee was slightly hyperextended when he made contact with the ball. That is not unusual for left-handed power hitters. But the torque on Mantle's knee was accentuated by the toe-in position of his lead foot, according to Amy Engelsman, P.T., a sports-certified specialist in Washington, D.C. who has treated many professional athletes. Worse, as he followed through on his swing, his right ankle turned outward, rolling over as you might in a bad sprain. All the weight was forced onto the outside of his foot. "There was no way he could stay completely balanced with his foot planted," Peavy said. "He would have ripped his knee apart."

Two months after that footage was taken, that knee was torn apart during the 1951 World Series, compromising ligaments and tearing insulating cartilage. "With an intact anterior cruciate ligament, your knee can remain stable even with that kind of swing," Engelsman said. "Without the ACL the shearing forces on the knee would have contributed to the breakdown of the damaged cartilage."

Over time, as the joint became more arthritic and less stable, "his swing would have atrophied," Peavy said, and his knee would have buckled as it did so often and so visibly late in his career.

Osteen saw that happen to other injured left-handed power hitters. "That knee won't let you spin the hips to bring your hands to the down and in position," he said. "Then they become vulnerable to being jammed. That's what happened to Mantle."

FINISH

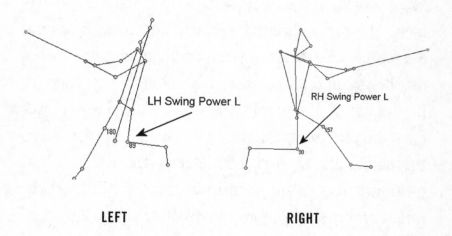

LH Swing Power L

RH Swing Power L

180

89

57

90

LEFT   **RIGHT**

Physicists define the perfect swing as one that facilitates the most complete transfer of energy from batter to bat. By that standard, Mantle was nearly perfect. Alan Nathan, an experimental nuclear particle physicist, studies high-speed collisions between subatomic particles and, in his spare time, collisions between bats and balls. "The ball is turned around in less than one-thousandth of a second," he said. "The collision generates a peak force of just under 10,000 pounds. The ball is compressed like a spring, maybe as much as an inch. It comes to a momentary stop and then expands."

At impact, Mantle's back leg formed what batting coaches sometimes call a "power L." Peavy tells his students that their shoelaces should face the pitcher and their heel should point to the sky. Mantle fit that configuration batting both right-handed and left-handed. His thigh and calf met in a perfect right angle. He looked like a supplicant, bowing to the pitcher's mound. Sometimes, his knee was no more than eight inches off the ground. "The back foot is like a gas gauge for hip rotation," Peavy said. "If you see the classic power L, then you can guarantee they have pulled the hip through as forcibly and completely as they can."

Mantle was nothing but forceful; he never compromised on his swing. Nowhere is this more visible than in an iconic photograph taken on October 8, 1961, when he had to leave game 4 of the World Series because of a gaping wound in his hip. The photograph reveals none of the particulars of the moment: not the inning (fourth), not the score (0–0), not the blood oozing through the protective padding on his hip. In the background the diffuse home crowd fills bunting-draped seats. Mantle occupies the foreground, frozen in time at the end of an empty swing. The futility of the effort is implicit in his posture and his facial expression. His chin is raised as if by a grimace. His jersey is creased by exertion, the full rotation of his hips and torso.

To a classicist, his form evokes Discobolos, the discus thrower of ancient Greece, probably the most famous athletic statue in history. Rendered originally in bronze in 460–450 B.C. by the sculptor Myron, the statue captures the fluidity of recent motion, an athlete who has just arrived at his current position. The similarity to the 1961 photograph is telling because it

suggests Mantle's classical form, one reason for his enduring hold on the imagination.

To a batting coach, the picture is evidence of the harnessing of all his potential power. "He wrung every pound of force out of that body," Peavy said.

To a pitching coach, it is a portrait of balance and depletion. "Whether he was fooled by a changeup and swung and missed or maybe popped one up, it's obvious that he was in pain," Osteen said.

But, he said, Mantle still aspired to "the perfect form that the picture shows. Normally when stars have an injury or pain, they change their form to alleviate the discomfort, but this picture shows that he was still trying to swing the way he always did."

APPENDIX 3: WHO'S BETTER?

MICKEY CHARLES MANTLE

Born: October 20, 1931, Spavinaw, Oklahoma
Died: August 13, 1995, Dallas, Texas
Buried at Sparkman-Hillcrest Memorial Park, Dallas,
    Texas (Mausoleum–St. Mark NE-N-C-13-A)

First Game: April 17, 1951; Final Game: September 28,
    1968
Bat: Both
Throw: Right
Height: 5' 11.5"
Weight: 195
Selected to the Hall of Fame in 1974
Named AL Most Valuable Player by Baseball Writers'
    Association of America (1956 to 1957 and 1962)

Named Major League Player of the Year by *The Sporting News* (1956)

Named AL Player of the Year by *The Sporting News* (1956 and 1962)

Named outfielder on *The Sporting News* Major League All-Star Team (1952 and 1956 to 1957)

Named outfielder on *The Sporting News* AL All-Star Team (1961 to 1962 and 1964)

Won AL Gold Glove as outfielder (1962)

Ejections as player: 1954 (1), 1957 (1), 1958 (2), 1964 (1), 1965 (1), 1968 (1). Total: 7

CAREER TOTALS: 18 YEARS

G	AB	R	H	2B	3B	HR	RBI	BB	IBB	SO	HBP	SH	SF	XI	ROE	GDP	SB	CS	AVG	OBP	SLG	BFW
2401	8102	1677	2415	344	72	536	1509	1733	144	1710	13	14	47	0	104	113	153	38	.298	.421	.557	71.8

Source: Retrosheet

When traditionalists compare Mays and Mantle, they use the old math: batting average (.302 vs. 298), RBIs (1,903 in 22 years vs. 1,509 in 18 years), and home runs (660 in 2,992 games vs. 536 in 2,401 games). Both hit .300 or better ten times; both hit more than 50 homers in a season twice; both finished their careers with a .557 slugging percentage; both fared well in Branch Rickey's measure of Isolated Power, with Mantle slightly higher than Mays (.256 vs. .259).

Very few boys on New York City street corners

bragged about how many more times The Mick walked. In ten seasons, he walked 100 times or more (Mays did that once). Or the number of times he grounded into double plays (113), half as many as Mays (251). But to the trained eye of a modern stat geek, walks are the key to Mantle's superior on-base percentage and the reason he fares so well in a preponderance of the new offensive metrics.

Mantle's lifetime batting average was much higher right-handed than left-handed (.329 compared to .275). But his on-base percentage was almost identical (.432 right-handed, .422 left-handed). His OPS is staggering from both sides of the plate—1.014 right-handed and .964 left-handed. (The major league average in 2008 was around .760.) By this standard, Mantle ranks twelfth in baseball history, ahead of Joe DiMaggio, Willie Mays, and Hank Aaron. In eighteen years in the major leagues, Mantle put 6,392 balls into play; 536 of them—or 8.4 percent—were home runs.

Bill James was a security guard at the Stokely Van Camp pork and beans factory in Kansas when he pioneered a formula for runs created (RC = total bases * [(hits + walks)/plate appearances]) that assessed credit for each run produced. Thirty years later, *Time* magazine named him one of the 100 most influential people in the world and the Boston Red Sox put him to work in their front office. In 2001, he unveiled a new formula for "win shares," an extrapolation of runs created that calculates a player's contribution to every victory. This system compares players at different positions as well as players of different eras, enabling fantasy baseball to expand into uncharted hypothetical territory. According to this calculus, Mantle should have been the Most Valuable Player nine times, not the three times he actually won the award. He led (or tied) the American League in win shares every year from 1954 to 1964, except 1963, when he played only sixty-five games. When Cyril Morong, an economist turned sabermatrician, extrapolated win shares per at-bat, Mantle finished second behind Ruth.

If the Bill James baseball abstracts are the sabermetric equivalent of the Old Testament, then Pete Palmer's 1984 *Hidden Game of Baseball* is the New Testament. Palmer's work began in the 1960s, when he stayed after work at the Raytheon Company, using the computer to develop a system of linear weights that assigned a value to each of the seven possible outcomes of an at-bat. It was a breakthrough that precipitated an entirely new way of assessing baseball performance. He pioneered total player rankings and, with co-author, John Thorn, devised a metric called on-base-plus slugging (OPS = SLG + OPB), which measured a player's ability to get on base and hit for power. Topps began putting OPS stats on the back of its baseball cards in 2004. Palmer's next evolutionary step was "batter-fielder wins" (BFW), a calculus for establishing the number of wins over (or under) what an average player would contribute to his team with his batting, baserunning, and fielding. (Retrosheet now includes BFW among other career totals.) Palmer credits Mantle with a total 71.8 BFW over the course of his career, meaning that he was responsible for nearly 72 additional wins beyond what a league-average performer at his position would have contributed. In the world of BFWs, two games per season are significant. In 1955 and 1956, Mantle is responsible for 8 BFW or better; in 1961, he is credited with 7.5 BFW. Three other times, he rated over 5 BFW. Mays, on the other hand, has a career total of 84.4 BFW, in part a reflection of his longevity. But he never had a single-season BFW rating over 7, though he was over 5 on eight other occasions.

Clay Davenport, a weather scientist at the National Oceanic and Atmospheric Administration, and chief statistician for Baseball Prospectus, spent a decade devising an equation for equivalent average, a metric that measures total offensive value per out with corrections for league offensive production, home ballpark, and team pitching. His most radical innovation was to translate his metrics into traditional baseball numbers, making the new math accessible to old fans. Mantle's translated EqA batting average in Davenport's system is .316, four points higher than Mays's.

	MANTLE	MAYS
EqA:	.340	.328
Black Ink:	65	57
OBP:	.421	.384
OPS:	12th all-time	30th all-time
OPS+:	6th all-time	19th all-time
RC:	2039	2368
RC/G:	9.3	7.9
RCAA:	7th all-time	11th all-time
RCAP:	6th all-time	10th all-time
TPRf:	7th all-time	9th all-time

BIBLIOGRAPHY

I am indebted to the research staff at the National Baseball Hall of Fame and Museum in Cooperstown, New York, for providing a copy of its Mantle file, containing newspaper and magazine stories, personal recollections, fan letters, poems, and cartoons. Some of the material was unattributed or undated or incomplete; some stories lacked volume and page numbers. I have used only those whose content I could verify. The list below, containing articles from multiple sources, includes only those I relied on extensively.

The following museums and archives provided invaluable primary source material: the Everett J. Ritchie Tri-State Mining Museum, Joplin, Missouri; the Heritage Center and Museum, Baxter Springs, Kansas; the Historical Society of Washington, Washington, D.C.; L.E.A.D. Agency, Inc., Vinita, Oklahoma; the LeDroit Park Civic Association, Washington, D.C.; the Library of Congress, Washington, D.C.; the *New England Journal of Medicine*; the New York Times Article Archive; the Oklahoma Geological Survey at the University of Oklahoma, Norman, Oklahoma; ProQuest; the University of Southern California film archive; the Washington Post Historical Archive.

Thanks to Maybell Bennett, the Howard University Community Association; Nick Dolin, HBO; Brad Garrett, Brad Garrett Investigations; C.C. Livingston, Howard University Hospital; Rhonda Schwartz, ABC News; Ted Spencer, Tim Wiles, and Russell Wolinsky, the National Baseball Hall of Fame; and Jason Zillo, New York Yankees.

Every game and career statistic for Mantle was verified by Dave Smith of Retrosheet.

BOOKS

Adair, Robert K. *The Physics of Baseball.* 3rd ed. New York: Perennial, 2002.

Allen, Maury. *Memories of the Mick.* Dallas: Taylor Publishing, 1997.

Anderson, Dave, ed. *The Red Smith Reader.* New York: Random House, 1982.

Arnold, Eve, et al. *The Fifties: Photographs of America.* New York: Pantheon, 1985.

Astor, Gerald. "Eyes Open, Mouth Shut." MS.

Baggelaar, Kristin. *The Copacabana*. Images of America. Charleston, S.C.: Arcadia Publishing, 2006.

Bahill, Terry A., and David G. Baldwin. "Mechanics of Baseball Pitching and Batting" in *Applied Biomedical Engineering Mechanics,* by Dhanjoo N. Ghista. Boca Raton, Fla.: CRC Press, 2009.

Baldwin, Dave. *Snake Jazz*. Philadelphia: Xlibris, 2007.

Barra, Allen. *Clearing the Bases: The Greatest Baseball Debates of the Last Century*. New York: Thomas Dunne Books, 2002.

———. *Yogi Berra: Eternal Yankee*. New York: W. W. Norton and Company, 2009.

Beckett, James. *Mickey Mantle*. Beckett Sports Heroes. New York: House of Collectibles, 1995.

Berkow, Ira. *Red: The Life and Times of a Great American Writer*. New York: Times Books, 1986.

Borelli, Stephen. *How About That!: The Life of Mel Allen*. Champaign, Ill.: Sports Publishing, 2005.

Bouton, Jim, and Leonard Schecter, ed. *Ball Four: The Final Pitch*. North Egremont, Mass.: Bulldog Publishing, 2000.

Breslin, Jimmy. *Can't Anybody Here Play This Game?* New York: Ballantine Books, 1970.

Brosnan, Jim. *The Long Season*. Chicago: Ivan R. Dee, 2002.

Burkard, Tom. *The Ultimate Mickey Mantle Trivia Book: A Citadel Quiz Book*. New York: Citadel Press, 1997.

Carrieri, Joseph R. *Searching for Heroes: The Quest of a Yankee Batboy.* Mineola, N.Y.: Carlyn Publications, 1995.

Castro, Tony. *Mickey Mantle: America's Prodigal Son.* Washington, D.C.: Brassey's Inc., 2002.

Ceresi, Frank, Mark Rucker, and Carol McMains. *Baseball in Washington, D.C.* Charleston, S.C.: Arcadia Publishing, 2002.

Click, James, and Jonah Keri. *Baseball Between the Numbers: Why Everything You Know About the Game Is Wrong.* New York: Basic Books, 2006.

Cohn, Beverly, and Laurie Cohn. *1951: What a Year it Was!* Los Angeles: MMS, 2000.

Courtois, Christine A. *Healing the Incest Wound: Adult Survivors in Therapy.* New York: W. W. Norton and Company, 1996.

———. *Recollections of Sexual Abuse: Treatment Principles and Guidelines.* New York: W. W. Norton and Company, 1999.

Cramer, Richard Ben. *Joe DiMaggio: The Hero's Life.* New York: Simon and Schuster, 2000.

Cramer, Richard Ben, and Glenn Stout, eds. *The Best American Sports Writing 2004.* The Best American Series. Boston: Houghton Mifflin, 2004.

Creamer, Robert W. *Mantle Remembered: Stories Excerpted from the Pages of* Sports Illustrated. New York: Warner Books, 1995.

————. *Stengel: His Life and Times.* New York: Simon and Schuster, 1984.

Duncan, David James. *The Mickey Mantle Koan.* Tuscaloosa, Ala.: Urban Editions, 2004.

Duren, Ryne, with Robert Drury. *The Comeback: The Story of Ryne Duren.* Dayton, Ohio: Lorenz Press, 1978.

Duren, Ryne, and Tom Sabellico. *I Can See Clearly Now: Ryne Duren Talks from the Heart about Life, Baseball, and Alcohol.* Chula Vista, Calif.: Aventine Press, 2003.

Einstein, Charles. *Willie's Time: Baseball's Golden Age.* Writing Baseball Series. Carbondale: Southern Illinois University Press, 2004.

Falkner, David. *The Last Hero: The Life of Mickey Mantle.* New York: Simon and Schuster, 1995.

Ford, Whitey, Mickey Mantle, and Joseph Durso. *Whitey and Mickey: A Joint Autobiography of the Yankee Years.* New York: The Viking Press, 1977.

Friends and Fans of Mickey Mantle. *Letters to Mickey.* New York: HarperCollins, 1995.

Gallagher, Mark, and Paul E. Susman, research associate. *Explosion: Mickey Mantle's Legendary Home Runs.* New York: Arbor House, 1987.

Gartner, Richard B. *Betrayed as Boys: Psychodynamic Treatment of Sexually Abused Men.* New York: Guilford Press, 1999.

Gershman, Michael. *Diamonds: The Evolution of the Ballpark.* Boston: Houghton Mifflin, 1995.

Gibson, Arrell M. *Wilderness Bonanza: The Tri-State*

District of Missouri, Kansas, and Oklahoma. Norman: University of Oklahoma Press, 1972.

Gifford, Frank, and Harry Waters. *The Whole Ten Yards.* New York: Random House, 1993.

Gittleman, Sol. *Reynolds, Raschi, and Lopat: New York's Big Three and the Great Yankee Dynasty of 1949–1953.* Jefferson, N.C.: McFarland and Company, 2007.

Gumbrecht, Hans Ulrich. *In 1926: Living at the Edge of Time.* Cambridge, Mass.: Harvard University Press, 1997.

Halberstam, David. *The Fifties.* New York: Villard Books, 1993.

——. *October 1964.* New York: Fawcett Books, 1995.

Hall, John G. *Majoring in the Minors: A Glimpse of Baseball in a Small Town.* Stillwater: Oklahoma Bylines, 2000.

——. *The KOM League Remembered.* Charleston, S.C.: Arcadia Publishing, 2004.

——. *Mickey Mantle: Before the Glory.* Leawood, Kans.: Leathers Publishing, 2005.

Hersh, Seymour M. *The Dark Side of Camelot.* Boston: Little, Brown, and Company, 1997.

Herskowitz, Mickey. *Mickey Mantle: An Appreciation.* New York: William Morrow and Company, 1995.

Herskowitz, Mickey, Danny Mantle, and David Mantle. *Mickey Mantle: Stories and Memorabilia from a Lifetime with The Mick.* New York: Stewart, Tabori and Chang, 2006.

Hirsch, James S. *Willie Mays: The Life, the Legend.* New York: Scribner, 2010.

Holtzman, Jerome. *No Cheering in the Press Box.* New York: Holt, Rinehart and Winston, 1974.

Honig, Donald. *Mays, Mantle, Snider: A Celebration.* New York: Macmillan Publishing Company, 1987.

James, Bill. *The Bill James Baseball Abstract, 1984.* New York: Ballantine Books, 1984.

———. *The Bill James Baseball Abstract, 1985.* New York: Ballantine Books, 1985.

James, Bill, and Jim Henzler. *Win Shares.* Morton Grove, Ill.: STATS Publishing, 2002.

Jamieson, Dave. *Mint Condition: How Baseball Cards Became an American Obsession.* New York: Atlantic Monthly Press, 2010.

Kaat, Jim, and Phil Pepe. *Still Pitching: Musings from the Mound and the Microphone.* Chicago: Triumph Books, 2003.

Kandel, Eric R. *In Search of Memory: The Emergence of a New Science of Mind.* New York: W. W. Norton and Company, 2006.

Kandel, Eric R., James H. Schwartz, and Thomas M. Jessell. *Principles of Neural Science.* 4th ed. New York: McGraw-Hill, Health Professions Division, 2000.

Kandel, Eric R., and Larry R. Squire. *Memory from Mind to Molecules.* Greenwood Village, Colo.: Roberts and Co., 2009.

Kravetz, Robert. *Where Have You Gone, Mickey Mantle?:*

A Collection of Stories About Times Past and America's Pastime. Pittsford, N.Y.: GK Creations, 1996.

Levi, Vicki Gold, Lee Eisenberg, Rod Kennedy, and Susan Subtle. *Atlantic City, 125 Years of Ocean Madness: Starring Miss America, Mr. Peanut, Lucy the Elephant, the High Diving Horse, and Four Generations of Americans Cutting Loose.* New York: Clarkson N. Potter, 1979.

Liederman, Bill. *Mickey Mantle's: Behind the Scenes in America's Most Famous Sports Bar.* Guilford, Conn.: Lyons Press, 2007.

Liederman, Bill, and Maury Allen. *Our Mickey: Cherished Memories of an American Icon.* Chicago: Triumph Books, 2004.

Lipsyte, Robert, and Peter Levine. *Idols of the Game: A Sporting History of the American Century.* Atlanta: Turner Publishing, 1995.

Lowry, Philip J. *Green Cathedrals: The Ultimate Celebration of Major League and Negro League Ballparks.* New York: Walker and Company, 2006.

Madden, Bill. *Bill Madden: My 25 Years Covering Baseball's Heroes, Scoundrels, Triumphs and Tragedies.* Champaign, Ill.: Sports Publishing, 2004.

———. *Pride of October: What It Was to Be Young and a Yankee.* New York: Warner Books, 2003.

Mantle, Merlyn, Mickey E. Mantle, David Mantle, and Dan Mantle. *A Hero All His Life.* New York: HarperCollins, 1996.

Mantle, Mickey. *The Education of a Baseball Player.* New York: Simon and Schuster, 1967.

———. *Mickey Mantle Auction Catalog.* New York: Guernsey's, 2003.

Mantle, Mickey, and Robert W. Creamer. *The Quality of Courage.* Lincoln, Neb.: University of Nebraska Press, 1999.

Mantle, Mickey, and Ben Epstein. *The Mickey Mantle Story.* New York: Henry Holt, 1953.

Mantle, Mickey, and Herb Gluck. *The Mick.* New York: Jove, 1986.

Mantle, Mickey, and Mickey Herskowitz. *All My Octobers: My Memories of Twelve World Series When the Yankees Ruled Baseball.* New York: Avon/Morrow, 2006.

Mantle, Mickey, and Phil Pepe. *My Favorite Summer, 1956.* New York: Island Books, 1991.

Marx, Jeffrey. *It Gets Dark Sometimes: My Sister's Fight to Live and Save Lives.* Washington, D.C.: JAM Publishing, 2000.

Mays, Willie, and Lou Sahadi. *Say Hey: The Autobiography of Willie Mays.* New York: Simon and Schuster, 1988.

Mazur, Gail. "Baseball." In *Zeppo's First Wife: New and Selected Poems.* Chicago: University of Chicago Press, 2005.

Neyer, Rob. *Rob Neyer's Big Book of Baseball Lineups: A Complete Guide to the Best, Worst, and Most Memo-*

rable Players to Ever Grace the Major Leagues. New York: Fireside, 2003.

Nieberding, Velma. *The History of Ottawa County.* Miami, Okla.: Walsworth Publishing Company, 1983.

Nuttall, David S. *Mickey Mantle's Greatest Hits.* New York: S.P.I. Books, 1998.

Okrent, Daniel, and Steve Wulf. *Baseball Anecdotes.* New York: HarperPerennial, 1990.

Parrott, Harold. *The Lords of Baseball: A Wry Look at a Side of the Game the Fan Seldom Sees—the Front Office.* Atlanta: Longstreet Press, 2001.

Podell-Raber, Mickey, and Charles Pignone. *The Copa: Jules Podell and the Hottest Club North of Havana.* New York: Collins, 2007.

Prudenti, Frank. *Memories of a Yankee Batboy 1956–1961.* Bronx, N.Y.: Pru Publishing, 2003.

Ritter, Lawrence S. *East Side, West Side: Tales of New York Sporting Life, 1910–1960.* New York: Total Sports, 1998.

———. *Lost Ballparks: a Celebration of Baseball's Legendary Fields.* New York: Viking Studio Books, 1992.

Robinson, Ray, and Christopher Jennison. *Yankee Stadium: Drama, Glamour, and Glory.* New York: Viking Studio Books, 1998.

Rosenfeld, Harvey. *Still a Legend: The Story of Roger Maris.* Lincoln, Neb.: IUniverse, 2002.

Rosner, David, and Gerald E. Markowitz. *Deadly Dust:*

Silicosis and the Politics of Occupational Disease in Twentieth-Century America. Princeton, N.J.: Princeton University Press, 1991.

Schaap, Dick. *Mickey Mantle, the Indispensable Yankee.* New York: Bartholomew House, 1961.

Shaffer, Benjamin. *Shoulder Problems in Athletes: An Issue of Clinics in Sports Medicine.* Philadelphia: Saunders, 2008.

Shannon, Bill, and George Kalinsky. *The Ballparks.* New York: Hawthorn Books, 1975.

Simon, Bryant. *Boardwalk of Dreams: Atlantic City and the Fate of Urban America.* New York: Oxford University Press, 2004.

Smalling, Jack. *The Baseball Autograph Collector's Handbook.* Durham, N.C.: Baseball America, 2001.

Smith, Curt. *The Voice: Mel Allen's Untold Story.* Guilford, Conn.: Lyons Press, 2007.

Smith, Marshall, and John Rohde. *Memories of Mickey Mantle: My Very Best Friend.* Bronxville, N.Y.: Adventure Quest, 1996.

Smith, Ralph Lee. *At Your Own Risk: The Case Against Chiropractic.* New York: Trident Press, 1969.

Smith, Red. *Red Smith on Baseball: The Game's Greatest Writer on the Game's Greatest Years.* Chicago: Ivan R. Dee, 2000.

———. *Out of the Red.* New York: Knopf, 1950.

Smith, Ron. *61* The Story of Roger Maris, Mickey Mantle*

and One Magical Summer. St. Louis: Sporting News, 2002.

Snider, Duke, and Bill Gilbert. *The Duke of Flatbush.* New York: Zebra Books, 1988.

Snyder, Brad. *Beyond the Shadow of the Senators: The Untold Story of the Homestead Grays and the Integration of Baseball.* Chicago: Contemporary Books, 2003.

Sokolic, William H., and Robert E. Ruffolo. *Atlantic City Revisited.* Charleston, S.C.: Arcadia Publishing, 2006.

Spatz, Lyle, ed. *The SABR Baseball List and Record Book: Baseball's Most Fascinating Records and Unusual Statistics.* New York: Scribner, 2007.

Stout, Glenn, and Richard A. Johnson, ed. *Yankees Century: 100 Years of New York Yankees Baseball.* Boston: Houghton Mifflin, 2002.

Suggs, George G. *Union Busting in the Tri-State: The Oklahoma, Kansas, and Missouri Metal Workers' Strike of 1935.* Norman, Okla.: University of Oklahoma Press, 1986.

Swearingen, Randall. *A Great Teammate: the Legend of Mickey Mantle.* Champaign, Ill.: Sports Publishing, 2007.

Sweet, Ozzie, and Larry Canale. *Mickey Mantle: The Yankee Years, the Classic Photography of Ozzie Sweet.* Richmond, Va.: Tuff Stuff Publications, 1998.

Thorn, John, ed. *Glory Days: New York Baseball 1947–1957.* New York: HarperCollins, 2008.

Vancil, Mark, and Alfred Santasiere. *Yankee Stadium: The Official Retrospective*. New York: Pocket Books, 2008.

Vincent, Fay. *We Would Have Played for Nothing: Baseball Stars of the 1950s and 1960s Talk about the Game They Loved*. New York: Simon and Schuster, 2008.

White, E. B. *Here Is New York*. New York: Little Bookroom, 1999.

Williams, Pete. *Card Sharks: How Upper Deck Turned a Child's Hobby into a High-Stakes, Billion-Dollar Business*. New York: Macmillan General Reference, 1995.

Wright, Doug, *I Am My Own Wife: Studies for a Play about the Life of Charlotte von Mahlsdorf*. New York: Faber and Faber, Inc., 2004.

ARTICLES/JOURNALS

Altman, Lawrence K. "Defending Tough Decisions in a Case Open to Hindsight." *New York Times*, August 15, 1995, Doctor's World section.

Andersen, Richard. "The Dust Fields Behind Us." *Elysian Fields Quarterly* 24, no. 1 (2007).

Anson, Robert S. "Playing with the Big Leaguers." *Philip Morris Magazine* (March/April 1989).

Astor, Gerald. "Mickey Mantle: Oklahoma to Olympus." *Look*, February 2, 1965.

Barringer, Felicity. "Despite Cleanup at Mine, Dust and

Fear Linger." *New York Times*, April 12, 2004, National section.

Barzlai, Peter, Stephen Borelli, and Gabe Lacques. "Yankee Stadium from Opening Day to the Last Day at the House that Ruth Built." *USA Today*, spring 2008, Sports Weekly Keepsake edition.

"Baseball Stars Top Poll of 'Trusted' Endorsers." *Advertising Age*, December 1973.

Bingham, Walter. "Assault on the Record." *Sports Illustrated*, July 31, 1961: 8–11.

———. "Double M for Murder." *Sports Illustrated*, July 4, 1960: 10–13.

———. "The Yankees' Desperate Gamble." *Sports Illustrated*, July 2, 1962: 10–13.

Broeg, Bob. "Muscle Plus Speed—Yankee Legend Mantle." *The Sporting News*, August 21, 1971.

Brooke, Holly. " . . . I Own 25% of Mickey Mantle . . . and There Were Times When He Was Mine—100%." *Confidential*, March 1957.

Callahan, Tom. "Willie, Mickey, and Nathan Detroit." *Time*, April 1, 1985: 86.

Cava, Pete. "The Perfect Day." *Elysian Fields Quarterly* 23, no. 4 (2006).

Chopra, Sanjiv. "Clinical Features and Natural History of Hepatitis C Virus Infection." UpToDate Inc.: http://www.uptodate.com/patients/content/topic.do?topicKey=~/mm QjpererXWw4 (accessed April 1, 2010).

Cinque, Chris. "Mantle Still Grappling with Old Knee Injuries." *Physician and Sportsmedicine* 17, no. 6 (June 1989).

Coleman, Jennifer, and Merrie Spaeth. "Transplanting the Mick's Liver." *Public Relations Strategist*, September 1996: 51–55.

Creamer, Robert W. "For the Want of a Warning a Pennant Was Lost." *Sports Illustrated*, June 17, 1963: 68–71.

———. "Mantle and Maris in the Movies." *Sports Illustrated*, April 2, 1962: 88–100.

———. "The Mantle of the Babe." *Sports Illustrated*, June 18, 1956: 11–14.

Daniel, Dan. "Blasé Broadway Buzzing over Maris, Mantle HRs." *The Sporting News*, August 23, 1961: 7.

———. "Broadway Busting Buttons over Bomber Thrill Show." *The Sporting News*, September 13, 1961: 9.

———. "Mantle Given Rare Salute on Yankee West Point Visit." *The Sporting News*, April 26, 1961: 9.

———. "Ushers Will Protect Mick from Mauling Fans." *The Sporting News*, June 8, 1960: 14.

Deford, Frank. "Hot Pitchmen in the Selling Game." *Sports Illustrated*, November 17, 1969: 110–20.

Donnelly, Joe. "Last of the Ninth for Mickey?" *Newsday* (New York), April 2, 1996.

Eisenhauer, Kelly. "The Complete Collectibles Guide, Mickey Mantle, A Portrait of a Young Mickey Mantle." *Sports Collectors Digest*, September, 18, 2009: 24–27, 30, 32–33.

Epstein, Ben. "What Manner of Man Is Mantle?" *Look,* July 24, 1956: 26–31.

Fair, Ray C., and Danielle Catambay. "Branch Rickey's Equation Fifty Years Later." Yale University. Cowles Foundation for Economic Research. http://cowles. econ.yale.edu/P/cd/d15a/d1529.pdf (accessed April 1, 2010).

Fay, Bill. "Cancer Quacks." *Collier's,* May 26, 1951.

Fimrite, Ron. "Mantle and Mays." *Sports Illustrated,* May 26, 1985: 70.

Frank, Stanley. "Boss of the Yankees." *Saturday Evening Post,* April 16, 1960: 31–34.

Furlong, William B. "Mickey, the Oklahoma Yankee." *True,* April 1953: 46.

Garvey, Alfred. "These Yankees Had a Ball." *Confidential,* September 1957.

Gentry, Layne O. "Osteomyelitis: Clinical Features." ACP Medicine Online (2008): http://www.acpmedicine.com/ acp/chapters/CH0716.htm.

Graham, Ed. "On the Trail of a Hero." *Sports Illustrated,* August 26, 1963: 50–55.

Graham, Jr., Frank. "The Trouble with Mantle." *SPORT,* October 1958: 6–7.

Gross, Milton. "Mickey Mantle: New Pride of the Yankees." *SPORT,* April 1953: 35.

———. "Nobody Tries Harder Than Mantle." *SPORT,* April 1961.

Hannon, Kent. "Beers with . . . Mickey Mantle." *SPORT*, February 1991.

Hano, Arnold. "Mickey Mantle: The Twilight of a Hero." *SPORT*, August 1965: 66–73.

Haupert, Michael J. "The Economic History of Major League Baseball." *EH.Net Encyclopedia*, edited by Robert Whaples. December 3, 2007: http://eh.net/ency clopedia/article/haupert.mlb.

Herskowitz, Mickey. "Mickey Mantle Is, Gulp, 50." *Gadsen Times*, August 15, 1982, Family Weekly section.

"Historic Homer." *Time*, April 27, 1953: 76.

Hoffer, Richard. "Mickey Mantle: The Legacy of the Last Great Player on the Last Great Team." *Sports Illustrated*, August 21, 1995: 18–30.

Holland, Gerald. "All Hail the Hero Mighty Mickey." *Sports Illustrated*, March 4, 1957: 52–60.

"If It Isn't One M It's Another." *Sports Illustrated*, July 8, 1963: 10–15.

Jerome, Richard. "Courage at the End of the Road." *People*, August 28, 1995.

Joiner, Andrea, ed. *Ottawa Co. Emporium Historical Newsletter*, June 1997.

Kahn, Roger. "Inside the Clubhouse: What the Yankees Think of Mickey Mantle." *SPORT*, June 1995: 16–21.

———. "Pursuit of No. 60: The Ordeal of Roger Maris." *Sports Illustrated*, October 2, 1961: 22–25, 70–72.

Katz, Harry. "Double Play." *Smithsonian,* October 2009: 7–8.

Keating, Frank. "Governor Frank Keating's Tar Creek Superfund Task Force Final Report." Office of the Secretary of the Environment. Oklahoma City, Okla.: October 1, 2000.

Keheley, Ed. "History of the Picher Mining Field." Reading, South-Central Section 40th Annual Meeting, Norman, Oklahoma, March 7, 2006. In *Abstracts with Programs.* 1st ed., vol. 38. Geological Society of America, 2006: 32.

Keheley, Ed, and Mary Ann Pritchard. "Report to Governor Keating's Tar Creek Superfund Task Force by the Subsidence Subcomittee." Report, 2000: http://www.deq.state.ok.us/lpdnew/Tarcreek/GovrTaskForce/SubsidenceFinalReport.pdf (accessed April 6, 2010).

King, Randall W., and David Johnson. "Osteomyelitis: eMedicine Emergency Medicine." EMedicine Medical Reference. http://emedicine.medscape.com/article/785020-overview (accessed April 6, 2010).

Kolata, Gina. "An Experimental Plan for Mantle." *New York Times,* June 8, 1995, Sports Desk section.

Koppett, Leonard. "Mays vs. Mantle: A Comparison; Injuries to Yankee over Years Give Edge to Giant." *New York Times,* June 12, 1965, Sports section.

Leavy, Jane. "Red: He Juggled for Us." *Village Voice,* November 20, 1978.

Linn, Ed. "If You Were Mickey Mantle." *SPORT,* August 1957.

―――. "The Last Angry Old Man." *Saturday Evening Post,* July 31, 1955.

Lipsyte, Robert. "How We Learned to Start Worrying and Hate the Bomb: Mickey Mantle, Barry Bonds, and the Bad Boys of Summer." TomDispatch.com. Web log entry posted May 22, 2007": http://www.tomdispatch. com/blog/174787/tomgram:_robert_lipsyte_on_the_ home run_wars.

"The *Look* All America Baseball Team." *Look,* October, 1954: 85.

"Look Who Switched to Natural Light." *Sports Illustrated,* May 5, 1980: 75.

Luza, Kenneth V. "A Study of Stability Problems Associated with Abandoned Underground Mines in the Picher Field, Northeast Oklahoma." Oklahoma Geological Survey/University of Oklahoma, Circular 88, June 8, 2005.

Mantle, Mickey. "The Bravest Man I've Ever Known." *SPORT,* December 1964: 22–25.

―――. "I Remember." *SPORT* , December 1967: 44–51.

―――. "A Year I'll Never Forget." Compiled by Charles Dexter. *SPORT,* December 1951: 8–9, 82–84.

Mantle, Mickey, and Gerald Astor. "Mickey Mantle's Decision." *Look,* March 18, 1969: 28–33.

Mantle, Mickey, and Jill Lieber. "Time in a Bottle." *Sports Illustrated,* April 18, 1994: 66–77.

"Mantle Fans Mays, Mays Fans Mantle." *Esquire,* August 1968: 46–47.

"Mantle of Greatness." *Time,* March 14, 1969: 52.

"Mantle, the Man and His Memorabilia." *Tuff Stuff's Sports Collectors Monthly,* December 2009.

"Math Muscles in on the Race Against Ruth." *Life,* August 18, 1961: 62–65.

Maule, Tex. "Yes, There Is a New Mantle: Healthy, Relaxed and at Last on Good Terms with the World." *Sports Illustrated,* April 17, 1961: 22–23.

McDermott, John R. "Last Innings of Greatness." *Life,* July 30, 1965: 46B–53.

Meany, Tom. "As Casey Stengel Sees Mickey Mantle." *Collier's,* July 20, 1956: 74–77.

———. "That Man Mantle." *Collier's,* June 2, 1951: 24.

———. "Wham! Woosh—Mantle's Away!" *Collier's,* July 4, 1953: 42–44.

Medical News. *Mickey Mantle Transplant: Battling Perceptions of Preferential Treatment.* Ardmore, Pa. Standish Publishing Company, 1996.

Meissner, Bill. "Mickey Mantle's Last Dream." *Elysian Fields Quarterly* 19, no. 1 (2002): 39.

"Mick and the Babe." *Time,* August 27, 1956: 52.

"Mickey Mantle Steps Up to Bat for Newly Released NSAI." *Sports Medicine News,* September 1988.

Millstein, Gilbert. "Case History of a Rookie." *New York Times Magazine,* June 3, 1951: 23.

Moss, Robert A. "Radio Days and the Boys of Summer."

Minneapolis Review of Baseball 9, no. 3 (Summer 1990).

"Music: Yankee Parsifal." *Time*, August 4, 1961: 78.

Nathan, Alan M. "Dynamics of the Baseball-Bat Collision." *American Journal of Physics* 68, no. 11 (November 2000): 979–90: http://webusers.npl.illinois.edu/~a-nathan/pob/AJP-Nov2000.pdf.

Neuberger, John S., and Joseph G. Hollowell. "Lung Cancer Excess in an Abandoned Lead-Zinc Mining and Smelting Area." *The Science of the Total Environment* 25, no. 3, (November 1982): 287–94: http://www.sciencedirect.com.

Neuberger, John S., and Stephen C. Hu et al. "Potential Health Impacts of Heavy-Metal Exposure at the Tar Creek Superfund Site, Ottawa County, Oklahoma." *Environmental Geochemistry and Health* 31, no. 1 (February 2009): 47–59: http://www.Springerlink.com.

Neuberger, John S., and Margaret Nuthall et al. "Health Problems in Galena, Kansas, A Heavy Metal Mining Superfund Site." *The Science of the Total Environment* 94, no. 3 (May 15, 1990): 261–72: http://www.sciencedirect.com.

Newman, Joel S. "Baseball Autographs." *Tax Notes*, vol. 116, no. 12 (September 17, 2007): Wake Forest University Legal Studies Paper No. 1084841: 1078–88.

O'Connell, T. S. "Mickey Mantle: The King of Cards: No Player Has Impacted the Card Hobby More Than Mick." *Sports Cards*, November 1995.

————. "The Mick Like No Other Mantle: The Man and His Memorabilia." *Sports Collectors Monthly*, December 2009.

Olsen, Jack. "The Week They Try to Catch The Babe." *Sports Illustrated*, September 11, 1961, 18–21.

"One More Trial for the Great Mick." *Life*, March 31, 1967: 50–53.

Povich, Shirley. "Mickey Mantle Incorporated." *Saturday Evening Post*, February 2, 1957: 19–25.

"A Prodigy of Power: Mickey Mantle Comes of Age as a Slugger." *Life*, June 25, 1956: 97–102.

Reichler, Joe. "Why It's Now or Never for Mickey Mantle." *Baseball Stars Magazine*, 1956: 12–14.

Rickey, Branch. "Goodbye to Some Old Baseball Ideas." *Life*, August 2, 1954: 78–89.

Riger, Robert. "Mantle: Seven Views of Genius." *Sports Illustrated*, March 4, 1957: 48–52.

Ritchie, Everett J., "Mining's Tragic Events in the Southwest Missouri Metal Mines, An Annotated Index." A publication of the Everett J. Ritchie Tri-State Mineral Museum, Joplin, Mo., 1994: 1–149.

Roosevelt, Margot. "The Tragedy of Tar Creek." *Time*, April 26, 2006: 42.

Rosenthal, Harold. "A Letter From Mickey." *Elks*, June 1962.

Schecter, Leonard. "Mantle: A Problem Child, First of a Series." *New York Post*, March 28, 1960.

————. "Problem Child: The Mantle Players Know, No. 2." *New York Post*, March 29, 1960.

————. "Problem Child: Mantle and His Temper, No. 3." *New York Post*, March 30, 1960.

————. "Problem Child: The Mantle Casey Knows, No. 4." *New York Post*, March 31, 1960.

————. "Problem Child: Mantle's Other Interests, No. 5." *New York Post*, April 1, 1960.

Shaffer, Benjamin, Frank Jobe, Marilyn Pink, and Jacquelin Perry. "Baseball Batting: An Electromyographic Study." *Clinical Orthopaedics and Related Research* 292 (July 1993): 285–93. Clinical Orthopaedic Practice: http://journals.lww.com/corr/Abstract/1993/07000/Baseball_Batting__An_Electromyographic_Study.38.aspx

Shah, Diane K. "A Lady Fan Has a Chat with Mickey Mantle." *National Observer*, September 7, 1970.

————. "Where Have You Gone, Mickey Mantle?" *New York*, April 21, 1980: 49–59.

Solotaroff, Paul. "Growing Up Mantle." *Men's Journal*, July 2003.

Souryal, Tarek and Kenneth Adams. "Anterior Cruciate Ligament Injury." eMedicine from WebMD.com, January 2, 2009.

Subsidence Evaluation Team, Ed Keheley, chairman. "Picher Mine Field, Northeast Oklahoma Subsidence Evaluation Report." Prepared for the U.S. Army Corps of Engineers, January 2006.

Susman, Paul E. "Mantle: All-Time King of Tape-Measure

Homers." *Baseball Digest,* June 1982: 46–53.

———. "Physical Ailments Took a Toll on Mickey Mantle's Career." *Baseball Digest,* June 1996: 50–53.

Szold, Lee. "Are the Yankees Heading for Trouble?" *True's Baseball Yearbook,* 1964.

Talese, Gay. "Diamonds Are a Boy's Best Friend." *New York Times Magazine,* October 1, 1961.

Telander, Rick. "The Record Almost Broke Him." *Sports Illustrated,* June 20, 1977: 60–68.

"That Leg Again: Is It Mickey's Fatal Flaw?" *Sports Illustrated,* July 16, 1956: 11.

"They All Went to Toots's." *Sports Illustrated,* July 27, 1959: 32–37.

Trimble, Joe. "The Yankees' Troubled Ace." *Saturday Evening Post,* April 18, 1953: 31–35.

U.S. Environmental Protection Agency. Region 6. Five–Year Review Tar Creek Superfund Site. Ottawa County, April 2000.

Vecsey, George. "Mickey Mantle on the Road." *SPORT,* October 1963: 18–24.

Watkins, Paul J. "Baylor Regional Transplant Reaches National Milestone." *Dallas/Fort Worth M.D. News,* February 2008.

Weil, Robert. "Interview: Mickey Mantle." *Penthouse,* September 1986.

Wetzsteon, Ross. "The Mick Hits 60." *New York,* September 30, 1991: 40–47.

"What Makes Mickey Tick?" *Sports Illustrated,* September 10, 1956: 60.

Wulf, Steve. "Superman in Pinstripes." *Time*, August 21, 1995: 72.

"Yankee Downfall." *Sports Illustrated*, May 28, 1962: 26–27.

Young, Dick. "Farewell to Mickey Mantle." *SPORT*, April 1969: 26–30.

"Young Man on Olympus." *Time* 61, June 15, 1953: 64–66.

FILM/VIDEO/AUDIO

Beesley, Bradley, and Julianna Brannum, and James D. Payne, producers and directors. *The Creek Runs Red*. DVD. Independent Lens, 2007.

Doniger, Walter, director. *Safe at Home!* Produced by Tom Naud. VHS. Columbia Pictures, 1962.

Major League Baseball. "Game 1." In *World Series Dodgers at Yankee Stadium*. Audiocassette. 1963.

———. *Mickey Mantle Memorable Moments*. Audiocassette. Miley Collection, 2001.

———. *World Series Highlight Film*. VHS. Major League Baseball Productions, 1963.

Mann, Delbert, director. Produced by Stanley Shapiro and Martin Melcher. *That Touch of Mink*. DVD. Universal, 1962.

Mantle, Mickey. Interview by Spencer Christian. *Good Morning America*. ABC, June 8, 1994.

———. Interview by Bob Costas. *NBC Now*. NBC, March 30, 1994.

——. Interview by Robert Lipsyte. *CBS Sunday Morning*. CBS, June 18, 1983.

——. "Mickey Mantle: A Self Portrait." Interview by Howard Cosell. ABC, August 19, 1965.

"Mickey Mantle." In *ESPN SportsCentury*. ESPN. August 11, 2000.

Mickey Mantle raw time-coded footage, reels 1-3. VHS. Major League Baseball Productions.

Nathan, Alan M. *The Physics of Baseball: How to Hit A Home Run, How a Physicist Thinks about Baseball.* Jefferson Science Series. http://www.hep.uiuc.edu/home/mats/Talks/Baseball/nathan_jlab.WMV

Ohlmeyer, Don, director. "A Comedy Salute to Baseball hosted by Billy Crystal." Produced by Perry Rosemond and David Israel. NBC, July 18, 1985.

Ross, Norman, producer. "Seasons in the Sun." In *SportsChannel Special.* AmericanLife. May 25, 1985.

Scorsese, Martin, director. *Good Fellas.* Produced by Irwin Winkler. DVD. Warner Home Video, 1997.

Senn, Jess, producer. *Trojan Review, Vol. 2, 1951, No. 1.* VHS. USC School of Cinematic Arts.

Sevush, Herb, director. *The 500 Home Run Club.* VHS. Chelsea Communications, Inc., Cabin Fever Entertainment, 1988.

WEB SITE RESOURCES

www.andrewclem.com

www.atvaudio.com

www.baseballanalysts.com

www.baseballdigest.com

www.baseballindex.org

www.baseballprospectus.com

www.baseballreference.com

www.batspeed.com

www.boston.com

www.bronxbanter.com

www.chicagotribune.com

www.digitalballparks.com

http://digital.library.okstate.edu

www.findlaw.com

www.hardballtimes.com

www.history@EOK.Lib.OK.US

www.hittracker.com

www.joplinglobe.com

www.latimes.com

www.leadagency.org

www.mikeepsteinhitting.com

www.nejm.com

www.newsLibrary.com

www.newspaperarchive.com

www.newyorktimes.com

www.NYDailynews.com

www.NYPost.com

http://ok.water.usgs.gov/tarcreek

http://optn.transplant.Hrsa.gov

www.organ.org

www.peavynet.com

www.retrosheet.org

www.sabr.org

www.sciencedirect.com

www.sportsjournalism.org

www.springerlink.com

www.TheMick.com

www.tulsaworld.com

www.unos.org

www.washingtonpost.com

www.webmd.com

THE NEW LUXURY IN READING

We hope you enjoyed reading
our new, comfortable print size and found it
an experience you would like to repeat.

Well – you're in luck!

HarperLuxe offers the finest in fiction and
nonfiction books in this same larger print size and
paperback format. Light and easy to read, HarperLuxe
paperbacks are for book lovers who want to see
what they are reading without the strain.

For a full listing of titles and
new releases to come, please visit our website:

www.HarperLuxe.com